CONVERSION OF SI UNITS TO ENGLISH UNITS

Lengths

Multiply By → From ↓	To → Inches	Feet	Yards	Miles
mm	3.94×10^{-2}	3.28×10^{-3}	1.09×10^{-3}	6.22×10^{-7}
cm	3.94×10^{-1}	3.28×10^{-2}	1.09×10^{-2}	6.22×10^{-6}
m	3.94×10^{1}	3.28	1.09	6.22×10^{-4}
km	3.94×10^{4}	3.28×10^{3}	1.09×10^{3}	6.22×10^{-1}

Area

Multiply By → From ↓	To → Square Inches	Square Feet	Square Yards	Square Miles
mm²	1.55×10^{-3}	1.08×10^{-5}	1.20×10^{-6}	3.86×10^{-13}
cm²	1.55×10^{-1}	1.08×10^{-3}	1.20×10^{-4}	3.86×10^{-11}
m²	1.55×10^{3}	1.08×10^{1}	1.20	3.86×10^{-7}
km²	1.55×10^{9}	1.08×10^{7}	1.20×10^{6}	3.86×10^{-1}

Volume

Multiply By → From ↓	To → Cubic Inches	Cubic Feet	Cubic Yards	Quarts	Gallons
cm³	6.10×10^{-2}	3.53×10^{-5}	1.31×10^{-6}	1.06×10^{-3}	2.64×10^{-4}
liter	6.10×10^{1}	3.53×10^{-2}	1.31×10^{-3}	1.06	2.64×10^{-1}
m³	6.10×10^{4}	3.53×10^{1}	1.31	1.06×10^{3}	2.64×10^{2}

Force

Multiply By → From ↓	To → Ounces	Pounds	Kips	Tons (short)
dynes	1.405×10^{-7}	2.248×10^{-6}	2.248×10^{-9}	1.124×10^{-9}
grams	3.527×10^{-2}	2.205×10^{-3}	2.205×10^{-6}	1.102×10^{-6}
kilograms	3.527×10^{1}	2.205	2.205×10^{-3}	1.102×10^{-3}
newtons	3.597	2.248×10^{-1}	2.248×10^{-4}	1.124×10^{-4}
kilonewtons	3.597×10^{3}	2.248×10^{2}	2.248×10^{-1}	1.124×10^{-1}
tons (metric)	3.527×10^{4}	2.205×10^{3}	2.205	1.102

(see over)

Pressure (or Stress)

From / To →	lb/in^2	lb/ft^2	$Kips/ft^2$	$Tons\ (short)/ft^2$	Feet of Water $(39.2°F = 4°C)$	Atmosphere
gm/cm^2	1.422×10^{-2}	2.048	2.048×10^{-3}	1.024×10^{-3}	3.281×10^{-2}	9.678×10^{-4}
kg/cm^2	1.422×10^{1}	2.048×10^{3}	2.048	1.024	3.281×10^{1}	9.678×10^{-1}
kN/m^2	1.450×10^{-1}	2.090×10^{1}	2.088×10^{-2}	1.044×10^{-2}	3.346×10^{-1}	9.869×10^{-3}
$ton\ (metric)/m^2$	1.422	2.048×10^{2}	2.048×10^{-1}	1.024×10^{-1}	3.281	9.678×10^{-2}

Torque (or Moment)

From / To →	lb-in	lb-ft	Kip-ft
gm-cm	8.677×10^{-4}	7.233×10^{-5}	7.233×10^{-8}
kg-m	8.6777	7.233	7.233×10^{-3}
kN-m	9.195×10^{3}	7.663×10^{2}	7.663×10^{-1}

Velocity

From / To →	ft/s	ft/min	m/h
cm/s	3.281×10^{-2}	1.9685	2.236×10^{-2}
km/min	5.467×10^{1}	3.281×10^{3}	3.728×10^{1}
km/h	9.116×10^{-1}	5.467×10^{1}	6.214×10^{-1}

1 mile = 1610 meters = 5282.152 ft

Unit Weight

From / To →	lb/in^3	lb/ft^3
gm/cm^3	3.613×10^{-2}	6.248×10^{1}
kg/m^3	3.613×10^{-5}	6.248×10^{-2}
kN/m^3	3.685×10^{-3}	6.368
$tons\ (metric)/m^3$	3.613×10^{-2}	6.428×10^{1}

Geotechnical Engineering

Geotechnical Engineering

John N. Cernica, P.E., Ph.D.
Youngstown State University

Holt, Rinehart and Winston

New York   Chicago   San Francisco   Philadelphia
Montreal   Toronto   London   Sydney   Tokyo
Mexico City   Rio de Janeiro   Madrid

Library of Congress Cataloging in Publication Data

Cernica, John N.
    Geotechnical engineering.

    Bibliography: p.
    Includes index.
    1. Soil mechanics.   I. Title.
TA710.C39   624.1′5136   81-7052
ISBN 0-03-059182-1   AACR2

Printed in the United States of America

1 2 3   144   9 8 7 6 5 4 3 2 1

CBS COLLEGE PUBLISHING
Holt, Rinehart and Winston
The Dryden Press
Saunders College Publishing

To my wife Patricia
and my six beautiful daughters:
Kathleen, Jude, Alice, Johanna, Tricia, and Sarah.

Preface

Although great strides have been made in assimilating knowledge related to soil behavior, one is still likely not to regard *Geotechnical Engineering* as an exact technology. Instead, it is perhaps more realistic to view soil mechanics and geotechnical engineering as a combination of engineering mechanics, engineering judgment and common sense. It is this view which is adhered to throughout this text.

While extensive use was made of the concepts presented in various solid and fluid mechanics courses in derivations of equations or in developing a point, a concerted effort was also made to expose their limitations and shortcomings when applied to geotechnical engineering. In this regard the author has drawn upon the background of his teaching various soil mechanics and foundation courses, both graduate and undergraduate, during more than two decades, and upon his experience as a consultant in the soil mechanics field on hundreds of projects during that time.

The first three chapters were intended to give students a brief introduction to the field of soil mechanics before exposing them to the more detailed evaluation of the basic soil properties. Chapters 4 through 10 formulate the basic techniques for the evaluations of these properties, and serve to develop the methodologies that make use of the basic data for analyzing various foundation problems common to the civil engineer. Such problems include the stability of slopes, retaining structures, shallow and deep foundations, as well as methods for site improvement. Chapters 11 through 16 are reflective of these problems.

The material coverage could be for either a one- or a two-semester course in soil mechanics and foundations. Chapters 1 through 10 are considered very basic for a soil mechanics course and are thereby viewed as essential if the coverage is limited to a one-semester course. The last six chapters deal more with "application" and with foundation analysis and, as such, may be incorporated as part of a two-semester course in soil mechanics and foundation analysis.

The format adopted for the presentation of the material was designed with the student in mind. For example, each chapter begins with an introductory section providing the student with a brief synopsis of the chapter's content as well as reflecting upon the relevance of that material to practical problems. A substantial number of examples were worked out in detail to illustrate procedures for working related problems. General statements were complemented by specific verbal descriptions designed to give more clarity to the statements and eliminate any apparent ambiguity. Furthermore, the example problems were divided into three component parts: *Given*, *Find*, and *Solutions*. The exposure of the student to such a systematic approach to problem-solving is deemed advisable and conducive to subsequent formulation of desirable work habits.

The basic units employed in this text are SI. However, the equivalent values in the English system were also provided, in parentheses, throughout the text in answers to problems. This was thought helpful during the current transition period from English to SI units.

In spite of the nearly exhaustive efforts made in eliminating errors, some errors are perhaps inevitable. In this regard, the author begs the indulgence of those using the text until such errors are corrected. Furthermore, bringing such errors to the attention of the author will be greatly appreciated.

The author is greatly indebted to numerous colleagues and students for their various contributions towards the preparation of this text and in particular to: Dr. P. Chan, New Jersey Institute of Technology; Dr. B. F. Cousins, University of Tasmania, Australia; Dr. K. Derucher, Stevens Institute of Technology; Dr. G. G. Meyerhof, Nova Scotia Technical College, Canada; Dr. T. F. Zimmie, Rensselaer Polytechnic Institute and Mr. S. Soleimanpour, graduate student in Civil Engineering at Youngstown State University. My special gratitude, however, must go to my extremely hardworking secretary, Miss Betty McBride, who did a most thorough job in typing and proofreading the manuscript. Thanks is also expressed to several of my daughters who helped in various ways during the research and preparation of the manuscript.

John N. Cernica

Contents

8 Stresses in Soils   **156**

9 Consolidation and Settlement of Structures   **199**

10 Shear Strength of Soil   **238**

11 Stability of Slopes   **276**

12 Lateral Earth Pressure   312

13 Flexible Retaining Structures   349

14 Bearing Capacity—Shallow Foundations   383

Symbols and Abbreviations

A	=	area; activity of clay
A_v	=	area of voids
$\mathring{A}$	=	Angstrom units
a	=	distance; area
a_v	=	coefficient of compressibility
B	=	width of base
b	=	width of footing or base; distance
c_α	=	coefficient of secondary compression
C_c	=	coefficient of compression
C_r	=	slope of rebound curve for overconsolidated clays
C_u	=	Hazen's uniformity coefficient
C_v	=	coefficient of consolidation
C_z	=	coefficient of curvature
c	=	cohesion; shape factor
$c_d, \bar{c}$	=	developed cohesion
c_e	=	effective or true cohesion
D	=	diameter; depth; dimension
D_r	=	relative density
D_{10}	=	Hazen's effective size diameter at which 10 percent of the soil is finer
D_{50}	=	diameter at which 50 percent of the soil is finer
d	=	depth; distance; diameter; dial reading
E	=	Young's modulus
e	=	void ratio; eccentricity
e_{max}	=	void ratio of soil in loosest state
e_{min}	=	void ratio of soil in densest state
e_0, e_i	=	initial void ratio of mass
F	=	friction force; factor of safety
f	=	coefficient of friction; friction factor
G	=	specific gravity
g	=	acceleration of gravity
H	=	height; hydrostatic head; thickness
h	=	hydraulic head; total head; height
h_c	=	height of capillary rise
I	=	indices; moment of inertia
I_p	=	plasticity index
I_f	=	flow index
I_t	=	toughness index

IV_i	= influence values
i	= hydraulic gradient; angle of inclination of slope with the horizontal
K	= coefficient of lateral pressure
K_a	= coefficient of active pressure
K_p	= coefficient of passive pressure
K_0	= Coefficient of earth pressure at rest
$K_{\text{f-line}}$	= line through p versus q
k	= permeability coefficient
k_e	= effective permeability
L	= distance; length
L.L.	= liquid limit
l	= length
M	= mass; moment
m	= distance ratio
N	= normal force; pressure index; various coefficients; standard penetration number
N_φ, N_q, N_γ	= bearing capacity coefficients
n	= number of equipotential drops; porosity; distance ratio; number
n_d	= number of equipotential drops
n_f	= number of flow paths
OCR	= overconsolidation ratio
P	= force; resultant pressure
P_a	= resultant active pressure
P_p	= resultant passive pressure
P.I.	= plasticity index
p	= pressure
Q	= rate of flow; concentrated force; pile capacity; bearing capacity
Q_p	= pile end-bearing resistance
Q_s	= pile frictional resistance
Q_u	= ultimate load
q	= rate of discharge per unit area; stress or pressure
q_a	= allowable bearing capacity
q_u	= ultimate bearing capacity
R	= radius; Reynolds's number; resultant force; resistance
r	= radius
S	= degree of saturation; distance; shear strength resultant
s	= shear strength per unit distance; distance; settlement
T	= time factor; tensile force
T_s	= surface tension
t	= time
U	= consolidation ratio; resultant neutral (pore-water pressure) force; unconfined compression

u	=	pore-water pressure; velocity
V	=	volume
V_s	=	volume of solids
V_v	=	volume of voids
V_w	=	volume of water
v	=	velocity
v_s	=	seepage velocity
W	=	weight
W_s	=	weight of solids
W_w	=	weight of water
w	=	water content
w_l	=	liquid limit
w_p	=	plastic limit
w_s	=	shrinkage limit
X, Y, Z	=	artesian coordinates
x, y, z	=	distances
$\bar{x}, \bar{y}$	=	centroidal distances
x_t	=	transformed x distance
Z	=	coordinate
z	=	depth
α	=	inclination of vectors; inclination of slopes; angle
α_m	=	maximum angle of stress obliquity
β	=	angle
γ	=	unit weight
γ_b	=	buoyant unit weight
γ_d	=	dry unit weight of soil
γ_s	=	unit weight of solids
γ_t, γ	=	total unit weight of mass
γ_w	=	unit weight of water
δ	=	angle; deflection; settlement
Δ	=	changes; increments
ε	=	strain
$\varepsilon_x, \varepsilon_y, \varepsilon_z$	=	strain in x, y, z directions, respectively
θ	=	angle
μ	=	Poisson's ratio; coefficient of viscosity
ρ	=	mass density; settlement
σ	=	normal stress; total stress
$\bar{\sigma}$	=	effective normal stress
$\sigma_1, \sigma_2, \sigma_3$	=	principal stresses
σ_t	=	combined or total pore-water stress and intergranular normal stress
$\sum$	=	sum
τ	=	shear stress
φ	=	friction angle

φ_d = developed friction angle
φ_e = effective friction angle
$\bar{\varphi}$ = friction angle based on effective stress
ψ = angle
ω = angle

1
Soil in Engineering

1-1 INTRODUCTION

Soil is man's oldest, perhaps most common, and probably the most complex construction material. Because of its function as the support for virtually all structures, soil becomes an indispensable component of construction and, therefore, in a broad sense, plays a most prominent role in civil engineering design. Hence, in spite of its complexity, we must work with it; we must determine its behavior under load; we must evaluate its interactions with the structure it supports. In the final analysis, we must fit it into its designated role such that a given design is economical and safe. These are rather demanding tasks which ordinarily require a working knowledge of soil mechanics, some relevant experience, and a great deal of engineering judgment.

Up to this point the typical civil engineering student has taken a series of courses in mathematics, analytical mechanics, or mechanics of materials in which the problem as well as the variables were relatively well defined. For example, the student may be able to readily solve for the reactions, moments, stresses, or deflections in a continuous-steel beam of a given size and length, which supports a designated load. The student may also know quite well that the modulus of elasticity for steel is a constant, that the material is reasonably homogeneous and a quality-controlled product. Hence, the behavior of steel under load can be predicted with reasonable accuracy. By contrast, the practicing civil engineer may be called upon to analyze, design, or construct a structure whose design parameters are not nearly as well defined. For example, let us imagine the problem of designing a dam, a dock, a tower, an airport, a retaining wall, a highway—all structures founded on soil or ledge. In their overall scope, these projects encompass much more than the idealized or hypothesized conditions and variables cited above. These are situations where one must employ the use of good judgment to solve the problem. In other words, the interaction of a reasonably "homogeneous" steel beam, supported by "elastic" columns which rest on rather "unpredictable" material such as soil poses a problem which is not clear-cut, and one which may or may not have a single and unique solution.

The term *soil* as used here will be taken to encompass the sediments and deposits of solid particles produced by the disintegration of rock, including all organic and inorganic materials overlying the solid and massive bedrock. The name *geotechnical*

1

engineering or *soil mechanics* is given to that part of science (and art) to which one applies the laws of mechanics and hydraulics to solve engineering problems associated with soils.

1-2 SOIL AND FOUNDATION PROBLEMS IN ENGINEERING

The word *foundations* usually refers to that part of a structure which transmits to the soil the dead and the live load of the superstructure. However, the term is sometimes used to denote the total general support system for a structure, including the soil or rock on which the structure rests.

Soil and foundation problems and solutions were certainly not limited to any one era, geographic location, or any particular people in history (15, 20, 25).* Ample evidence exists of successful solutions to foundation problems faced by some of the early builders. For example, some earth dams in India have been storing water for more than 2000 years; the successful construction of many aqueducts, bridges, roads, etc., by the Romans 2000 years ago provides some proof of their mastery of the art in foundation design (22); the many projects in China and Egypt are further evidence of the ability of the ancient civilizations to master some of the problems associated with a suitable foundation.

Needless to say, the ancient designers experienced failures as well as successes—perhaps many more failures than successes. These, however, are not as historically traceable as the successes. Hence it is perhaps a little more than conjecture at this point for us to explain the nature and degree of their understanding of the interaction of the structures with the soil supporting them. In all probability, their knowledge was derived through experience, through trial and error, and through common sense approaches.

Most frequently, in the design of a foundation, one is confronted with a problem of *settlement*—indeed a most important design consideration. Perhaps of even greater importance is the difference in settlement between parts of a structure, commonly referred to as *differential settlement* (5, 16). The extent of settlement tolerable in a design usually depends on the type and function of the structure. On the other hand, the magnitude of settlement depends on such factors as the type and intensity of the loads transmitted to the soil, and the general physical and mechanical characteristics of the soil on which the structure rests.

Another concern often confronted by a designer is that of *stability*. Among the more common problems related to stability, one might include the sliding of an embankment (17), the tipping of a retaining wall, excessive penetration of a footing into the ground (19, 26, 28, 33, 36), collapse of an excavation, cracking of a dam, soil creep, etc. These are reflected upon in greater detail in subsequent sections.

Many soil-related failures are sudden and rather obvious (2, 3, 5, 7, 9, 12–14, 21); others are more gradual and less evident during a relatively short span. For example,

---

* The numbers in parentheses designate references from the bibliographical list located at the end of each chapter which the student may find of interest for further reading.

a hill slide may be sudden, either partial or total. On the other hand, many of the structures in Mexico City experience progressive and cumulative deformation or consolidation of the soil, and subsequent settlement of the structure—perhaps several feet during the life of the structure (33, 36). The Tower of Pisa (19, 26) is leaning not only from the total deformation of the soil under it, but also from the results of differential settlement. In this case the stability of the structure as well as the functional aspects related to the deformation are obviously interrelated.

Safety and economy, in that order, are perhaps the most important aspects of foundation design. The designer must strive for a stable and functional design at minimum cost. To do so, engineers must first understand the problem, systematically evaluate various available data, and then proceed toward a solution. That is, they must understand the interaction of the superstructure with the foundation, then, based upon careful evaluation of available information in the field of soil mechanics, experience, and sound judgment, they subsequently proceed to develop a reasonable solution to the problem.

With time, students will find that soil and foundation problems are not nearly as well defined as most others which they confront; they will find their solutions less concise and/or uniquely oriented. They will also be impressed with the fact that geotechnical engineering permits much more intuitive judgment than most other forms of mechanics. At this point in time, the field is perhaps as much of an art as it is a science (30, 31).

1-3 SOIL MECHANICS—A BRIEF HISTORICAL REVIEW

Remnants of many notable structures built by the Romans, Egyptians, Chinese, and others provide evidence that some knowledge existed during ancient civilizations of the interaction of a superstructure with the soil supporting it. The Great Wall of China, the Pyramids of Egypt, the many buildings and durable roads constructed by the Romans, and the mastery of dam building displayed by the Indians are parts of such evidence. There is insufficient evidence, however, to suggest that these ancient people had a "systematic" approach to the solution of their foundation-related problems. In all probability, their basic knowledge of soil mechanics was rather skimpy, quite probably unstructured, limited in scope, and perhaps confined to the geography of a given region. During this era, lack of transportation and writing proved to be major obstacles to the dissemination of knowledge and ideas over wide areas, thereby imposing limitations to the propagation of any basic information. Hence it is quite probable that if a structure did not perform satisfactorily, it would be replaced by a new one that would, and so on.

The Egyptians, the Romans, and the Chinese appeared to be particularly knowledgeable of foundation-related problems even before Christ. In fact, they appeared to be more aware of the importance of a suitable foundation design than were generations several centuries later. Unfortunately, clear and specific details of their knowledge are lacking for an accurate and comprehensive evaluation of their state of the art. However, as mentioned above, their many and varied building accomplishments are testimonials to their ability and provide implicit proof that they were capable builders of their time. Perhaps they ought to be credited as the beginners of a more basic approach to the solution of problems in soil mechanics and foundations.

An appreciable decrease in interest and knowledge in problems related to soils appeared from the fall of the Roman Empire until about the fifteenth century. It was not until about the seventeenth and eighteenth centuries that real attention and interest were again revived and special attention focused on soil mechanics and foundation engineering.

The analysis of earth pressures on retaining walls was one of the first branches in the field of soils and foundations to generate a renewed interest. Among the first to formulate guidelines in this area—indeed a pioneer in the branch—was a French military engineer named Marquis Sebastian le Prestre de Vauban (1633–1707). Approximately 300 years ago he provided integral guidelines and regulations for the construction of structures for retaining the soil behind these structures. Several other contributors followed. The most notable early contribution to this phenomenon was that of Charles Augustin Coulomb (1736–1806). He, too, was a French military engineer who is generally credited with the first basic and scientific approach for calculating the stability of retaining walls. Another important contribution related to earth pressures and the stability of retaining walls was made by William John Macquorn Rankine (1820–1872), a Scottish engineer and physicist, best known for his research in molecular physics, and one of the founders of the science of thermo-dynamics.

Rankine and Coulomb are perhaps the two best known nineteenth-century contributors to the study of lateral pressures on retaining walls. Among many others who contributed to this phase of soil mechnics was Jean Victor Poncelet (1788–1867), a French mathematician and engineer, and one of the founders of modern projective geometry. His contribution focuses on a graphical method of solving earth pressures and retaining-wall problems. Karl Culmann (1821–1881), a German engineer whose method of graphostatics has been used extensively in many problems of engineering and mechanics, further improved on the graphical solution for retaining walls.

Joseph Valentin Boussinesq (1842–1929) provided us with a tool for analyzing and estimating stresses in a soil stratum from loads applied at the surface, an offshoot from the subject of strength of materials. Otto Mohr (1835–1918) in 1882 proposed a graphical device to analyze stress at a point. His "Mohr circle," as it is commonly known, may be used when evaluating the strength of soils.

Sir Benjamin Baker (1840–1907), an English civil engineer responsible for designing many bridges in England and the United States, as well as tunnels and embankments, provided valuable reflections on the characteristic of slope failures.

By the turn of the twentieth century much of the knowledge in the soil mechanics field was getting disseminated through various forms of publications, planned meetings, and organized presentations of experimental findings via periodicals published on a regular basis by professional societies. Particularly noteworthy is the American Society of Civil Engineers (1). This was, very probably, the real formidable beginning of a new science and the evolution of a genuine interest on the part of many civil engineers and groups in the broad and comprehensive field of soil mechanics. Among many of these groups were engineers both from this country and from abroad, particularly Europe.

Early in the twentieth century much research on the subject was done by the U.S. Bureau of Public Roads, the U.S. Army Corps of Engineers, the U.S. Bureau of Reclamation, the Portland Cement Association, various state highway departments, as well as many universities. Most notable were the efforts in the United States, with significant contributions made by others in Germany and Sweden.

Perhaps the greatest and truly profound influence in the field of soil mechanics was exerted by Karl Terzaghi (1882–1963). He is generally credited with having established soil mechanics as one of the basic subjects of civil engineering science. As a result of his research during the period of 1919 through 1925, and his genuine interest, his ingenuity, and his immense contributions to the field, he provided the unifying concepts that made soil mechanics a science. His publication of *Erdbaumechanik* formulated what is generally accepted as the basis for today's soil mechanics. This work, coupled with his more than 200 scientific papers and his worldwide influence as a consultant, lecturer, and contributor to scientific efforts, elevates him to the level of recognition as probably the most fruitful contributor to the field of soil mechanics in history. He is frequently regarded as the "father of soil mechanics."

Since Terzaghi's famous publication to the present time, numerous works have appeared in this regard. Among the many who left an indelible mark in the field of soil mechanics are the following, listed alphabetically, together with the subject matters with which they are perhaps most frequently associated:

Bishop, A. W.	Pore pressure, slope stability, shear strength
Bjerrum, L.	Shear strength, stability of slopes
Casagrande, A.	Soil classification, seepage, shear strength
Crawford, C. B.	Consolidation and settlement
Lambe, T. W.	Studies of engineering behavior of fine-grain soils, textbook on soil mechanics
Meyerhoff, G. G.	Bearing capacity
Peck, R.	Pressures in clay, Chicago subway problems, textbook on soil mechanics
Richart, F. E.	Dynamically loaded foundations, textbook on soil dynamics
Rowe, P. W.	Lateral earth pressures
Seed, H. B.	Earthquake studies
Skempton, A. W.	Pore pressure in clays, effective stress-bearing capacity, slope stability
Taylor, D. W.	Consolidation, shear strength, slope stability, textbook on soil mechanics
Vesic, A. S.	Bearing capacity of deep foundations
Whitman, R. V.	Dynamically loaded foundation, textbook on soil mechanics

Some particularly good sources for information related to soil mechanics are:

Journal of Geotechnical Engineering, American Society of Civil Engineers, New York
Canadian Geotechnical Journal, Ottawa, Canada
Geotechnique, Institution of Civil Engineers, London, England

1-4 ORIGIN OF SOILS

Soils are the products of disintegration of rock. On the other hand, some rock is formed by cementation of loose soil particles into a hard, solid substance that we call rock. Common examples are shale, sandstone, siltstone, etc. A more detailed evaluation of such processes is given in Chapter 2. At this time the discussion will be limited to a brief survey of the factors related to the origin of soil.

There has been much speculation about the origin of the solar system. Also, many theories about the origin of the planets have been propounded and many rejected.

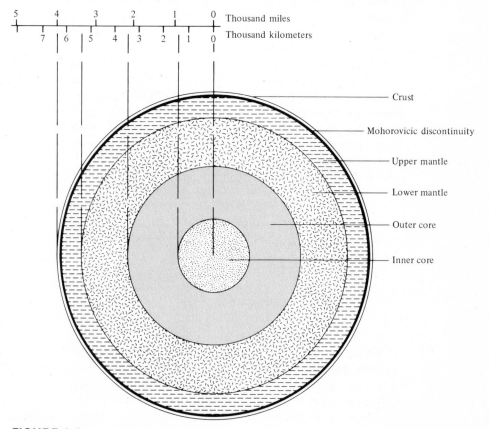

FIGURE 1-1
Significant layers of the earth.

It is possible that the formation of the sun occurred much earlier than that of the planets. It is generally accepted that the planetary system developed out of a disk-shaped mass of gas believed to be surrounding the sun. From all available evidence it appears that the overwhelming part of original cosmic matter consisted of hydrogen and some helium, and that these gases have been lost in space from the disk at an early stage. The less volatile matter condensed, primarily in the form of silica and metallic iron—the prime constituents of the interior of the earth.

Earlier speculations viewed the earth as an original ball of molten lava. Current-day geophysicists question, however, that the earth was ever completely molten. At some time during its formation the earth must have been reasonably soft to permit separation into layers. These are generally divided into four major parts: the outer crust, the mantle, the outer core, and the inner core, Figure 1-1 shows this division. The thickness of the crust varies from about 8 km (5 mi) under the oceans to about 35 km (22 mi) under the continents. There is a boundary between the earth's crust and the mantle, called the Mohorovicic discontinuity. The earth's mantle is solid rock, and it goes down about 2900 km (1800 mi). It is speculated that the earth's outer core is about 2250 km (1400 mi) thick and that it is made of melted iron and nickel. The inner core of the earth is about 1300 km (800 mi) thick and probably consists of mostly iron and nickel.

It is estimated that within the deepest part of the crust the earth's temperature may be 870° C (about 1600° F). Where the mantle meets the outer core, the temperature is believed to be about 2200° C (about 4000° F). The temperature in the inner core is probably about 5000° C (about 9000° F).

As engineers we are concerned almost exclusively with the outer portion of the earth, generally referred to as the *crust*. The greatest percentage (by weight) of the crust is formed by the following elements:

Oxygen	49.2%
Silicon	25.67%
Aluminum	7.5%
Iron	4.71%
Calcium	3.39%
Sodium	2.63%
Potassium	2.40%
Magnesium	1.93%
	97.43%

Various processes are continually acting on the earth's surface. They subsequently affect not only the configuration of the surface but also the material of which it is composed. *Gradation* is a process whereby the physical topography (mountains, valleys, etc.) may be altered by the action of water, air, ice, or other weathering factors. *Diastrophism* designates a process wherein portions of the earth move relative to other portions. *Vulcanism* refers to the action of the molten mechanic rock, both on the earth's surface and within the earth.

The above-mentioned processes result in a general disintegration and decomposition of the rock. Although these changes may result in a transformation of old rock into new rock, part of the product, and particularly near the surface of the earth, is that which we generally classify as *soil*. It is formed from rock by *mechanical disintegration* or *chemical decomposition*, or both. Disintegration is related to freezing and thawing, the action of water, glaciation, etc. Decomposition is related to oxidation or hydration. The combination of the mechanical and chemical processes is called *weathering*.

Briefly, the rocks and minerals of the earth's surface were the starting material from which soils originated. Exposure to atmospheric conditions and volcanic and tectonic action has subsequently transformed these rocks and minerals into a more or less unconsolidated blanket over the earth's surface which we call *regolith*. With time this separated into an orderly sequence of layers or horizons. These chemically and biologically differentiated top layers of the regolith are the soil. Figure 1-2 shows a schematic of this development.

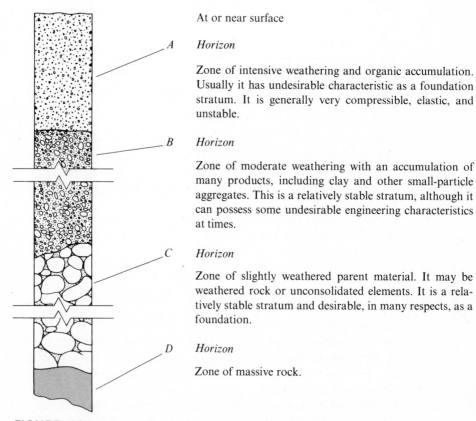

At or near surface

A Horizon

Zone of intensive weathering and organic accumulation. Usually it has undesirable characteristic as a foundation stratum. It is generally very compressible, elastic, and unstable.

B Horizon

Zone of moderate weathering with an accumulation of many products, including clay and other small-particle aggregates. This is a relatively stable stratum, although it can possess some undesirable engineering characteristics at times.

C Horizon

Zone of slightly weathered parent material. It may be weathered rock or unconsolidated elements. It is a relatively stable stratum and desirable, in many respects, as a foundation.

D Horizon

Zone of massive rock.

FIGURE 1-2
A typical soil profile in the outer portion of the regolith.

1-5 SOIL MECHANICS AND SOIL ENGINEERING

During the past century, and particularly during the last 50 years, monumental strides have been made in our accumulation and dissemination of information and in our systematic organization of this information in solving problems in soil mechanics.

Theoretical and experimental methods and procedures have been developed and are available for solving many of the structural problems in a systematic manner. For example, we have formulated procedures for estimating the settlement of a structure due to the consolidation of the soil supporting it, for estimating the pressures behind a retaining wall, for estimating stresses in a soil stratum from surface loads, for determining the seepage loss through a dam, etc. These established and accepted procedures have resulted in significant changes in the design and construction practices in the field of foundations, and have established an uncompromising place for the soil engineer in the field of structural design.

Although the emergence of *geotechnical engineering* or *soil mechanics* has greatly reduced the try-and-miss or rule-of-thumb approaches which have prevailed in the past, in most instances the methodology for solving problems still includes many assumptions and empirical procedures. Although geotechnical engineering embraces many components of established sciences, laws of mechanics and hydraulics, it does not establish any single or unique approach to the solution of a foundation problem. Instead, it provides some basic tools with which the soils engineer must work. With these "tools" plus diligence, common sense, and careful evaluation of available data, the engineer labors a well-developed solution to a foundation problem. In that respect, the term *soil mechanics* or *geotechnical engineering* is a relatively young field. It is one of the fastest growing disciplines in civil engineering today. Few, if any, of present-day structures of any significance are designed or built without thorough and detailed information regarding the soil on which they rest, or without the careful scrutiny and advice of a qualified soils engineer. In fact, many of the major engineering or architectural firms carry, as part of their staff, personnel and laboratory facilities for the express purpose of evaluating the foundation material. Indeed, many firms have specialized in this field to the degree that most, if not all, of their efforts are expanded within this speciality.

1-6 ADAPTATION OF THEORY TO PRACTICAL USE

Virtually all theories used in geotechnical engineering owe their origin to certain basic assumptions. Hence their accuracy and their dependability are only as reliable as the assumptions on which they are based.

Many of the assumptions used in developing theoretical expressions are based on the physical properties of the soil in question. When one recognizes that the physical characteristics of the soil vary extensively in both the horizontal and the vertical directions, it becomes readily clear that the determination of these properties is not an easy task; quite to the contrary, these efforts could be monumental tasks indeed.

For example, one may proceed to predict the settlement of a structure from the behavior of a small sample of soil subjected to a consolidation test. The question which must ultimately be resolved is: to what extent are these data representative of the overall behavior of the actual soil stratum? Inherent in the above question are a series of related questions: to what extent was the tested sample disturbed during extraction, preparation, etc.?; how representative is that sample of the total stratum?; what is the variation of the stratum characteristics for a given site? Obviously the answers to these and other questions are not easy ones. Hence it is important that careful scrutiny be made of the experimentally obtained results before they are to be used in calculations and subsequent conclusions.

It is equally important to note that many of the theoretical considerations are derived on the basis of assumptions which frequently include homogeneity, elasticity, uniformity in stratum distribution, and many more. Yet in all probability very few, if any, of these characteristics are found in a typical soil stratum. Hence the accuracy of an analysis is an inseparable combination of theoretical considerations and sound engineering judgment. It is imperative that one recognize the limitations of theory and place the results in their proper perspective. It is in this light that many of the practitioners regard the application of soil mechanics as a composite of science and art.

It is important to recognize that a good theoretical background, coupled with sound engineering, is a desirable combination; on the other hand, decisions based on only theory or only judgment are highly undesirable—perhaps even disastrous.

PROBLEMS

1-1.   Describe the characteristics which make *soils* such a complex construction material.

1-2   What is encompassed by the term *soil* as used in engineering?

1-3   Define the following terms: (a) Settlement. (b) Differential settlement. (c) Stability as related to foundation design.

1-4.   How does geotechnical engineering differ from other forms of mechanics, such as statics and dynamics, fluid mechanics, etc.?

1-5.   Name at least two early contributors to the study of lateral pressures on retaining walls.

1-6   How and what did Karl Terzaghi contribute to the field of soil mechanics?

1-7.   Name at least four contributors, and state briefly their contributions to the field of soil mechanics.

1-8.   What is generally believed to be the origin of the planetary system?

1-9.   Name the four major parts of the cross section of the earth.

1-10. Describe the role that soil engineering assumes in present-day foundation design.

1-11. Why must information obtained from soil borings and laboratory tests undergo much careful and detailed scrutiny?

1-12. Why are judgment and experience so important in foundation engineering?

1-13. What part of the regolith is of the most interest to the foundation designer? Why is it not always advisable to place the foundation in that part of the regolith where stable support is virtually guaranteed?

1-14.   How does mechanical disintegration differ from chemical decomposition?

1-15.   How does geotechnical engineering resemble other forms of mechanics? How does it differ?

BIBLIOGRAPHY

1.   American Society of Civil Engineers, "Selected Bibliography on Soil Mechanics," *Man. Eng. Practice*, no. 18, 1940.

2.   Baracos, A., "The Foundation Failure of the Transcona Grain Elevator," *Eng. J.*, vol. 40, 1957.

3.   Bazynski, J., "Studies of the Landslide Areas and Slope Stability Forecast," *Bull. IAEG*, Krefeld, West Germany, no. 16, 1977.

4.   Bell, R. A., and J. Iwakivi, "Settlement Comparison Used in Tank-Failure Study," *ASCE J. Geotech. Eng. Div.*, Feb. 1980.

5.   Belloni, L. A., A. Garassini, and M. Jamiolkowski, "Differential Settlements of Petroleum Steel (Tanks)," *Conf. Settlement of Structures*, Brit. Geotech. Soc., Cambridge, England, Apr. 1974.

6.   Bishop, A. W., "The Strength of Soils as Engineering Materials," 6th Rankine Lecture, *Geotechnique*, vol. 16, no. 2, 1966.

7.   Bjerrum, L., and A. Overland, "Foundation Failure of an Oil Tank in Fredrikstad, Norway," *4th Int. Conf. Soil Mech. Found. Eng.*, Oslo, Norway, 1957.

8.   Bjerrum, L., "Geotechnical Problems Involved in Foundation of Structures in the North Sea," *Geotechnique*, vol. 23, no. 3, 1973.

9.   Brown, G. D., and W. G. Paterson, "Failure of an Oil Storage Tank Founded on Sensitive Marine Clay," *Can. Geotech. J.*, vol. 1, no. 4, Nov. 1964.

10.   Casagrande, A., and R. E. Fadum, "Applications of Soil Mechanics in Designing Building Foundations," *Trans. ASCE*, 1944.

11.   Casagrande, A., "Role of the Calculated Risk in Earthwork and Foundation Engineering," *ASCE J. Geotech. Eng. Div.*, vol. 91, no. SM4, 1965.

12.   Clark, J. S., "Survey of Oil Tank Failures," *Ann. Inst. Belge du Petrol*, no. 6, 1969.

13.   Duke, C. M., and D. J. Leeds, "Response of Soils, Foundations, and Earth Structures to the Chilean Earthquakes of 1960," *Bull. Seismol. Soc. Am.*, vol. 53, no. 2, 1963.

14.   Eden, W. J., and M. Bozozuk, "Foundation Failure of a Silo on Varved Clay," *Eng. J.*, vol. 45, no. 9, Sept. 1962.

15.   Forbes, R. J., *Man the Maker, A History of Technology and Engineering*, Henry Schuman, New York, 1950.

16.   Green, P. A., and D. W. Hight, "The Failure of Two Oil Storage Tanks Caused by Differential Settlement," *Conf. Settlement of Structures*, Brit. Geotech. Soc., Cambridge, England, Apr. 1964.

17.   Jakobson, B., "The Landslide at Surte on the Gota River," *R. Swed. Geotech. Inst.*, vol. 5, 1952.

18.   Legget, R. F., "Geology and Geotechnical Engineering," *ASCE J. Geotech. Eng. Div.*, vol. 105, Mar. 1979.

19.   Mitchell, J. K., V. Vivatrat, and T. W. Lambe, "Foundation Performance of Tower of Pisa," *ASCE J. Geotech. Eng. Div.*, vol. 103, Mar. 1977.

20.   Painter, S., *A History of the Middle Ages 284–1500 A.D.*, Knopf, New York, 1953.

21.   Peck, R. B., and F. G. Bryant, "The Bearing Capacity Failure of the Transcona Elevator," *Geotechnique*, vol. 3, 1953.

22. Platner, S. B., *The Topography and Monuments of Ancient Rome*, Allyn and Bacon, Boston, 1904.

23. Rominger, J. F., and P. C. Rutledge, "Use of Soil Mechanics Data in Correlation and Interpretation of Lake Agassiz Sediments," *J. Geol.*, vol. 6, no. 2, Mar. 1954.

24. Skempton, A. W., "Soil Mechanics in Relation to Geology," *York. Geol. Soc.*, vol. 29, 1953.

25. Szechy, C., *Foundation Failures*, Concrete Publications Ltd., London, 1961.

26. Terracina, F. "Foundations of the Tower of Pisa," *Geotechnique*, vol. 12, 1962.

27. Terzaghi, K., *Erdbaumechanik*, Franz Deuticke, Vienna, 1925.

28. Terzaghi, K., "Die Ursachen der Schiefstelling des Turmes von Pisa," *Bauingenieur*, no. 1/2, 1934.

29. Terzaghi, K., "Relation between Soil Mechanics and Foundation Engineering," presidential address to *1st Int. Conf. Soil Mech. Found. Eng.*, Cambridge, Mass., vol. 3, 1936.

30. Terzaghi, K., "Soil Mechanics—A New Chapter in Engineering Science," *J. Inst. Civil Eng.*, London, England, vol. 12, 1939.

31. Terzaghi, K., "Ends and Means in Soil Mechanics," *Eng. J.*, Montreal, Canada, vol. 27, no. 12, Dec. 1944.

32. Terzaghi, K., "The Influence of Modern Soil Studies on the Design and Construction of Foundations," *Build. Res. Congr.*, London, England, vol. 1, 1951.

33. Thorney, J. H., G. B. Spencer, and P. Albin, "Mexico's Palace of Fine Arts Settles 10 ft," *Civ. Eng.*, vol. 25, no. 6, June 1955.

34. Tschebotarioff, G. P., "A Case of Structural Damages Sustained by One-Storey High Houses Founded on Swelling Clays," *3rd Int. Conf. Soil Mech. Found. Eng.*, Zurich, Switzerland, vol. 1, 1953.

35. Williams, J. H., and J. D. Vineyard, "Geological Indicators of Catastrophic Collapse in Karst Terrain in Missouri," *Transp. Res. Rec.*, no. 612, 1976.

36. Zeevaert, L., "Foundation Design and Behavior of Tower Latino Americana in Mexico City," *Geotechnique*, vol. 7, no. 3, Sept. 1957.

2

Formation of Soil and Soil Deposits

2-1 INTRODUCTION

As briefly indicated in Section 1-4, the rocks and minerals of the earth's crust were the parent materials from which soils originated. Exposure to weathering agents, volcanic action, and stresses induced by an ongoing deformation of the earth's crust are factors directly tied to the disintegration of these rocks and minerals. Soils were formed in this process. Thus the type as well as the characteristics of the soil are inherently related to its origin or its parent material.

The nature of the soil stratum is continually transformed (7). Both the rate of transformation and the resulting product are dependent on and influenced by a number of factors. Among these factors are the following:

1. *Climate*, particularly temperature and humidity
2. *Hydrology* of the area (such as rainfall, run-off conditions)
3. *Topography* of the area, including the rate and amount of percolation of water, exposure to the sun or winds, etc. (3)
4. *Organisms*, animal and plant life
5. *Time*

The disintegrating effects on the rock and the subsequent soil transformation are most evident near the surface. However, this is a continuous process, to various degrees, for appreciable depths below the surface as well. In the process of decomposing rock and the gradual transformation of soil, the end result may be a new type of rock. For example, sandstone is a rock formed by sedimentary deposits of sand; shale is a rock which is formed by the consolidation of clay, mud, or silt; slate is a rock formed from shale, etc. This cyclical transformation, and some of the factors related to this process, are described in the following sections.

2-2 THE GEOLOGICAL CYCLE

Geologists tell us that there is a continuous slow process whereby rocks are decomposed into soil, while some of the soil goes through various stages of conversion back into rock (10, 14). The transformation whereby rocks are converted into soil, and vice versa, generally occurs over millions of years and through complex chemical and physical processes. This phenomenon is described generally as the *geological cycle* and is schematically depicted by Fig. 2-1.

We may begin with the cycle where the molten magma cools and thus produces *igneous rock*. These rocks thus become subject to attack, particularly near or at the earth's surface, by environmental agents, such as water, oxygen, carbon dioxide, and temperature changes. Subsequently the rock decays and disintegrates into an initial state of *residual soils*. Additional factors, such as chemical action, organisms, wind, water, and ice, may move the loose materials. The transported materials will ultimately be dropped in a form generally classified as *sedimentary deposits*. A portion of such deposits may cement and consolidate into *sedimentary rock*, while others precipitate during transportation into sedimentary beds of fragmented materials. Still other

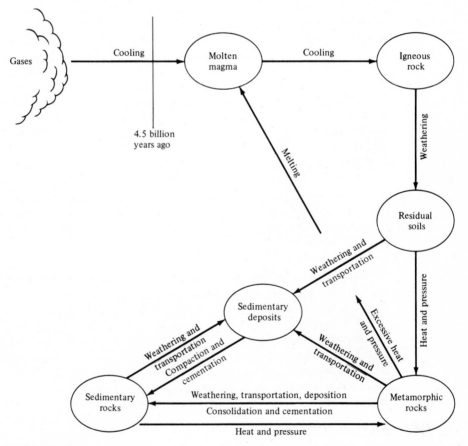

FIGURE 2-1
The geological cycle.

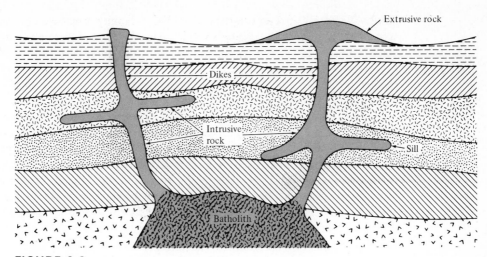

FIGURE 2-2
Section through dikes and sills.

parts of the transported sediments may remain in a form of solution, as evidenced by the salts in the oceans and seas.

Subjected to new environmental impositions of pressure and heat, some of the sedimentary deposits will be transformed into *metamorphic rocks*. Other parts of the sedimentary deposits may be subjected again to weathering and disintegration, transportation, and redeposition, forming a new generation of secondary rocks and fragmented materials and soils.

Excessive pressures and heat may result in melting of virtually all rocks, resulting in new igneous rocks. Hence a new geological cycle begins again as an ongoing process. Indeed, volcanic action provides us with rather direct proof that the interior of the earth is still in a molten state and under high heat and pressure. Cracks and faults in the earth's crust may permit this molten magma, commonly referred to as *lava*, to permeate upward into the crust formation and/or at the surface in the form of volcanic eruptions. (The eruptions of Mount Saint Helens in the state of Washington in 1980 are among the more recent such occurrences, although volcanic eruptions in Japan, Italy, and Hawaii are more renowned.)

The magma (or lava) which cools at the surface forms *extrusive* igneous rock; that formed within the crust is commonly referred to as *intrusive*. During the upward flow from the earth's interior to the surface, through faults in the earth, some of the molten mass will cool within such conduits to form a crystalline formation known as *dike rock*, which is usually considered to belong to the intrusive group of rocks. This is shown in Fig. 2-2.

2-3 THE MAJOR TYPES OF ROCK

Geologists place rocks into three major groups: *igneous*, *sedimentary*, and *meta-morphic* rocks. Their formations have already been briefly described in the preceding section. From a geotechnical point of view it may be relevant at this time to identify some of the subgroups within the major categories and to identify some of their characteristics.

Igneous Rocks

These rocks may vary considerably in their mineral composition, texture, and mode of occurrence. Slow cooling of the molten magma tends to yield large crystals, whereas rapid cooling generally results in finer grains. Hence extrusive rocks are usually fine grained, while the intrusive rocks have much coarser grain crystal development. The main texture classifications designate the relative sizes of the mineral grains, grouped as coarsely crystalline, finely crystalline, cryptocrystalline, or glassy; glassy is a non-crystalline texture. The strongest rocks are those possessing interlocking crystal grains; the weakest are those that do not have interlocking crystals. Unweathered igneous rocks are very strong and durable, and are usually quite satisfactory for most engineering needs, including concrete aggregates. Extrusive rocks, however, are subject to appreciably more variation in properties and should be evaluated with more scrutiny. Examples of some igneous rocks are given in Table 2-1.

Sedimentary Rocks

The sedimentary rocks are generally classified into two major sediment types, *chemical* and *clastic* (mechanical). Within the chemical group fit the sediments formed by materials that have been transported in solution and later precipitated. The clastic, or mechanical, sediments evolved from materials that were transported and deposited primarily by mechanical means. Examples of sedimentary rocks of both types are given in Table 2-2.

Many of the chemical sediments develop crystalline textures varying from microscopic in size (cryptocrystalline) to individual crystals readily visible to the eye. The texture of the clastic variety usually consists of fragments of a variety of shapes and sizes. The range may stretch from the various sandstones with different grain sizes and deposition patterns to the pyroclastic variety (e.g., tuff and volcanic ash) which resemble pieces of broken pottery.

Many of the engineering properties, particularly strength, of the clastic sediments are closely related to their state of consolidation. Most shales will tend to slake and disintegrate, and will subsequently revert to clay or silt, etc., if subjected to any prolonged exposure to atmospheric elements. On the other hand, the coarser grained rocks are usually more stable, and usually less compressible than shales and clay. In fact, in a cemented and confined state, the coarse-grained clastic rocks (e.g., sandstones) are virtually incompressible as far as most engineering requirements are concerned. Most well-cemented fault-free sandstone formations, for example, provide a good foundation support; that is, they provide good strength and permit minimum settlement. On the other hand, sandstones are perhaps the most porous of the clastic group.

TABLE 2-1
Some Common Igneous Rocks

Name	Usual		Color	Mineral Composition	
	Grain Texture	Classification		Essential	Accessory
Granite	Coarse	Intrusive	Light gray	Quartz, orthoclase	Biotite, muscovite hornblende, magnetite
Diorite	Coarse	Intrusive	Intermediate	Plagioclase, amphiboles	Hornblende, quartz
Gabbro	Coarse	Intrusive	Dark	Plagioclase, pyroxenes	Biotite, magnetite
Peridotite	Coarse	Intrusive	Dark	Pyroxenes, magnetite	Chromite
Syenite	Coarse	Intrusive	Light to medium	Orthoclase, plagioclase	Biotite, hornblende
Rhyolite	Fine	Extrusive	Light	Quartz, orthoclase	Plagioclase, mica
Andesite	Fine	Extrusive	Intermediate	Plagioclase, amphiboles	Biotite, magnetite
Basalt	Fine	Extrusive	Dark	Plagioclase, pyroxenes	Biotite, magnetite

TABLE 2-2
Examples of Sedimentary Rock

Name	Group Type	Grain Texture	Brief Description
Limestone	Chemical	Usually fine	The primary constituent is calcium carbonate ($CaCO_3$) in the form of calcite or aragonite. Usually relatively soft (Moh's hardness scale), it is readily soluble in acidic solutions (e.g., dilute hydrochloric acid solutions); organically or inorganically precipitated.
Dolomite	Chemical	Fine	The primary constituent is calcium magnesium carbonates [$(CaMg)(CO_3)_2$]. It closely resembles limestone, although less affected by, say, dilute hydrochloric acid solutions; organically or inorganically precipitated.
Salinastone	Chemical	Fine	This category includes several mineral types precipitated from sea water; anhydrite ($CaSO_4$); gypsum (hydrated calcium sulfate compound, $CaSO_4 \cdot 2H_2O$); chlorides ($NaCl$, $CaCl_2$). Most are fairly soluble in water; inorganically precipitated.
Conglomerate	Clastic	Coarse (over 2 mm size)	The primary constituents consist of particles, roundstones or sharpstones, which may vary in size from 2 mm to perhaps boulders over 200 mm in size.
Sandstone	Clastic	Intermediate (0.1–2 mm)	The primary constituents are pressure-cemented particles of sand; feldspar, hornblende, volcanic matter, and other minerals may also be present. Rather porous; sandrock mass is relatively easily crushed into smaller particles.
Shale	Clastic	Fine (0.001–0.1 mm)	The primary constituents are clay minerals and fine particles of silica; various oxides and colloidal and organic matter may also be present. *Cementation* shale is composed of particles bonded together primarily by cementation materials (e.g., silica and calcium carbonates). It is rather stable when exposed to limited weathering. *Compaction* shales are held together primarily by molecular attraction of their particles. This shale is generally less stable than the cementation type, but both types may slake or break down when exposed to prolonged atmospheric conditions.

Metamorphic Rocks

Metamorphism is a general term for a process whereby sedimentary rocks, and to a lesser extent igneous rocks, are altered via heat, pressure, and solutions into metamorphic rocks. Two categories of metamorphism have been identified, *contact* metamorphism and *dynamic* metamorphism. Contact metamorphism occurs primarily as a result of temperature increases, high hydrostatic pressures, and various solutions. It appears mostly about the boundaries of intrusive rocks, but may also occur at lower contact zones of lava flows. Dynamic metamorphism results primarily from the action of differential pressure, commonly related to major earth movements and deformations.

TABLE 2-3
Examples of Metamorphic Rocks

Name	Texture	Mineral Arrangement	Brief Description
Slate	Fine	Foliated	A metamorphized shale, composed mostly of quartz and secondary mica, is a dense rock characterized by well-developed tabular cleavage or splitability. It is widely used in "slate" roofs, floors, mantels, etc.
Schist	Medium to coarse	Foliated	Contains readily visible slaty cleavages. Mica and chlorite are common minerals, with feldspar in lesser amounts. It is rather weak and not widely used.
Gneiss	Medium to coarse	Poor	The rock is characterized by alternating bands of different colors and highly contorted shapes. Its common minerals are mica, feldspar, hornblende, and quartz.
Quartzites	Fine	Nonfoliated	It may be formed by both contact and dynamic metamorphism, or from cementation of silica. It is very strong and durable, with the principal mineral being quartz. It is an excellent crushed aggregate in concrete and in glass making. It is white, when pure, but may have pink, red, yellow, or gray tints, depending on its impurities (e.g., red from iron oxide presence).
Marble	Medium to coarse	Nonfoliated	It is formed by the recrystallization of limestone and dolomite. It is used as a source for lime and as material for building or decorative purposes (e.g., walls, columns, blocks, etc.).
Hornfels	Fine	Nonfoliated	Sometimes referred to as "trap" rock, hornfels is dense, tough, and strong contact rock formed from the baking and silicification of shales. It produces excellent crushed aggregate.

TABLE 2-4
Order of Relative Rock Hardness

Mineral									
Diamond	Corundum	Topaz	Quartz	Feldspar	Apatite	Fluorpar	Calcite	Gypsum	Talc

Hardest   Softest

Metamorphism may be resolved into four separate processes, although in nature their existence is generally concurrent: (1) *plastic deformation*, the permanent or nonelastic change in shape; (2) *granulation*, the crushing of rock; (3) *recrystallization*, the regrouping of various elements into new crystals; and (4) *metasomatism*, solution and precipitation of original minerals, replaced by or altered into other minerals.

Metamorphic rock may be grouped into one of two texture groups, *foliated* and *nonfoliated*. Foliation occurs during dynamic metamorphism as some rock mass undergoes elongation and shortening and orientation of the minerals into platy shapes, roughly comparable to pages of a book, etc. Some of the metamorphic rocks, including some born from dynamic metamorphism, depict a much less directional orientation and foliation. These are generally classed as nonfoliated. Examples of both types are given in Table 2-3.

Table 2-4 lists the ten minerals shown in the order of their relative hardness, from the hardest to the softest.

2-4 CLAYS

Produced primarily from the weathering of feldspathic rock, clays are fine-grained soils consisting of mostly hydrous aluminum silicates. They consist of particles which are, for the most part, less than 0.002 mm in size, although some of several classification systems regard as clay particles up to 0.005 mm in size. Hence one may group into the clay category two possible classes: (1) the deposits derived from various minerals, also referred to as *argillaceous* matter; and (2) the materials that fit into the range of particle sizes designated as clay. An example of the latter are colloids, or rock flour (e.g., most glacial clays) produced through abrasion of rock. In nature virtually all clay contains clay *minerals* as well as various contaminants.

Clays are common constituents in most foundation soils. Generally they are mixed with other forms of soils, such as silts, sands, gravels, or rock formations. That is, most foundation soil formations consist of a heterogeneous mixture of both cohesive (clays and silts) and cohesionless (sand, gravel, etc.) materials. In the final analysis, the evaluation of the foundation material must be viewed in terms of the total mixture. However, at this time we shall focus only on the various characteristics of clay.

Clay Minerals

With the aid of the electron microscope, differential analysis, and X-ray diffraction techniques the study of clay minerals has been enhanced significantly. Hence it is now known that clay minerals are essentially aluminum silicates and/or iron and magnesium, with some containing forms of alkaline materials as essential components.

The majority of clay minerals are insoluble in acids. They appear to have appreciable affinity for water, are elastic when wet, water retentive, and coherent when dry. Most clay minerals are crystalline, with sheetlike or layered structures of two varieties: *silica sheets* and *alumina sheets*.

Clay minerals are quite small, less than 2 μm,* and electrochemically very active. For example, minute clay particles carry similar electric charges which induce mutual repulsion. Neutralization of these charges, say through electrolytes, can bring about coagulation and subsequent precipitation of the floccules of clay. Furthermore, as the size of the particles decreases below 2 μm, the electrical charges on the particles increase with the decrease in size. Hence it would indeed be useful to the civil engineer to be able to manipulate the cation-exchange feature toward a desired goal when confronted with a flocculation situation.

The silica sheet is made of tetrahedrons, each tetrahedron being bounded by four triangular plane surfaces, with four equally spaced oxygen atoms at the vertices and a silicon atom within the interior, equally spaced from the oxygen atoms. The basic unit arrangement is shown in Fig. 2-3a. The symbolic equivalent of this unit is shown in Fig. 2-3b. The tetrahedrons are combined into hexagonal units, in a repetitious manner, to form a lattice of the mineral.

The alumina sheet has two-row units as shown in Fig. 2-4. One aluminum atom is at the center of the octahedron, with oxygen atoms of hydroxyl (OH) units at the vertices of alternative rows, respectively.

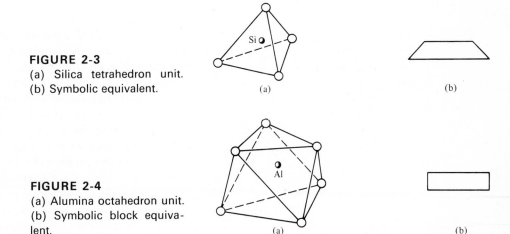

FIGURE 2-3
(a) Silica tetrahedron unit.
(b) Symbolic equivalent.

(a)   (b)

FIGURE 2-4
(a) Alumina octahedron unit.
(b) Symbolic block equivalent.

(a)   (b)

* 1 micron (μm) $= 1 \times 10^{-6}$ meters.

The clay minerals are divided into three main groups, with the lattice structures of the minerals serving as the basis for their classification. The groups are the *kaolinites*, the *montmorillonites*, and the *illites*.

Kaolinite. The kaolinite minerals are formed of units consisting of a single tetrahedral silica and a single octahedral alumina sheet. These units may repeat themselves indefinitely to form a *lattice* of the mineral. Figure 2-5 gives a symbolic arrangement of the kaolinite minerals. Their general chemical composition is expressed by the formula

$$(OH)_8 Al_4 Si_4 O_{10}$$

Kaolinite is the most abundant constituent of residual clay deposits, derived mostly as a by-product of the weathering of rock and/or certain clay minerals, and is commonly intermixed with illites in sedimentary clays. Kaolinites are very stable, possess a tight cohesive structure which resists the penetration of water into the lattices, and, generally, are not subject to expansion when saturated. Also, the coefficient of internal friction is somewhat higher than that of most other clay minerals.

Halloysites are minerals that belong to the kaolinite family. They possess a round or flattened tubelike shape. Some other members of the kaolinite groups are *nacrite* and *dickite*. The halloysites are similar to the kaolinites in the chemical composition, with halloysites being distinguished by one additional water molecule to the basic kaolinite unit. This is given by the formula

$$(OH)_8 Al_4 Si_4 O_{10} \cdot 4H_2 O$$

When wet, halloysite masses have a tendency to creep or flow horizontally. Thus they may be viewed as potentially unstable, less than desirable as materials for embankments, etc. On the other hand, both kaolinites and halloysites are common materials in the pottery industry.

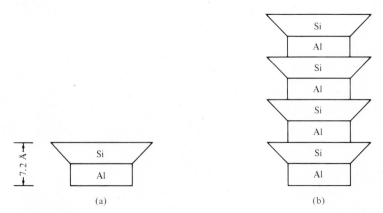

FIGURE 2-5
The kaolinite mineral. (a) Basic kaolinite unit. (b) Lattice of kaolinite mineral. (1 Angstrom $= 1 \times 10^{-6}$ meters).

Montmorillonites.　　The chemical composition of this group is expressed by the formula

$$(OH)_4Al_4Si_8O_{20} \cdot nH_2O$$

The montmorillonites are made up of sheetlike units comprising an alumina octahedral sheet between two silica tetrahedral sheets, as shown in Fig. 2-6. The bonding of these sheets is rather weak, resulting in a rather unstable mineral, especially when wet. In fact, montmorillonites display a significant affinity for water, with subsequent swelling and expansion. Conversely, upon drying a saturated montmorillonite, the result is appreciable shrinkage and cracking. In practical terms, such characteristics may be of significance to the engineer. For example, the expansion of the clay may mean lifting of slabs, excessive lateral thrusts on retaining structures, endangering the stability of slopes, etc. On the positive side, such expansion may be beneficial as described below.

　　Bentonite is part of the montmorillonite clay family, usually formed from the weathering of volcanic ash. It is noted for its expansive properties in the presence of water. As such, it was found to have beneficial uses as a general grout in preventing leakage from reservoirs, for plugging leaks in tunnel construction, and as a drilling mud in connection with soil borings and oil and gas wells. It prevents flocculation and facilitates the removal of the drill cuttings of the rotary drill.

Illites.　The illites are somewhat similar to montmorillonites in the structural units, but are different in their chemical composition. The chemical composition of illites is expressed by the formula

$$(OH)_4K_y(Al_4 \cdot Fe_4 \cdot Mg_4)(Si_{8-y} \cdot Al_y)O_{20}$$

where y varies from 1 to 1.5. The symbolic structure of illites is shown by Fig. 2-7.

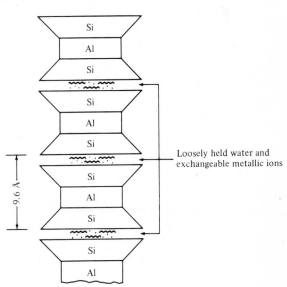

FIGURE 2-6
The montmorillonite minerals.

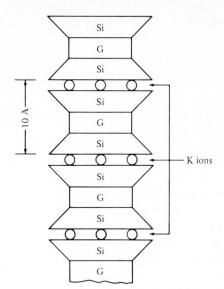

FIGURE 2-7
The illite clay mineral.

The basic structure of the illite unit consists of a gibbsite octahedral sheet between two silica tetrahedral sheets. Unlike montmorillonite particles, which are extremely small and have a great affinity for water, the illite particles will normally aggregate and thereby develop less affinity for water than montmorillonites. Correspondingly, their expansion properties are less. Also, their angle of internal friction is higher than that for montmorillonites.

Flocculation and Dispersion

The minute clay particles carry small electrical charges of the same sign, a condition that causes mutual repulsion of the particles. For example, clay particles in distilled water will not attract or collide; instead, the negative charge of the particles will cause interparticle repulsion. Hence subject to rather insignificant gravitational pull, these particles may remain in a state of colloidal suspension, or *dispersed*, for some time and vibrate back and forth in the solution (Brownian movement) with others settling, however slowly. Compounds used to maintain colloidal particles dispersed are known as *peptizers*. A number of organic acids may be used for this purpose.

If electrolytes or oppositely charged colloids are introduced in the "*colloidal*" solution, positive ions are attracted to the negatively charged particles, thereby causing coagulation into *flocs*, which may grow large enough to precipitate to the bottom rather rapidly under the influence of gravity. This process is referred to as *flocculation*. Salt water, for example, from oceans may cause flocculation of colloidal suspension carried by the streams discharging into the oceans; waters from limestone deposits may result in flocculation of colloidal suspension with which it comes in contact.

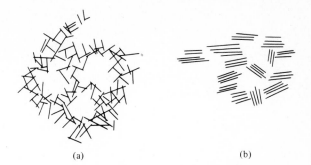

FIGURE 2-8
Structural arrangement of clay
sediments. (a) Flocculated.
(b) Dispersed.

(a)　　　　　　　(b)

The sediment structure may consist of an *edge-to-edge* arrangement, as shown in Fig. 2-8a, or of a *parallel, face-to-face* arrangement shown in Fig. 2-8b, or both. These soil structures are termed *flocculated* and *dispersed*, respectively. Honeycomb formations may be developed by flocs that tend to form a particle chain around voids which are larger than the flocs themselves. Figure 2-9 shows this diagrammatically.

Generally a flocculated clay has higher strength and is more permeable and less compressible than the same clay of the same porosity in a dispersed state. Furthermore,

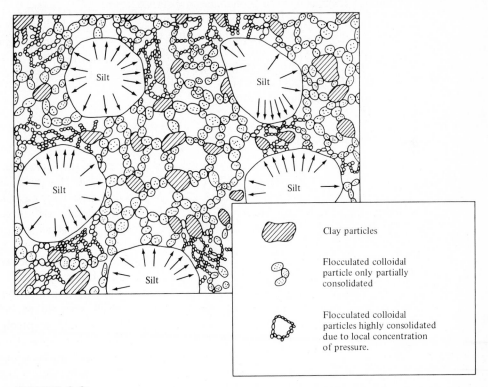

Clay particles

Flocculated colloidal
particle only partially
consolidated

Flocculated colloidal
particles highly consolidated
due to local concentration
of pressure.

FIGURE 2-9
Structure of undisturbed marine clay. (After A. Casagrande.)

mechanical remolding of flocculated clay results in a certain degree of dispersion as particles tend to develop a parallel array and, thus, alter some of these characteristics. For example, driven piles (to be discussed in Chapter 15) may penetrate with relative ease in soft, saturated clay due to remolding effects during driving. Later that same clay develops an appreciable increase in strength, and, correspondingly, the pile capacity is significantly improved. Similarly a given load may compress a remolded sample several times the amount the same load would have compressed the clay sample, at the same void ratio, in the undisturbed state.

Absorbed Water

We generally think of water as but a liquid composed of water molecules, H_2O. However, some of the molecules disassociate themselves from the compound into hydrogen ions (H^+) and hydroxyl ions (OH^-). Where water, pure or impure, comes in contact with a clay particle, the negative electrical charge by the mineral attracts the cations, including the hydrogen ions (H^+) disassociated from the water, to the surface of the mineral. Thus we have a case of *absorbed water*.

The absorbed water molecules are more intense near the clay particle, with the intensity decreasing with an increase in distance from the clay particles. This is shown qualitatively in Fig. 2-10. The combination of the negatively charged mineral surface and the positively charged space around the mineral forms what is known as the *electric double layer* or just the *double layer*. In addition to the attraction of water molecules to the mineral surface via the exchangeable ions, hydrogen bonding and van der Waals forces (i.e., neutral molecules attract) are additional causes believed to be affecting this phenomenon.

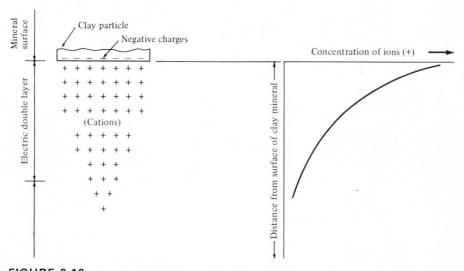

FIGURE 2-10
Intensity of cations with distance from surface to clay mineral.

While the attraction of the water molecules appears to be particularly strong immediately adjacent to the mineral surfaces, two adjacent clay particles experience repulsion exerted by the double layer of each of these particles, That is, each clay particle carries a net negative charge, thereby creating an electrical force that repels the particles, similar in nature to that of two like poles of a magnet (e.g., two negative or two positive poles) which are pushed toward each other.

The ions absorbed on the soil particles may exchange places with another ion within a diffuse double layer. For example, if we were to add lime to a wet clay mixture which contains sodium, the calcium ions will replace the sodium ions. This transformation is known as a *base exchange*, and the ions involved in the transformation are known as *exchangeable ions*. The nature of the transformation can have a significant effect on the properties and behavior of a clay mass. For example, the thickness of the absorbed water layer may be reduced as a result of an ion exchange, as described above, and thus the relative movement of adjacent particles will be reduced. In turn, this permits the soil to deform plastically without cracking.

Compressibility and Expansion

As the particles of clay are forced closer together, the soil is said to *consolidate*. Consolidation is directly related to the decrease in thickness of the diffuse double layer between the particles as the water content, and perhaps air, is reduced. The change in the clay particle size is relatively insignificant. The causes for such a volume reduction are commonly attributed to external loads, but evaporation and/or changes in the diffuse double layer due to ion transformation or base exchange may be additional causes.

Not all clay masses deform equally under comparable external influences, since wide variations exist in particle properties and structural arrangements for various clay types. Among the factors influencing compressibility are: (1) the amount of water absorbed and retained by different minerals; (2) the ion exchange capacity of different clay minerals (e.g., montmorillonites have a high exchange capacity, kaolinites have a low one); (3) the structural arrangement and the clay particle orientation (e.g., flocculated or dispersed); and (4) the type and duration of these external influences.

Swelling is the rebound of a clay soil, and thus, in a sense, it is the opposite of consolidation. Like consolidation, however, it is related to a number of clay properties: (1) the affinity of the clay mineral for water; (2) the base exchange behavior and electrical repulsion; (3) the expansion of entrapped air within the mass; and (4) the mineral type. Both phenomena are indeed complex and beyond the scope of this text. For those interested in more detail regarding clay properties, a good source may be reference (13) in the Bibliography.

2-5 ROCK WEATHERING

The process of rock disintegration and decay resulting from exposure to and the influence of the atmospheric agents is known as *rock weathering* (6). This is further divided into two groups, *mechanical weathering* and *chemical weathering*. Although the end result for both processes is a breakdown of the massive rocks into smaller sizes, the primary causes for this breakup are different. The weathering agents associated with the respective subdivision are as follows:

Mechanical Weathering	*Chemical Weathering*
1. Temperature changes	1. Oxidation
2. Freezing and thawing	2. Carbonation
3. Splitting action of plant roots	3. Hydration
4. Abrasive movements	4. Vegetation

The essence of mechanical weathering is the disintegration of the rock, while that of chemical weathering is the decomposition of the rock. A large variation in *temperature* may cause a rock to disintegrate either because of fatigue due to cyclic stresses of compression and tension, or because of thermal expansions of minerals within the rock, or both. *Freezing* and *thawing* may widen the crack in the rock by the expansion of moisture in the crack, or may create cracks in a sound rock if the pores (voids) of the rock are filled with moisture. *Roots* of vegetation may penetrate rock pores or existing cracks to further increase their sizes. *Abrasive* action due to mass movements by means of wind, water, or ice may cause an erosion and disintegration effect on the rock.

Oxidation is an agent in the decomposition process whereby oxygen ions combine with some minerals in the rock which subsequently decomposes in a manner similar to the rusting of steel. *Carbonation* is a form of decomposition where carbon dioxide and water form carbonic acid which decomposes many minerals containing iron, sodium, calcium, etc. Limestone, for example, is readily dissolved in this manner. *Hydration* is the process of the chemical addition of water to the minerals which subsequently convert into new minerals. For example, the carbonation and hydration of the mineral feldspar in granite may produce a clay mineral, kaolinite. Decaying *vegetation* may be a factor in the production of organic acids, carbon dioxide, and oxygen which, when mixed with water penetrating through the rock, may extract certain chemical elements from the rocks. Silica, for example, may be extracted from the silicate minerals in this manner.

2-6 WATER-TRANSPORTED SOILS

Each of us has undoubtedly observed on many occasions soil being carried away by water, or the erosion that water creates in the earth in this process. We might have also observed that swift running water, such as after a heavy rain, carries more soil and creates more and larger channels or gullies than does a slower moving stream. Generally we have observed that water movement erodes the hills and deposits into the valleys, and that the higher the velocity, the greater the amount of transported soil.

The transport of soil may be in the form of suspended particles in the flowing water, or larger particles which are rolled and/or pushed over the bottom of the stream.

The swift water carries with it sediment which ranges in size from sand grains to coarse gravel, perhaps even boulders. The scrubbing of this material over the stream bed causes abrasion of sharp edges and tends to reduce the particles themselves, but it also wears on the floor of the stream, including any bedrock. Furthermore much of the bedrock, particularly that which has laminations (shale, for example), may be loosened and dislodged by the force of the stream to become part of the transported mass.

With decreasing velocity, the coarse particles are the first to be deposited. The finer particles remain in suspension until the velocity decreases further and are therefore deposited at a much farther point downstream. This may be illustrated by the effects of a typical river which discharges into a large body of water such as a sea or the ocean. For example, in the high hills or the mountains where the slope of a stream is steep, the channels of erosion are correspondingly pronounced, generally deep and narrow. As the gradient of the stream decreases (in valleys, deltas, etc.) the larger size particles are left at the mouth of the valley, and the finer particles are carried further downstream. Figure 2-11 gives a schematic illustration of this process.

River deltas are formed in this manner. They are produced by the deposition of alluvium where a stream enters a body of quiet water and where the decrease in the velocity permits the stream to drop its load of transported soil. Most deltas are very flat, with the highest point being less than 20 m above the sea level. Because of the relatively flat floor at such points, rivers develop several discharge channels called tributaries, which are generally only used during flood periods. Such floods usually contribute to the gradual widening of the valley and the continuous widening of the flood plane.

The mode of transportation of soil by water may be generalized or divided into three parts: (1) that due to the rolling along the river bed of the heaviest particle; (2) the bumping and rolling of a smaller size particle; and (3) the suspension of the finest particles of soil carried by the river.

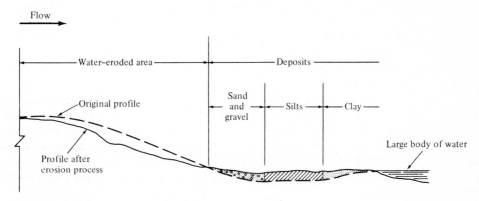

FIGURE 2-11
Process of water erosion, transportation, and deposits.

The phenomenon of erosion and subsequent deposition of a transported soil may be of some special interest to a civil engineer in some specific instances (2). For example, the engineer may be concerned with the removal of such deposition (dredging) if the river is used for navigational purposes. Or the overall effects such erosions may have on the downstream portion of a dam, around a dam, or perhaps through a dam may be of interest. Also, the engineer may be interested in the stability of a bank, or in the overall outcome if a structure were to be placed on such deposits. Mexico City, for example, poses rather difficult foundation conditions because it is founded on an unusually thick stratum of transported soil.

The Mississippi delta in the Gulf of Mexico is one of the largest in the world. Some other particularly large deltas are those of the Nile in Egypt, the Po in Italy, the Rhône in France, the Danube near the Black Sea, and the Ganges in India.

Many of the great deltas encompass rather vast exposed surfaces. For example, in round figures, the surface area of the Ganges delta is approximately 130,000 km² (about 50,000 mi²), that of the Nile is about 24,000 km² (close to 9200 mi²). In fact, the total surface area from the largest few deltas in the world approaches nearly 1 million km² (400,000 mi²). Likewise the amount of sediment associated with these deltas is indeed enormous. For example, the Mississippi delta averages about 2 million tons per day, all predominantly fine-grained soil (e.g., clay, silt, and fine sand).

The layers of the deposits are not uniform. In fact, due to a continuous change of stream flow and various changes in sea levels (subsidence or rise of the sea level) these deposits frequently mix with marine sediments to create various complex mixtures and formations.

The typical characteristics of the fine-grained deposits (clays, silts, and fine sands) generally found in the deltas include the following: (1) their bearing capacity is generally very low; (2) this stratum is rather compressible and thereby permits a high degree of settlement of a structure; (3) it is generally good agricultural soil, poor bearing capacity, particularly when wet (5, 12).

2-7 WIND-TRANSPORTED SOILS

The transportation of soil by wind is an important geological process which produces sedimentary deposits of widespread occurrence. As in the case of water, wind will erode, transport, and deposit fine-grained soils. Also, the finer particles are carried for longer distances, while the coarser and heavier particles will be the first to settle. Unlike water, wind will generally carry the sediment from lower to higher ground.

The transportation of soil by winds occurs predominantly in windy, dry, sandy, and silty areas, generally unprotected by vegetation or natural obstacles. Prior to becoming airborne, the sand and silt exert an abrasive effect on other soil or rocks in their path, thereby causing further erosion and sculpturing of surfaces. Deposits are the end product of this process. Figure 2-12 is a graphical representation of this behavior.

Portions of the lower elevation are generally eroded by a two-pronged process: (1) the loose, fine-grained soil may be picked up and transported by the wind to a

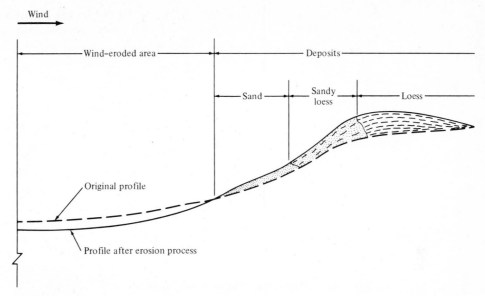

FIGURE 2-12
Process of wind erosion, transportation, and deposits.

higher level; and (2) some of the exposed bedrock may be eroded by the abrasive action of the soil being transported.

The first particles to be deposited are the coarser materials, perhaps a mixture of sand and small gravel. The next to be dropped is the small-grain material; the last to be deposited is the very fine-grain soil such as fine silt and clay, generally categorized as *loess*. Generally loess is of low density, highly compressible, and of poor bearing capacity, particularly when wet (5, 12).

Large sand dunes are formed from wind-transported soil deposits. They are rather common occurrences in deserts of Africa, Asia, and the United States; they may be found in a number of states, including Texas, the Dakotas, Wyoming, Kansas, Colorado, and California. While many of the dunes are formed in dry areas, such as the deserts, appreciable erosion and subsequent transportation of the soil could take place across the temporarily dry beds of streams or deltas. On the other hand, the erosive effects of the wind are frequently canceled out by the washing-in of new material by water.

The effects of erosion, transportation, and final deposition of wind-transported soil can be quite devastating. Roads, fields, forests, buildings, and villages have been known to be inundated in the process. Furthermore, under particularly dry conditions much of the fertile soil from ground surfaces can be removed by the wind, leaving behind a barren and unproductive land. For example, during the middle 1930s, as a result of extreme drought, repeated crop failures, and continued pulverization by farm implements, the soil in the western Great Plains became loose and powdery over wide areas and was subsequently picked up by the prevailing winds and carried away, leaving behind a major economic problem to the region. Frequently erosion of soil could be checked and/or reasonably well controlled by various forms of vegetation, such as trees, plants, and a variety of ground cover.

2-8 GLACIERS

Glaciers are formed by the compaction and recrystallization of snow; that is, if a region receives more snow in the winter than melts in the summer, the weight of the accumulated snow over the years causes melting and recrystallization of the snow into ice.

Glaciers cover approximately 10 percent of the earth's surface, with almost all concentrated in Greenland and Antarctica. Furthermore it is estimated that there is enough ice in the glaciers to raise the sea level nearly 60 m (about 200 ft) if the ice were to melt. The thickness of some of the Greenland ice sheet was measured to be over 3300 m (almost 11,000 ft), while some of the ice sheets in Antarctica are in excess of 2500 m.

As the thickness of the snow and ice continues to increase, the pressure, as a result of the accumulated weight, also increases. Once this pressure exceeds the strength of the ice, the mass becomes rather plastic and lateral flow begins. Depending on a number of factors, including the temperature of the ice, the slope of the land surface, etc., the thickness of the ice mass varies rather widely. Masses of thicknesses of only 20 to 25 m have been known to move in this manner.

As a general rule, glaciers move extremely slowly, perhaps a few centimeters per day being typical, but velocities of up to 50 m per day have been recorded. Furthermore the movement is greater near the center than near the edge of the glacier.

The glacial movement generally deforms and scours the surface and the bedrock over which it passes. In the process it carries with it soil varying in size from fine grains to huge boulders. When the glacier melts, all of the material is deposited to form what is generally referred to as *till*. At its terminus a melting glacier drops its load in the form of ridges, commonly referred to as *terminal moraine*. Of course much of this material is eventually reworked by the action of water and wind, as described in the preceding two sections.

It is estimated that there were at least four major periods of time in the earth's history when ice sheets covered large portions of the land. These are commonly referred to as the ice ages, and they may last several million years each. The earliest ice age came during the Precambrian time, which had its beginning more than 600 million years ago. The next ice age came during the Cambrian period, about 600 million years ago. The third occurred during the Carboniferous-Permian era between 350 and 230 million years ago. The most recent ice age happened during the Pleistocene epoch, which began about 1.75 million years ago and lasted until about 10,000 years ago. However, it is estimated that during the Pleistocene ice age between 6 and 20 glaciations (formation of a glacier) took place, with the typical glaciation lasting between 40,000 and 60,000 years. The last major ice advance in North America appears to have culminated about 20,000 years ago, when the last ice sheet began to retreat. It is estimated that the ice-covered area of the earth during the early ice age or the Pleistocene epoch was several times its present size. During this time so much water was turned into ice that the level of the oceans dropped, perhaps in excess of 100 m.

When the ice on the glaciers melted, some of the water flowed back into the oceans, filling them to the present level. However, not all of the water flowed into the ocean. Some of the water filled the depressions which were dug out by the glaciers as they receded. Lakes Superior, Erie, Ontario, Michigan, and Huron were formed in this manner as a result of the scouring glacier movements caused 250,000 years ago, and they were filled less than 20,000 years ago as a result of ice melting.

The effects of glaciers were not only in carving mountains, rivers, or lakes, but also in their final deposits. For example, the soil which supported the enormous weight of the ice sheets was subjected to great pressures and preconsolidation. Preconsolidation information is quite useful in estimating the bearing capacity and the potential settlements of structures. Similarly, the knowledge of the type of deposit (e.g., glacial till) which resulted from glacial movements is sometimes a useful piece of information when a building site is evaluated (1, 19, 21, 28, 30).

2-9 RESIDUAL SOILS

Disintegrated material above the rock crust which has not been transported by water, wind, and glacial movements is generally classified as *residual soil*. This soil is the product of rock weathering, but its characteristics differ significantly from those of transported soils. Furthermore the degree of disintegration may vary appreciably over the thickness of its stratum. Its profile generally shows a gradual transition from soil into rock, rather than a distinct line of separation of soil and rock. Also, because of its partially disintegrated condition, it offers relatively weak resistance to further weathering and other disintegrating factors such as abrasive action, dynamic forces, or other destructive effects. In this regard it is difficult to establish the actual size of the grains since they are subject to the process of further disintegration.

Shales convert to residual soil if the shale disintegrates back into clay or silt. In this respect the stability of shale becomes of particular concern as a foundation material. That is, an already weathered and perhaps unstable shale becomes even more weathered and therefore less stable when exposed to the atmospheric conditions. Typical consequences from such exposure is delamination, expansion, and overall decomposition of the shale layer, particularly near the surface of the layer where exposure is generally most severe. In turn, shale expansion may have adverse effects on foundations, particularly lightly loaded footings and/or slabs. If the use of such sites could not be avoided, the usual corrective approach may be to either excavate beyond the scope of the foundation and back-fill with an inert material such as sand and gravel, or minimize the exposure time to the atmospheric conditions, or both.

Limestone is particularly vulnerable to the effects of solvents and weathering, forming a rather irregular and highly plastic residual stratum. Some of the disintegration of the limestone takes the form of vertical erosion, forming what are known as *sinkholes* and *caves*. In some instances the vertical erosion and general subsidence of the overlying strata could occur in a surprisingly short period of time. Although the usual occurrence of sinkholes is not uniform or continuous over a site, sometimes a series of sinkholes can be detected. Hence a detailed subsurface soil investigation and careful scrutiny are indeed advisable whenever limestone formations are detected or even suspected. A solution to this problem is sometimes the bridging of the site with a properly designed slab so as to "cap" the site in a floating manner.

Sandstone, which ordinarily is composed of predominantly cohesionless particles, will form a weathered surface which is generally nonplastic and rather stable. Sandstone mixed with any significant amount of clay particles may be rather unstable and with time may convert to a rather clayey and plastic material. This may then be poorly resistant to an accelerated weathering process and erosion. Soil stabilization through chemical or cement grouting may sometimes be used to stabilize the stratum and prevent further disintegration.

2-10 SEDIMENTARY DEPOSIT

Some of the products of weathering are subject to erosion and possible transportation via water, wind, glacial movements, etc.; the transported mass will eventually be deposited. The material moved and deposited in this manner is referred to as *sedimentary deposit*.

Water and air are the transportation agents responsible for most of the sedimentary deposits. Glacier movements, earthquakes, and gravitational forces are generally less significant in this regard. As the forces of these transporting agents lessen, the heavier particles of the soil will be released, and the finest will be carried to the point where their velocity is checked. In a sense they are sorted with remarkable efficiency into layers of reasonable homogeneity, according to particle weight, size, and shape. Although a heavier particle may be the first to reach the bottom of a river, for example, it may or may not remain in that spot permanently. Depending on the flow of the river, it may again be moved further down along the river and eventually deposited in a new place. Hence although the bottom of a river and the stable bed of the river generally consists of coarser material, with the finer being in suspension in the moving stream, some lack of homogeneity in the stratification may be expected. Nevertheless, for the most part a significant degree of uniformity and stratification is the result of water and wind transportation.

The type and degree of sedimentation is indeed a factor relevant in many of the civil engineering designs. For example, a stream discharging into a reservoir may eventually "silt up" on the upstream face of the dam. That is, as the velocity of the stream decreases once it hits the reservoir water, the transported load gets deposited to form a sediment on the reservoir bottom. With a continuous increase in the silting layer, the amount of water storage by the reservoir will obviously decrease, thereby significantly affecting or limiting the useful life of the reservoir. Dredging of these deposits is usually a viable but also a costly solution to the problem.

Sedimentation deposits in harbors are a problem sometimes faced by marine engineers. Although tides are usually the cause, river and/or underground currents may also be responsible for the silting of the harbor. Regardless of the cause, however, the silt buildup may impose a serious limitation to the proper function and usefulness of the harbor as a navigational tool. Again, dredging may be a feasible solution to the problem.

While sedimentary deposits related to marine structures may be very important, many wind- or water-transported sedimentary deposits may be equally important and perhaps more frequently encountered. For example, many sedimentary deposits were formed by rivers which changed their course or by lakes which are no longer in existence. Now suppose that some of these become building sites. It is readily apparent that the characteristics of such deposits are also very important. Generally such deposits consist of fine-grained soil such as clay or silt, which may be compressible and of a low bearing capacity. In that case the amount of settlement, the type of foundation, and the overall feasibility of the site become of paramount importance in the overall planning and design.

PROBLEMS

2-1.   What is the geological cycle? Give a brief description.

2-2.   Differentiate between intrusive and extrusive rocks.

2-3.   Name several factors that are directly tied to the disintegration of rocks.

2-4.   Name and describe briefly the origin of three main groups of rocks, and give three examples of each.

2-5.   Describe metamorphism.

2-6.   Name and describe briefly the three main clay minerals.

2-7.   Name a beneficial feature of bentonite.

2-8.   Differentiate between flocculation and dispersion.

2-9.   How does clay absorb water?

2-10.   What constitutes consolidation of clay soil?

2-11.   How does mechanical weathering differ from chemical weathering?

2-12.   Define: (a) Oxidation. (b) Carbonation.

2-13.   Describe briefly how soil is transported and deposited by water.

2-14.   How does the transportation by wind resemble that by water? How does it differ?

2-15.   What is the general range of speeds of glaciers?

2-16.   What is residual soil? What is sedimentary deposit?

2-17.   Give two examples where sedimentary deposits are of concern to engineers.

2-18.   How is sandstone formed? How is shale formed? Discuss their relative merit, stability, and overall reliability as foundation materials.

2-19.   How are river deltas formed? Name and give the size of a few of the largest deltas in the world.

2-20.   What is loess? How is it formed? How does it rate in quality as a foundation base? Explain.

2-21.   How does the formation and melting of glaciers affect the water levels in oceans and lakes? Explain.

2-22.   How might a building site be preconsolidated during the geological cycle? during a glacier era?

BIBLIOGRAPHY

1. Adams, J. I., "Tests on Glacial Till," *14th Can. Soil Mech. Conf.*, Tech. Memo. 69, Assoc. Comm. Geotech. Res., N.R.C., Ottawa, Canada, 1960.
2. Boswell, P. G. H., *Muddy Sediments*, W. Heffer and Sons, Cambridge, England, 1961.
3. Branner, J. C., "Geology in Its Relations to Topography," *Trans. ASCE*, vol. 39, 1898.
4. Casagrande, A., "The Structure of Clay and Its Importance in Foundation Engineering," *J. Boston Soc. Civ. Eng.*, 1932.
5. Clevenger, W. A., "Experiences with Loess as Foundation Material," *Trans. ASCE*, vol. 123, 1938.
6. Dearman, W. R., "Weathering Classification in the Characterization of Rock for Engineering Purposes in British Practice," *Bull. IAEG*, Krefeld, West Germany, no. 9, 1974.
7. Dunbar, C. O., *Historical Geology*, Wiley, New York, 1960.
8. Eckel, E. C., "Engineering Geology and Mineral Resources of the T.V.A. Region," *Geol. Bull.* no. 1, Tennessee Valley Authority, Knoxville, Tenn., 1934.
9. Flint, R. F., *Glacial Geology and the Pleistocene Epoch*, Wiley, New York, 1947.
10. Flint, R. F., and B. J. Skinner, *Physical Geology*, Wiley, New York, 1974.
11. Fookes, P. G., "Geotechnical Mapping of Soils and Sedimentary Rocks for Engineering Puposes," *Geotechnique*, vol. 19, 1969.
12. Gibbs, H. J., and W. Y. Holland, "Petrographic and Engineering Properties of Loess," *Eng. Monograph* no. 28, U.S. Bureau of Reclamation, Denver, Colo., 1960.
13. Grim, R. E., *Clay Mineralogy*, McGraw-Hill, New York, 1953.
14. Holmes, A., *Physical Geology*, 2nd ed., T. Nelson and Sons, London, England, 1965.
15. Jennings, R. A. J., "The Problem Below," *Quart. J. Eng. Geol.*, London, England, vol. 9, no. 2, 1976.
16. Legget, R. F., and M. W. Bartley, "An Engineering Study of Glacial Deposits at Steep Rock Lake, Ontario, Canada," *Econ. Geol.*, vol. 48, no. 7, Nov. 1953.
17. Legget, R. F., *Geology and Engineering*, 2nd ed., McGraw-Hill, New York, 1962.
18. Legget, R. F., *Cities and Geology*, McGraw-Hill, New York, 1973.
19. Leggett, R. F., "Glacial Till: An Inter-Disciplinary Study," *R. Soc. Can. Spec. Publ.* no. 12, Ottawa, Canada, 1976.
20. Legget, R. F., "Geology and Geotechnical Engineering," *ASCE J. Geotech. Eng. Div.*, Mar. 1979.
21. Linell, K. A., and H. F. Shea, "Strength and Deformation of Various Glacial Tills in New England," *ASCE Res. Conf. Shear Strength of Cohesive Soils*, Boulder, Colo., 1960.
22. MacClintock, P., "Glacial Geology of the St. Lawrence Seaway and Power Project," pamphlet from New York State Museum and Science Service, Albany, N.Y., 1958.
23. MacClintock, P., and D. P. Stewart, "Pleistocene Geology of the St. Lawrence Lowland," *Bull.* 394, New York State Museum and Science Service, Albany, N.Y., 1965.
24. McQuillin, R., and D. A. Ardus, *Exploring the Geology of Shelf Seas*, Graham and Trotman, London, England, 1977.
25. Moum, J., and I. T. Rosenqvist, "On the Weathering of Young Marine Clay," *Proc. 4th Int. Conf. Soil Mech. Found. Eng.*, vol. 1, 1957.
26. Moye, D. G., "Engineering Geology of the Snowy Mountain Scheme," *J. Inst. Eng.*, Sydney, Australia, vol. 27, 1955.
27. Newberry, J., "Engineering Geology in the Investigation and Construction of the Batang Padang Hydroelectric Scheme, MI Malaysia," *Quar. J. Eng. Geol.*, vol. 3, 1970.
28. Skempton, A. W., "A Study of the Geotechnical Properties of Some Post-Glacial Clays," *Geotechnique*, vol. 1, 1948.

29.   Skempton, A. W., "Soil Mechanics in Relation to Geology," *York. Geol. Soc.*, vol. 29, 1953.

30.   Skempton, A. W., and D. J. Henkel, "The Post-Glacial Clays of the Thames Estuary at Tilbury and Shellhaven," *3rd Int. Conf. Soil Mech. Found. Eng.*, Zurich, Switzerland, vol. 1, 1953.

31.   Sowers, G. F., "Engineering Properties of Residual Soils Derived from Igneous and Metamorphic Rocks," *2nd PanAm Conf. Soil Mech. Found. Eng.*, Brazil, vol. 1, 1953.

32.   Upson, J. E., E. B. Leopold, and M. Rubin, "Postglacial Change of Sealevel in New Haven Harbour, Connecticut," *Am. J. Sci.*, vol. 262, no. 1, Jan. 1964.

33.   White, G. W., "Engineering Implications of Stratigraphy of Glacial Deposits," *24th Int. Geol. Congr.*, Montreal, Canada, sec. 13, 1972.

34.   Wilson, A. E., "Geology of the Ottawa–St. Lawrence Lowland, Ontario and Quebec," *Memoir 241*, Geol. Surv. of Can., Ottawa.

3

Subsurface Exploration

3-1 INTRODUCTION

An adequate subsurface investigation should be conducted for the purpose of determining the foundation conditions for a proposed structure prior to its design. A well-planned exploration program may provide:

1. Information regarding the feasibility of the project as influenced by the subsurface conditions
2. Data for establishing design criteria and parameters
3. Information to formulate a basis for estimating the effects subsurface conditions may have on the proposed and/or nearby facilities

This chapter introduces some of the methodologies and techniques employed in the more routine subsurface exploration efforts.

Generally subsurface exploration provides a reasonable depiction of the stratigraphy (soil profile) and physical characteristics of the soil strata under the proposed building site. More specifically, items obtained from a typical subsurface investigation usually include:

1. The nature of the underlying material
2. The extent (e.g., thickness, width, and length) of the strata
3. Information regarding ground water (e.g., level, seasonal fluctuation, contaminants)
4. Information regarding depth and nature of bedrock, if encountered
5. Samples for visual identification and laboratory tests
6. The relative stiffness characteristics of the soil strata

When establishing the extent of the subsurface investigations (including the degree of sampling and laboratory analysis), the soil engineer usually takes into account the following:

1. The type, size, and function of the proposed structure
2. The magnitude and effects of the loads transmitted to the soil stratum by the structure

3. The effects the soil reaction (e.g., settlement) may have on the structure
4. The effect the proposed structure may have on the adjoining structures or facilities

Generally the soil engineer determines the spacing of the borings or pits, the type of exploration, the type of samples, and the frequency and/or degree of sampling. Furthermore, it is up to the engineer to designate the extent of laboratory testing considered necessary for the determination of the properties of the soil strata to the degree that these properties affect the design and function of the proposed structure (1, 28, 29, 67).

It is frequently advisable to run preliminary explorations prior to more detailed evaluations. Relatively speaking, such an exploration is a rather inexpensive and expedient method for early or preliminary design estimates. It is also a rather sound basis on which to develop the more detailed exploration program. Depending on these early findings and various project needs, the preliminary subsurface evaluation may be all that is necessary for the project. However, most frequently a detailed exploration is necessary for a more reliable evaluation of the site conditions.

3-2 SCOPE OF EXPLORATION PROGRAM

There are no hard and fast rules regarding the type and extent of the soil exploration program (22, 52, 65). For example, it may be apparent even to one not versed in geotechnical engineering details that the scope of exploration for a parking garage whose column loads may exceed 800 tons is likely to be significantly different from that for, say, an ordinary lightweight office building of one or two stories; or that the scope of exploration for a large dam is likely to be quite different from, say, that for a road project. Hence the scope of the program should be viewed relative to:

1. The proposed "building" details (e.g., type, weight, size, function, type of construction)
2. Specified or anticipated needs (e.g., design data and parameters, settlement restrictions, construction and/or site limitations)
3. Cost (e.g., cost of exploration compared to that of the total project, anticipation of abandonment, and/or subsequent evaluation of alternate building sites)
4. Available information (e.g., previous boring data, performance history and design parameters of adjoining buildings, and other general or specific information of the site, such as type of soil, water table, depth of rock, existence of coal mines)
5. Nature of soil involved

The character and extent of the exploration program must be sufficient to provide information of the nature and dimensions of the soil strata under a given site. The procedure usually encompasses conducting soil borings or test pits and extracting samples for visual and laboratory analysis. This is generally complemented, to various degrees, by a site reconnaissance and various field in situ tests. Although the scope of the program is usually established or formulated prior to the start of "drilling" and sampling, it is common and advisable to evaluate the findings as the work progresses, and to make adjustments in the program as warranted. For example, the boring depth, the type and depth of samples, the number of borings, etc., may be adjusted as deemed necessary. On the other hand this should be done under careful scrutiny, and in conjunction with and with approval of other responsible parties (e.g., structural engineer, architect, owner) if these changes reflect an increase in cost.

Depth of Exploration

The exploration should extend through the unsuitable material and to a stratum where the material in that stratum or below is "stable" against potential damaging deformation (e.g., settlement). A rule of thumb sometimes used for this purpose is to extend the boring to a depth where the additional load resulting from the proposed building is less than 10 percent of the average load of the structure, or less than 5 percent of the effective stress in the soil at that depth (1). Chapter 8 provides some commonly used methods for determining this stress. Where piles are anticipated, the depth of exploration is to at least the anticipated pile tip; the author prefers a 20 to 30 percent increase in depth beyond the anticipated tip of the pile.

Artesian pressures and artesian flow may be a source of problems for low-lying areas or for deep excavations. This is sometimes the case in connection with sewage-treatment plants, since the sites are normally at low elevations, the water table is usually high (shallow depth), and some plant components are rather deep (e.g., pits, storage tanks). Dewatering construction details (e.g., shoring, sheet-pile installations) and design considerations (e.g., hydrostatic pressures, uplift potential) demand fairly detailed information regarding the presence, depth, and magnitude of aquifers (water-bearing stratum) as well as pressure data and the type and character of the soil below the excavation. Heave, or boiling (see Section 7-3), is closely tied to these conditions. Again, as a rule of thumb, the depth of exploration is usually a minimum of 1.5 times the depth of excavation.

Rock exploration may be necessary if the foundation rests directly on rock, if excavation of the rock is necessary during construction (i.e., the cost for rock excavation may be many times that for ordinary soil), if the rock formation may be subject to weathering effects during construction, or if irregularities are likely (e.g., flaws or fractures, variation in characteristics and formations, elevations, variations, etc.). The sampling is commonly done via a core drill (discussed later in this chapter) for a minimum depth of 3 m. The depth may vary, however, and greater depths are common if much variation is detected in the rock formation within the site during the coring operations, or if voids (e.g., coal-mine voids, limestone sinkholes) may be present. (A giant limestone sinkhole about 200 m in diameter and 50 m deep swallowed homes, vehicles, etc., in Winter Park, Fla., in May 1981.) In such cases the strength and soundness of the strata above voids may be of paramount importance in the assessment of the ability of the rock formation to bridge over these voids and support the superstructure.

Spacing of Borings

It is normally difficult to determine the spacing and number of borings prior to the commencement of the drilling work because so much of such planning is tied to the underlying soil conditions, which are unknown at the time of planning. Hence it is common practice to proceed with a rather skimpy preliminary investigation, and then follow up with a more structured and better planned effort.

The preliminary borings usually lack detailed sampling. Instead, drill cuttings or disturbed samples (see Section 3-5) and water table information usually suffice. The followup borings are planned as complements to the preliminary ones, but

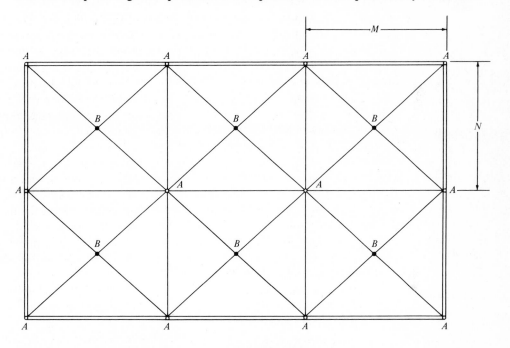

Column or Wall Loads	Depth [m (ft)]		Spacing [m (ft)]	
	Borings A	Borings B	M	N
Light	3–6 (10–20)	6–7.5 (20–25)	30 (100)	30 (100)
Medium	6–7.5 (20–25)	9–12 (30–40)	25–30 (80–100)	25–30 (80–100)
Heavy	9–12 (30–40)	15–25 (50–80)	15–25 (50–80)	15–25 (50–80)

FIGURE 3-1
General guidelines for boring layout and sampling used by the author

include specific requirements regarding the type, method, depth, and degree (or frequency) of sampling. For relatively light structures the preliminary phase may be sufficient if the soil strata are good and appear uniform throughout the site. Frequently a comparison of the design data and performance history of an adjoining structure, if one is available, becomes a useful tool for establishing the need for a more detailed investigation effort. On the other hand, for heavy and/or important structures, and for cases where the stratification information is doubtful or inconclusive, there is little choice but to proceed with the more comprehensive survey. This should be done with due care to extract only samples and perform only tests as required. For example, it is fruitless, if not irresponsible, to extract samples and perform tests which are not relevant to the design needs, or to perform tests if it becomes convincingly clear that the site is totally unsuitable for the intended purpose.

As is apparent by now, there are many factors related to the formulation of such a program. However, after a careful weighing of all the factors, the final decision on the part of the soil engineer is a rather empirical one, usually based on judgment and experience. Figure 3-1 shows a grid pattern for a typical boring layout used by the author on numerous projects. This particular layout assumes that the shape of the building is a rectangle, but the basic pattern could be improvised for nonrectangular buildings by merely adding grid blocks of M by N dimensions, as required by the shape of the building. This arrangement is rather convenient for plotting the stratification or profile of a series of borings in various directions (vertically, horizontally, and/or diagonally). It facilitates evaluating the soil profile from a number of angles with perhaps a minimum of distortion of data. Again, it is merely a guide, intended to give the student an idea of some commonly used values for boring depth, spacing, and sampling for some rather general groups of structures. It depicts a wide range of values, provides only general ranges, and is very approximate in nature. In this regard, one cannot overemphasize the importance of judgment in the overall program. Seldom are two sites identical, and even more seldom are two programs found to be completely identical in scope or needs.

3-3 SITE RECONNAISSANCE

Preliminary exploration is perhaps the most substantive basis for formulating the type and extent of a more detailed exploration and testing program. In fact, such "early" information may serve as the deciding factor for the feasibility of the site, or its rejection. For example, a preliminary subsurface exploration program conducted by the author, consisting of ten borings spread over several acres, revealed coal mines (voids) under the building site for a proposed new high school in a western Pennsylvania community. After some deliberation, and after reflecting on the preliminary findings, the local school board authorized a more detailed investigation and directed the design of a foundation as required to place the school over these mines. Obviously, in the judgment of the school board the location of the school was important enough to warrant the additional expenditures not only for the much more detailed surface exploration, but also for the anticipated increase in design and foundation costs.

Sometimes the preliminary subsurface investigation may serve as the catalyst for a change in building design. For example, the author recalls a preliminary investigation which revealed a 2-m-thick layer of undesirable fill. The original design called for a basementless construction. However, after reviewing the cost of removing the undesirable fill and replacing it with controlled (compacted) fill, it was decided to incorporate a basement in the final design.

While indeed useful, a preliminary exploration should be regarded only as such. It is a useful intermediate step and one that is generally recommended whenever the time schedule permits. It is not intended to replace a more detailed and thorough evaluation of the site.

3-4 SUBSURFACE EXPLORATION

Within the efforts of evaluating a given site, one may reflect on one or any combination of the following considerations:

1. Environmental data
2. Relevant information on the behavior of adjoining structures
3. Electrical or geophysical testing
4. Soil drilling and sampling

In addition, the study is generally extended to include laboratory testing and evaluation and, perhaps, additional in situ (in-place) field tests (49).

Environmental Considerations

Geology. The study of the geology of an area may provide useful information on some or perhaps most of the following points:

1. The general soil profile (e.g., groundwater conditions, flooding, erosion, metastable soil formation)
2. The state of the mass-rock formations (e.g., fractures or faults, formations, voids)
3. Areas of seismic activity

Sources for geological information may include the U.S. Geological Survey publications, state geological surveys of the Bureau of Mines, the U.S. Department of Agriculture, state highway departments, geological departments of universities, local well drillers, mining companies, and local libraries.

Seismic Zones. Zones of potential seismicity should be identified in terms of both occurrence and intensity of seismic activity. This information is necessary if structures are to be properly designed, protected against potential damage from such occurrences, and the loss of life therefrom is to be minimized.

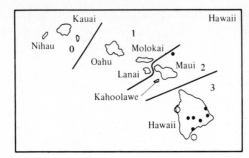
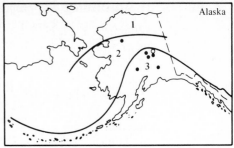
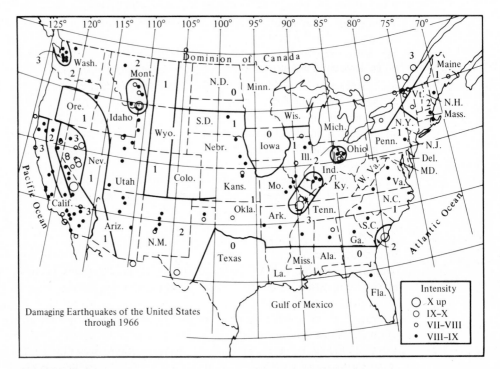

FIGURE 3-2

Location of damaging historic earthquakes through 1966. Zone 0 represents minimum risk, while zone 3 represents maximum risk. Similar maps exist in building codes. (*Courtesy of Prentice-Hall, Inc.*)

Many of the more active seismic zones are identified in Fig. 3-2. Most building codes governing within a known affected area limit or stipulate design criteria for the zones in question. A good source for information regarding zones of seismic activity is the U.S. Coast and Geodetic Survey, Environmental Sciences Service Administration, Superintendent of Documents, no. 41-1, pts. I and II, Washington, DC.

Behavior of Adjoining Structures

Whenever available, information regarding the design criteria (e.g., type of foundations, design parameters), the behavioral history (e.g., settlement, problems during construction or later), and general soil information (e.g., available boring logs, conversations with persons acquainted with the general site and/or specific soil conditions) becomes an invaluable basis for formulating plans related to the new construction.

Underground facilities such as tunnels, pipes, or cables must be located and not interfered with either during the drilling and exploration or from the imposed building loads. Similarly, such installations or other buildings must be viewed in terms of the effects and changes in soil stresses and water conditions imposed by the new structure either during construction, or during its life, or both. For example, some common questions may be: "What is the effect of the additional building load on the settlement of the existing structure?" "What may be the effects on the existing structure from excavations, dewatering, pile driving, etc.?" or, "What is the effect of the construction work and/or new building loads on an existing underground water main, sewer, or tunnel?" In this regard it is advisable that a careful inventory be taken of the conditions prior to construction of adjoining "structures" and a methodical assessment made of the potential effects the new conditions might impose on these structures. A final comparison should be undertaken to evaluate any subsequent changes of conditions and to ascertain potential damages, if any. Many lawsuits are instituted in this area of activity.

Geophysical Testing

Information regarding the topography of underlying rock and the water table is sometimes obtained by an indirect method known as *seismic refraction*. Briefly, the method consists of measuring the velocity of the compression wave that propagates through the soil. The waves are usually induced by an explosion (say, dynamite cartridge) at the ground surface, and are picked up by detectors located at different known distances from the point of explosion. The velocity may vary: (1) in soils from 150 to 2500 m/s; (2) in water, 1440 m/s; and (3) in rock from 1800 to 8000 m/s. Hence some estimates of the stratification, the degree of weathering and hardness, and the depth of the hard rock may be made prior to a more reliable testing program (such as borings).

The *electric resistivity* method measures the change in the electric resistivity of the soil via electrodes placed at the surface. For example, hard and dense rocks have high resistivity, while softer rocks and soils are less resistant. Application of the method might also include the detection of cavities within the rock strata (e.g., mine voids, sinkholes, faults). It might further be a means for establishing the extent of a certain formation by comparative evaluations with certain quantities already established (say, from boring information), or to obtain groundwater information such as water tables and aquifers.

3-5 BORINGS

The drilling of holes, commonly referred to as *test borings*, and subsequent soil sampling at varying depths is one of the most widely used methods for subsurface investigations. The "drill" is usually advanced into the soil vertically, but may be inclined if one suspects faulty formations, say, vertical rock fractures or joints, which might be missed by vertical drilling. The soil samples are taken at specified intervals and/or at changes of strata.

In nonrock formations (or in soft rock) the drilling is advanced by either augering or chopping the soil with a bit attached to a hollow drill rod through which water (sometimes mud, usually a bentonite clay mud described in Section 2-4) is injected into the hole under pressure and transports the loosened soil out. This is referred to as a *wash boring*. The use of water should be avoided in soils such as loess whose properties may be altered by such water. A steel casing, pushed or pounded downward with the advancing depth of the boring, is normally used to prevent a potential collapse, particularly in cohesionless soils, of the hole during this operation. Samples may be obtained from the soil transported out, or via a sampling spoon or a thin-wall tube which is pushed or driven to the stratum at the bottom of the hole. (The latter two sampling devices are discussed in more detail below.)

Virtually all present-day augering is done by mechanically powered drills. The typical drilling rig generally consists of a power unit, usually mounted on a truck, and is equipped with various types of augers, bits, core drills, hydraulic tools, and paraphernalia (e.g., pumps, reservoirs, hoses). Other accessories which may be needed for drilling, sampling, preparation and preservation of samples, and recording of information include split-spoon samplers, thin-wall tubes, sealing wax, sampling containers, and log sheets. Power units are sometimes mounted on skids to faciliate the drilling and sampling in places such as steep hills or inside low-clearance buildings which are rather inaccessible to truck-mounted types.

Augers consist of flutes welded to a solid or hollow (pipe) stem. Their tips are usually sloped in order to facilitate the downward advancement of the auger. They are normally less than 20 cm (8 in) in diameter for routine drilling, but sizes large enough for a person to enter and observe the formations within a drilled hole may be employed for special needs. Samples may be obtained from the flutes as the augers are extracted, or from the bottom of the hole. For samples from the bottom of the holes the auger must be extracted in the case of solid-stem augers. Samples may be obtained through the stem without extracting the augers in the case of hollow-stem types.

Augering through soils which contain large cobbles, boulders, or rock may be quite difficult. In that case the drilling is usually done by fracturing these boulders or rocks via percussion drills or by means of cable-tool drilling. In the latter case a steel rod, which is equipped with jagged teeth and weighs several hundred kilograms, is raised and dropped repeatedly by steel cables, thereby crushing the boulders or rock. Water and/or mud may be injected into the hole to facilitate the removal of the cuttings and subsequently extracted via a bailer at various intervals, as required.

Disturbed Samples

Disturbed samples encompass a rather wide array of soil specimens. Included are *wash-boring samples* which are transported out by water and subsequently deposited in a tub or other container, sometimes on the ground. As mentioned previously, these samples have relatively limited value and are seldom kept for any laboratory analysis. The process obviously permits mixing of the various strata in the boring, and therefore one may be provided with information of an average or general nature and texture of the total deposit. Continuous *auger boring samples* are somewhat more valuable. They may provide reasonable data for stratum delineation and sample identification with depth, some information regarding moisture or the water table, etc., but they are greatly lacking in information regarding the in-place characteristics of the soil such as stiffness, density, shear strength, or compressibility. Of those falling in the category of disturbed samples, the *split-spoon sample* is by far the most reliable. These samples are obtained by driving a steel tube into the undisturbed stratum and extracting the sample. The sampler is shown in Fig. 3-3. Briefly, the assembly consists of a short tube with a cutting edge (cutting shoe) on one end and threads on the other. A split tube threads to the shoe and to a head assembly which is attached to the drilling rod, as detailed in Fig. 3-3. When unscrewed from the shoe and head assembly, the split barrel can actually be opened into two equal segments for visual inspection of the sample or for removing part of the sample for preservation or future analysis.

The split-spoon sample is obtained by driving the sampler a total of 450 mm (18 in) with a 64-kg (140-lb) hammer falling for a distance of 760 mm (30 in). A record is made of the number of blows required to drive each 150-mm (6-in) segment. The number of blows required to drive the sampler for the last 300 mm (12 in) is an indication of the relative density of the material and is generally referred to as the *standard penetration resistance* (SPR). A more detailed description of such sampling and the relevant factors is given by ASTM D-1586-67 and other sources (1, 12, 18, 23, 25, 40–42, 47, 58–61).

Split-spoon samples are generally taken at every change of soil stratum and/or at specified intervals of depth, usually every 1.5 m (approx. 5 ft) or at every change of

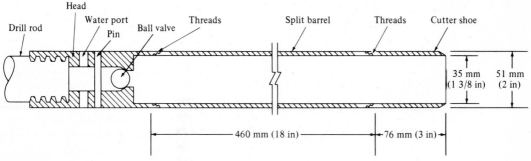

FIGURE 3-3
Split-barrel (split-spoon) sampler.

stratum detected by the driller. The samples are preserved in properly labeled sample bottles (e.g., project, date, boring number, sample depth). Occasionally some of these samples are coated with a paraffin wax for the purpose of preserving the moisture in the material as well as the sample shape, however deformed it may be.

The blow count from the standard penetration test is frequently used as a measure of the relative density of sand or of the stiffness of the stratum in which the split-spoon sample is taken. The method is not recommended to measure comparable characteristics of formations that contain gravels, particularly large sizes, or for cohesive soils.

Undisturbed Samples

An undisturbed sample is a somewhat more expensive and more time-consuming sample, but it is considerably more valuable (30, 52, 54, 67, 72). As might be expected, the grain structure of particle arrangement of many soils may be sensitive to disturbance. Hence if the samples have experienced considerable disturbance during extraction, the test results and the predictions based on such results may be seriously in error. Yet total duplication of in situ conditions is virtually impossible. That is, the mere extraction of a sample changes the pressure of the sample from in-place conditions to atmospheric, etc. (20, 32, 33, 38, 44, 46, 62). Quite obviously, therefore, due care should be exercised to obtain samples least disturbed and to account for whatever disturbances have occurred during the interpretation of the test results. A thin-walled tube, sometimes referred to as a *Shelby tube* (Shelby Tube Company was among the first to manufacture thin-wall tubing) is one of the most widely used devices for in situ or "undisturbed" soil sampling.

Figure 3-4 shows a typical Shelby-tube sampler. The tube is thin wall, about 1.6 mm thick, and anywhere from 50 to 100 mm in diameter. It is recommended, however, that the diameter be 76 mm or larger. In order to minimize friction between soil and tube, the cutting tip of the sample is slightly beveled inward (about 0.5 mm), as shown in Fig. 3-4. Furthermore it is recommended that the ratio of the peripheral cross section (encompassed by the wall thickness and beveling) to that of the soil should not exceed 10 percent. For example, if we assume the diameter of the specimen

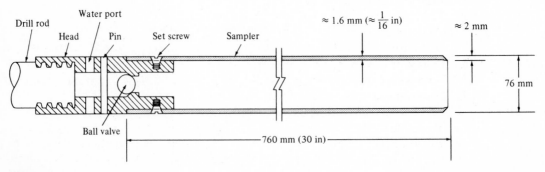

FIGURE 3-4
Thin-wall Shelby-tube sampler.

to be 73 mm (2.875 in) and a beveled thickness (wall thickness plus inward protrusion) of 1.8 mm, the ratio is $(74.8)\pi \times 1.8 \times 4/\pi(73)^2 = 0.10$ (or 10 percent).

The samples are usually obtained by pushing, smoothly and continuously, the Shelby tube into the soil. (Some engineers claim less disturbance by driving the sample into some of the more plastic soils.) Once extracted, the sample is left in the tube, with both ends of the tube waxed to prevent moisture escape. The tube with the sample inside is taken to the laboratory where it is cut and the sample is extracted and perhaps tested. The procedure for performing the various tests may be found in the ASTM standards or in most laboratory manuals dealing with the subject.

Core Boring in Rock

For most rock drilling and sampling the same drilling rig could be used as for soil sampling. On the other hand, most of the sampling tools and implements used for drilling in the soil are not normally adequate for rock sampling. For example, a split-spoon sampler or auger bit may penetrate some soft rock, but usually these tools are limited to a relatively negligible portion of the upper segment of the massive rock formation. Coring the rock layer is perhaps the most reliable method of sampling. The typical tool used for rock coring is the *core barrel*. In essence it is a hardened steel tubing, 5 to 10 cm in diameter and 60 to 300 cm long (2 to 4 in in diameter and 2 to 10 ft long), equipped with a cutting bit which contains tungsten carbide or commercial diamonds at its cutting end. During sampling the bit and core barrel rotate, while a steady stream of water or air is pushed down through the hollow rods and barrel into the bit. The water and air serve as coolants and as transporting agents in the process of bringing the cuttings up to the surface. A number of such tools are available commercially, and a more detailed description of their characteristics is deemed unnecessary at this time.

The typical rock sample consists of a cylindrical core cut from the rock formation. Through careful analysis, one may extract much useful and relevant information about the rock:

1. Type (e.g., shale, sandstone, limestone)
2. Texture (e.g., fine or coarse grain, mixtures)
3. Compressive strength (e.g., compression test results)
4. Orientation of formation (e.g., bedding planes, vertical, horizontal variations)
5. Degree of stratification (e.g., laminations)
6. Soundness (e.g., weathering, fissures, faults, degree of fracture)
7. Miscellaneous (clay seams, coal formations, stability)

By noting the ratio of the length of core obtained to the distance drilled, commonly referred to as *core recovery*, one is able to formulate an opinion regarding the consistency and soundness of the rock. For example, a shale stratum laminated with a number of clay seams will generally result in a low percentage recovery since much of the clay may be "washed away" by the water or air during the drilling process.

Similarly, a layer of broken sandrock may show a relatively small percentage recovery, with a percentage of the rock stratum being carried away by the water or air.

The small-diameter core samples are generally sufficient for the exploration of the rock strata for most jobs. Occasionally larger diameter cores are taken for evaluating the rock strata not only from the extracted cores, but also from more detailed observations via actual visual inspection of the natural formation, perhaps through mirrors or personal inspection from within the hole. Such explorations and investigations are frequently coupled with coring for caisson installations.

Strengths as determined from compression tests of core samples may be of rather limited value in predicting the strength of the in situ rock. Strength and deformation characteristics of the in situ rock are greatly affected by discontinuities (e.g., joints, bedding planes, fractures, weathered seams) which are too complex to account for during the strength evaluation from the rock core compression test.

3-6 TEST PITS

Another method of subsurface exploration is via an *open pit*, usually dug with a backhoe or power shovel. An ordinary construction backhoe with a reach of 3 to 4 m (10 to 13 ft) is usually adequate for this type of exploration. It is one of the most dependable and informative methods of investigation, since it permits a most detailed examination of the soil formation for the entire depth. For example, the stiffness of the strata (via penetrometers), the texture and grain size of the soil, detailed sampling, in situ testing, moisture evaluations, etc., are among some of the items of information which might be rather conveniently and reliably obtained from such explorations.

For deeper holes the excavation is frequently shored to protect against collapse, so that an observer may be able to view the stratification from within the hole and/or perhaps take samples of the undisturbed soil from the walls or bottom of the hole. One may obtain relatively undisturbed samples by entering the pit itself. One sampling method is to press a thin-wall steel tube into the bottom or sides of the excavation and extract a portion of the stratum by means of the tube. Another sampling method consists of carving a relatively undisturbed soil sample from the sides or bottom of the pit. The sample is then waxed to preserve its moisture during the interim between sampling and laboratory testing.

The open-pit method of exploration has both advantages and disadvantages. Some of the positive features are:

1. It provides a vivid picture of the stratification.
2. It is relatively fast and inexpensive.
3. It permits rather reliable in-place testing and sampling.

Some of the disadvantages associated with the open-pit method of explorations are:

1. It is usually practical for only relatively shallow depths, generally 4 to 5 m.
2. A high water table may prohibit or at least limit the depth of excavation.

3. If shoring and/or extraordinary safety requirements are confronted during excavation, the cost of such exploration may be unacceptably high.
4. The back-filling of the holes, generally under controlled compaction conditions, may produce a series of nonuniform stratum characteristics over the site.

3-7 FIELD TESTS

The reliability of laboratory results to represent the in situ soil properties remains a significant concern to the geotechnical engineer. The change in the environmental conditions (e.g., pressure, moisture) and the disturbance that the soil sample undergoes during extraction and subsequent handling and testing may greatly influence the test data. Furthermore, some valuable data may be obtained only through actual "field" testing. Several of the more commonly used in situ tests are discussed in this section.

Penetrometer Tests

A number of penetrometer-type tests have been used by engineers to determine relative density, stiffness, strength, or bearing capacity of the soil strata. In essence, they all relate the force or energy required to push or drive a probe a certain distance through a soil. A cone-type penetrometer has been developed in Holland and is widely used in Europe. The Standard Penetration Test (SPT) is a more widely used method in the United States, although the cone penetrometer has experienced increased use in the United States as well.

The details for performing the Standard Penetration Test (ASTM D-1586-67) have already been briefly mentioned in Section 3-5 and will therefore not be repeated here. The test is used as a measure of the relative density of sands and noncohesive soils, excluding cobbles, boulders, or very coarse gravels; it is not recommended for cohesive soils. The blow count increases with depth and with increasing coarseness of the cohesionless material. Figure 3-5 shows the relationship between the relative density of sand or gravelly sands and the corresponding blow count for various effective overburden pressures (or confining stresses).

Cone penetrometers come in a variety of designs (56). Basically they all have a cone-shaped tip which is used to facilitate the downward advancement. Some are solid rods, others are a composite of rod and casing. Some are pushed down (static), while others are driven (dynamic) into the soil. Generally they are limited in accuracy and should be used relative to and in conjunction with information from other methods and evaluations. The Begemann friction cone (4–7) is a rather sophisticated widely used tool, which appears relatively reliable for measuring the relative density and bearing capacity of strata in a continuous fashion.

Vane Shear Tests

Vane shear tests (ASTM D-2573-72) are used with increasing acceptability to determine the shear strength of soils in situ. Although reportedly first used in Sweden

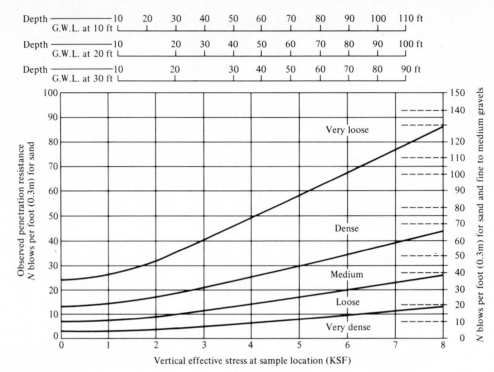

FIGURE 3-5
Relative density of sand from Standard Penetration Test. (*Courtesy of ASCE.*)

almost 50 years ago, their use did not gain significant acceptance until their extensive use and development from 1947 to 1949 by the Royal Swedish Geotechnical Institute. There are a number of different versions of the instrument, based on essentially the same principle. In essence the instrument consists of a rod with radial vanes, as shown in Fig. 3-6a. Once pushed into the soil, the vane is rotated by an applied torque. The resistance to the applied torque T in Fig. 3-6a is provided by the shearing forces on the two ends of the vane and on the circumferential plane. Figure 3-6b shows a commonly used ratio of the vane; Fig. 3-6c shows an assumed shear stress distribution on the ends of the vane, similar to that used in round bars subjected to torsion; Fig. 3-6d shows a uniform shear stress around the vertical cylindrical surface of the rotating vane.

The torque for the dimension shown in Fig. 3-6b is resisted by T_1 and T_2 (Fig. 3-6c and d, respectively). If both ends of the vane are "submerged" in the soil stratum, and if the maximum shear stress is τ for all shear surface, then

$$T = 2T_1 + T_2 = 2\tau(\pi R^2)(\tfrac{2}{3}R) + \tau(2\pi R)(4R)R$$

or

$$T = \tfrac{28}{3} \cdot \tau \cdot \pi R^3$$

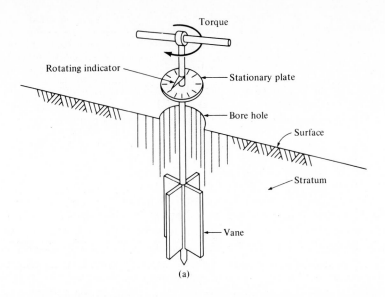

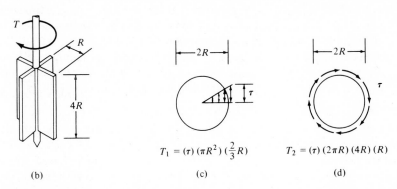

$$T_1 = (\tau)(\pi R^2)(\tfrac{2}{3}R)$$

$$T_2 = (\tau)(2\pi R)(4R)(R)$$

(b) **(c)** **(d)**

FIGURE 3-6
Basic features of a shear vane tester. (a) Schematic of shear vane apparatus. (b) Common dimensions of shear vane. (c) Assumed shear stress distribution at ends of vane. (d) Assumed shear stress distribution on vertical surface.

from which,

$$\tau = \frac{3T}{28\pi R^3} \tag{3-1a}$$

If the vane penetration is only to the top of the vane, only one end plane develops shear resistance. Therefore

$$T = \tau(\pi R^2)(\tfrac{2}{3}R) + \tau(2\pi R)(4R)R$$

or

$$\tau = \frac{3T}{26\pi R^3}$$ (3-1b)

The vane shear test is adapted to clay soils, particularly to soft sensitive clays. Generally speaking, for such clays these tests give results which are somewhat superior to those obtained from unconfined compression or direct shear tests.

Load Tests

Load tests are sometimes run at the surface or on the bottom of an excavation in order to determine the behavior of the soil stratum under load, e.g., shear strength or settlement. However desirable it may be to have the test conditions simulate the "actual" loads and footer sizes, this approach is rather impractical. That is, the footer sizes, loads, duration of loading, etc., are normally not reasonable to duplicate. Instead loads are generally transmitted through relatively rigid bearing plates, usually less than 1 m² in area, bearing on the soil stratum at the anticipated footer elevations. The resulting pressures from the plates to the soil may be anywhere from 100 to 300 percent of the expected pressures from the footing. Likewise, the time of the sustained pressures may vary from a few hours to perhaps several weeks. Deformation readings (rough approximations of settlement and/or shear strength) during the time of the sustained load are made at regular intervals. Also, a reading of the net settlement with the load removed is made after completion of the test. Load-settlement readings may be used to determine the *coefficient of subgrade reaction*. Figure 3-7 is a schematic of such a load test.

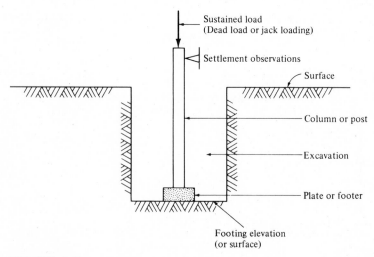

FIGURE 3-7
Schematic of a field load test.

Observation Wells

Special installations are sometimes used for the purpose of determining the ground-water location and seasonal fluctuations. Although the level of the water table is generally recorded during the drilling of bore holes, the information may be misleading; the reading may represent water from capillary saturation, surface infiltration, or a perched water table. And, of course, such short-time observations do not indicate the seasonal fluctuations that the water table may undergo.

A commonly used approach for this purpose, serving for both granular and fine-grained strata, consists of using perforated plastic pipe. The perforated pipe is lowered into the bore hole, which is held open by a steel casing. The space between the perforated pipe and the steel casing is filled with a sand and gravel mix, and then the steel casing is pulled out. In order to minimize surface infiltration, the top of the bore hole is sometimes sealed with a concrete cap, with only a small opening left to serve as a vent. Figure 3-8 shows a schematic installation of such a well. One notes that the purpose of the sand and gravel is to act as a filter; that is, it is rather easy to plug the perforations via fine-grained particle infiltration, thereby reducing or perhaps completely destroying the effectiveness of the well.

In Situ Permeability Tests

A rough estimate of in situ permeability can be obtained by pumping or bailing water out of a boring. Another way is to observe the rate of infiltration of water poured into a hole. (Percolation tests run in connection with septic systems employ rather shallow and small holes dug near the surface.) Another method entails observations of the time required for an identifiable liquid (e.g., dye) injected into the soil to reach a well at a given distance from the point of injection. Still another method is described in

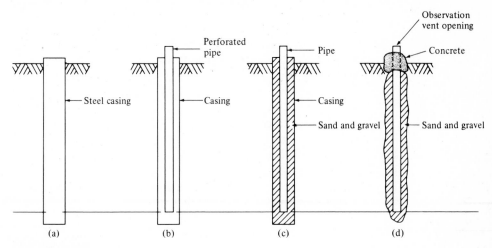

FIGURE 3-8
Schematic of common, long-term observation well.

Section 5-6. *The Earth Manual* of the U.S. Bureau of Reclamation provides procedures and sample calculations for this purpose.

Pile Load Tests

The purpose of the pile load test is to verify the capacity of the pile. Pile tests provide acceptably reliable information regarding load capacity, but are not generally reliable for long-term settlement data since the pile loads are normally sustained for only relatively short periods (e.g., 3 to 4 days). The procedure for performing the tests is described in Section 15-10, as well as in ASTM D-1143.

3-8 BORING LOGS

Data obtained from drilling the bore holes (or test pits or core samples) must be recorded accurately, completely, and at the time the data become available—not later. That is, as the drilling progresses and information regarding the strata becomes available, either through visual observations of the material brought up by the auger, air, or water, or from samples taken by the split-spoon or Shelby-tube samplers, the information is immediately recorded. One does not depend on memory to record this at completion of drilling. In fact, samples which are saved for future evaluations in the laboratory (Shelby-tube samples, split-spoon or rock cores) are likewise properly labeled on the container in which they are preserved (perhaps a jar, a Shelby tube, or a core box). Simultaneously that information is also recorded in the boring log.

Typically a driller's boring log will have the following information:

1. Name, address, and telephone number of the drilling company
2. Name and address of the project
3. Type and number of boring
4. The date of drilling
5. The type of drilling rig and equipment
6. Ground elevation at the boring location
7. The driller's name
8. The water level at the time of drilling and at a given later time (frequently 24 hours)
9. The sample number
10. The sample depth
11. The resistance to penetration of the split-spoon sampler
12. The description of the sampler
13. A complete description of the strata, coupled with pertinent notes that the driller or field technician deemed appropriate

Figure 3-9 shows a typical boring log the driller used to record the above-mentioned data. The correlation chart in the upper right-hand corner of the figure is a rather empirical guide for the sake of uniformity and classification of the strata

Driller's name:

Address:

Phone number:

Name _____ Address _____ Location _____

Boring no. _____ Page no. _____ of _____

Type boring _____ Date _____

Rig _____ Elevation _____

Casing size _____ Driller _____

				Standard Penetration Test correlation chart
				Sampler 51 mm OD; 35 mm ID
				Hammer 64 kg; 760 mm fall

Water level				Soil–Clay	Blows	Sand and Silt	Blows
Time				Very soft	2	Loose	0–10
Date				Soft	3–5	Medium	11–30
				Medium	6–15	Dense	31–50
					16–25	Very dense	Over 50
				Hard	Over 25	Stiff	

Sample number	Sample Depth	Blows Each 150 mm	Casing Depth	ID. of Sampler	Length of Sampler	Shelby	S.S.	Depth in Feet or Meters	Description — Soil Type, Water, Firmness, Drive Notes, Remarks
								0	X X X X X X X X X X X X X
								1	
								2	
								3	
								4	
								5	
								6	
								7	
								8	
								9	
								10	
								11	
								12	
								13	
								14	
								15	
								16	
								17	
								18	
								19	
								20	
								21	
								22	
								23	
								24	
								25	
								26	
								27	
								28	
								29	
								30	

FIGURE 3-9
Typical boring log form.

from job to job. It is not a universally accepted standard, but it is widely accepted as a reasonable guide for general classification purposes in the field.

From the information provided from the field log and that obtained from laboratory samples, a new log is drawn which somewhat facilitates the stratification interpretation. Figure 3-10 shows a typical log.

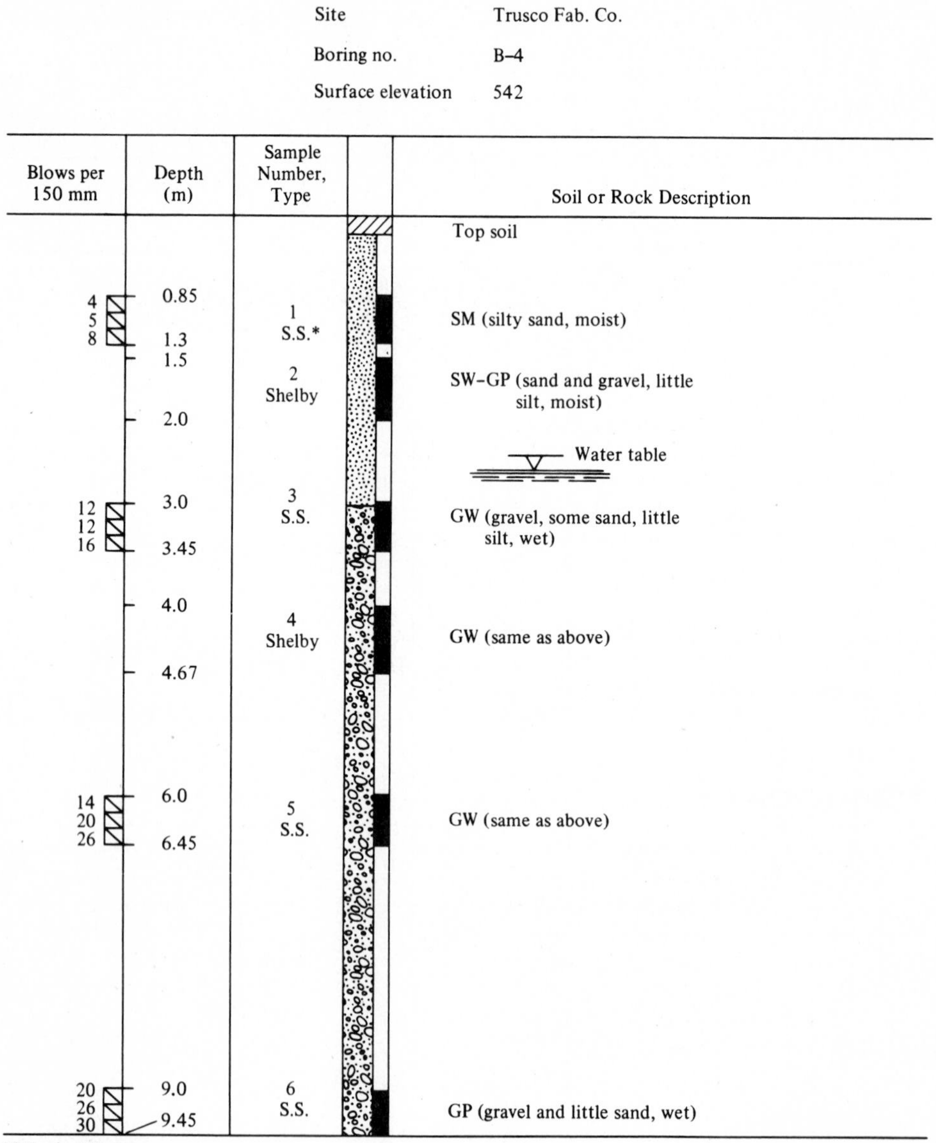

Site	Trusco Fab. Co.
Boring no.	B–4
Surface elevation	542

Blows per 150 mm	Depth (m)	Sample Number, Type		Soil or Rock Description
				Top soil
4 5 8	0.85 1.3 1.5	1 S.S.*		SM (silty sand, moist)
	2.0	2 Shelby		SW–GP (sand and gravel, little silt, moist)
				Water table
12 12 16	3.0 3.45	3 S.S.		GW (gravel, some sand, little silt, wet)
	4.0 4.67	4 Shelby		GW (same as above)
14 20 26	6.0 6.45	5 S.S.		GW (same as above)
20 26 30	9.0 9.45	6 S.S.		GP (gravel and little sand, wet)

*S.S. = split spoon

FIGURE 3-10
Typical boring log.

PROBLEMS

3-1. Name several sources for information and preliminary evaluation of a building site.

3-2. Explain briefly the difference between a *preliminary* exploration and a *detailed* exploration program; suggest relevant features for each of the two scopes.

3-3. Name several pieces of information which are obtained from a typical subsurface investigation.

3-4. Name several factors which are relevant to the planning for a well-balanced exploration program.

3-5. How does one go about planning the depth of the boring, the boring layout, and the type of samples?

3-6. What is a test pit? Give some of the negative and positive aspects of a test pit.

3-7. What is a test boring? How does it differ from a test pit?

3-8. What is a disturbed soil sample? What is an undisturbed soil sample? How is each obtained?

3-9. Describe the features of a Shelby-tube sampler. How does this differ from the split-spoon sampler?

3-10. What is a core sample? How is it obtained? What information can be obtained by evaluating this sample?

3-11. What is a vane shear test? Describe the apparatus.

3-12. What are boring logs?

3-13. Describe the basic construction and features of observation wells.

3-14. Can a split-spoon sampler penetrate a typical rock formation? Can a flight auger penetrate a rock formation? Explain.

3-15. Shelby tubes are usually pushed into the strata. However, some practitioners regard driving the tube an acceptable approach. How might the disturbance be affected by the two methods for each type listed below?
(a) Very soft clay.
(b) Hard clay.
(c) Fine sandy silt.
(d) Sandy soil.

3-16. What is drilling mud? What is bentonite?

BIBLIOGRAPHY

1. American Society of Civil Engineers, "Subsurface Investigation for Design and Construction of Condition of Building," *Manual no. 56,* 1976.

2. Adam, J., discussion of V. F. B. deMello, "The Standard Penetration Test," *4th PanAm. Conf. Soil Mech. Found. Eng.,* vol. III, 1971.

3. Alperstein, R., and S. A. Leifer, "Site Investigation with Static Cone Penetrometer," *ASCE J. Geotech. Eng. Div.,* vol. 102, May 1976.

4. Begemann, H. K. S., "A New Method for Taking of Samples of Great Length," *5th Int. Conf. Soil Mech. Found. Eng.,* vol. 1, 1961.

5. Begemann, H. K. S., "The New Apparatus for Taking a Continuous Soil Sample," *LGM Meded.,* vol. 10, no. 4, 1966.

6. Begemann, H. K. S., "Soil Sampler for Taking Undisturbed Sample 66 mm in Diameter and with a Maximum Length of 17 m," *4th Asian Conf. Int. Soc. Soil Mech. Found. Eng.,* Speciality Session Quality in Soil Sampling, 1971.

7. Begemann, H. K. S., "The Delft Continuous Soil Sampler," *Eng. Geol.,* no. 10, 1974.

8. Berre, T., K. Schjetne, and S. Sollie, "Sampling Disturbance of Soft Marine Clays," *7th Int. Conf. Soil Mech. Found. Eng.*, Speciality Session 1, Mexico, 1969; also *Norw. Geotech. Inst. Publ.* no. 85.

9. Bishop, A. W., "A New Sampling Tool for Use in Cohesionless Soils below the Water Level," *Geotechnique*, vol. 1, 1948.

10. Bishop, A. W., D. L. Webb, and P. I. Lewin, "Undisturbed Samples of London Clay from the Ashford Common Shaft, Strength–Effective Stress Relationships," *Geotechnique*, vol. 15, 1965.

11. Broms, B. B., and A. Hallen, "Sampling of Sand and Moraine with the Swedish Foil Sampler," *4th Asian Conf. Int. Soc. Soil Mech. Found. Eng.*, Specialty Session Quality in Soil Sampling, 1971.

12. Broms, B. B., "Soil Sampling in Europe, State-of-the-Art," *ASCE J. Geotech. Eng. Div.*, Jan. 1980.

13. Brown, R. E., "Drill Rod Influence on Standard Penetration Test," *ASCE J. Geotech. Eng. Div.*, vol. 103, no. GT11, Proc. Paper 13313, Nov. 1977.

14. Burghignoli, A., and G. Calabresi, "A Large Sampler for the Evaluation of Soft Clay Behavior," *9th Int. Conf. Soil Mech. Found. Eng.*, papers presented at Speciality Session 2, Soil Sampling, vol. 1, 1977.

15. Cass, J. R., "Subsurface Explorations in Permafrost Areas," *ASCE J. Geotech. Eng. Div.*, vol. 85, Oct. 1959.

16. Christian, J. T., and F. Swiger, "Statistics of Liquefaction and SPT Results," *ASCE J. Geotech. Eng. Div.*, vol. 101, no. GT11, Proc. Paper 11701, Nov. 1975.

17. Cooling, L. F., discussion of "Site Investigation Including Boring and Other Methods of Sub-Surface Exploration," *J. Inst. Civ. Eng.*, vol. 32, 1949.

18. De Mello, V. F., "The Standard Penetration Test," *4th PanAm. Conf. Soil Mech. Found. Eng.*, San Juan, Puerto Rico (published by ASCE), vol. 1, 1971.

19. Drnevich, V. P., and K. R. Massarsch, "Effect of Sample Disturbance on Stress–Strain Behavior of Cohesive Soils," presented at the 1978 ASCE Speciality Session on Soil Sampling and Its Importance in Dynamic Laboratory Testing, Chicago, Ill., 1978.

20. Ducker, A., "Method for Extraction of Undisturbed Frozen Cores," *7th Int. Conf. Soil Mech. Found. Eng.*, Specialty Session 1, Soil Sampling, Mexico, 1969.

21. Durante, V. A., J. L. Kogan, V. I. Ferronsky, and S. I. Nosal, "Field Investigations of Soil Densities and Moisture Contents," *Int. Conf. Soil Mech. Found. Eng.*, London, England, vol. 1, 1957.

22. Fahlquist, F. E., "New Methods and Technique in Subsurface Explorations," *Contribution to Soil Mechanics 1941–1953*, Boston Soc. Civ. Eng., 1941.

23. Fletcher, G. F., "Standard Penetration Test, Its Uses and Abuses," *ASCE J.. Soil Mech. Found. Eng. Div.*, vol. 91, 1965.

24. Friis, J., "Sand Sampling," *5th Int. Conf. Soil Mech. Found. Eng.*, vol. 1, 1961.

25. Gibbs, H. J., and W. G. Holtz, "Research on Determining the Density of Sands by Spoon Penetration Testing," *4th Int. Conf. Soil Mech. Found. Eng.*, London, England, vol. 1, 1957.

26. Helenelund, K. V., and C. Sundman, "Influence of Sampling Disturbances of the Engineering Properties of Peat Samples," *4th Int. Peat Congr.*, Helsinki, Finland, vol. 2, 1972.

27. Holm, G., and R. D. Holtz, "A Study of Large Diameter Piston Samplers," *9th Int. Conf. Soil Mech. Found. Eng.*, papers presented at Speciality Session 2, Soil Sampling, vol. 1, 1977.

28. Hvorslev, M. J., "Subsurface Exploration and Sampling of Soils for Civil Engineering Purposes," Rep. Comm. Sampling and Testing, ASCE Soil Mech. Found. Div., 1948.

29. Hvorslev, M. J., "Subsurface Exploration and Sampling of Soils for Civil Engineering Purposes," *Rep. Res. Project ASCE*, U.S. Army Eng. Exp. Stn., Vicksburg, 1949.

30. Idel, K. H., H. Muhs, and P. von Soos, "Proposal for 'Quality-Classes' in Soil Sampling in Relation to Boring Methods and Sampling Equipment," *7th Int. Conf. Soil Mech. Found. Eng.*, Specialty Session 1, Soil Sampling, Mexico, 1969.

31. Jakobson, B., "Influence of Sampler Type and Testing Method on Shear Strength of Clay Samples," *Swed. Geotech. Inst.*, no. 8, 1954.

32. Kallstenius, T., and W. Kjellman, "A Method of Extracting Long Continuous Cores of Undisturbed Soil," *2nd Int. Conf. Soil Mech. Found. Eng.*, vol. 1, 1948.

33. Kallstenius, T., "Mechanical Disturbances in Clay Samples Taken with Piston Samplers," *Swed. Geotech. Inst.*, no. 16, 1958.

34. Kallstenius, T., "A Standard Piston Sampler Prototype," *Swed. Geotech. Inst.*, no. 19, 1961.

35. Kallstenius, T., "The Soil Mechanics Aspects of Soil Sampling in Organic Soils," *7th Int. Conf. Soil Mech. Found. Eng.*, Specialty Session 1, Soil Sampling, Mexico, 1969.

36. Kezdi, A., I. Kabai, E. Biczok, and L. Marczal, "Sampling Cohesive Soils," *Period. Polytech. Civ. Eng.*, Budapest, Hungary, vol. 18, no. 4, 1974.

37. Kirkpatrick, W. M., and I. A. Rennie, "Stress Relief Effects in Deep Sampling Operations," *Underwater Constr. Conf.*, University College, Cardiff, Wales, Apr. 1975.

38. Kjellman, W., "A Method of Extracting Long Continuous Cores of Undisturbed Soil," *2nd Int. Conf. Soil Mech. Found. Eng.*, vol. 1, 1948.

39. Kjellman, W., T. Kallstenius, and O. Wager, "Soil Sampler with Metal Foils," *Swed. Geotech. Inst.*, no. 1, 1950.

40. Kovacs, W. D., *et al.*, "A Comparative Investigation of the Mobile Drilling Company's Safe-T-Driver with the Standard Cathead with Manila Rope for the Performance of the Standard Penetration Test," School of Civ. Eng., Purdue University, Lafayette, Ind., 1975.

41. Kovacs, W. D., J. C. Evans, and A. H. Griffith, "Towards a More Standardized SPT," *9th Int. Conf. Soil Mech. Found. Eng.*, vol. 2, 1977.

42. Kovacs, W. D., "Velocity Measurements of Free-Fall SPT Hammer," *ASCE J. Geotech. Eng. Div.*, vol. 105, Jan. 1979.

43. Landva, A., "Equipment for Cutting and Mounting Undisturbed Specimens of Clay in Testing Devices," *Norw. Geotech. Inst. Publ.*, no. 56, 1964.

44. La Rochelle, P., and G. Lefebvre, "Sampling Disturbance in Champlain Clays," *ASTM Spec. Tech. Publ.* no. 483, Am. Soc. Test Mater., Philadelphia, Pa., 1970.

45. Lundstrom, R., "Influence of Sample Diameter in Consolidation Tests," *Swed. Geotech. Inst.*, no. 19, 1961.

46. McGown, A., L. Barden, S. H. Lee, and P. Wilby, "Sample Disturbance in Soft Alluvial Clyde Estuary Clay," *Can. Geotech. J.*, vol. 11, no. 4, 1974.

47. McLean, F. G., A. G. Franklin, and T. K. Dahlstrand, "Influence of Mechanical Variables on the SPT," *Specialty Conf. In-Situ Meas. Soil Prop.*, ASCE, vol. 1, 1975.

48. Mohr, H. A., "Exploration of Soil Conditions and Sampling Operations," *Harvard Bull.*, no. 208, 1962.

49. Mulilis, J. P., C. K. Chan, and H. B. Seed, "The Effects of the Method of Sample Preparation on the Cyclic Stress–Strain Behavior of Sands," Rep. 75–18, Earthquake Eng. Res. Center, University of California, Berkeley, Calif., July 1975.

50. Noorany, I., and I. Poormand, "Effect of Sampling on Compressibility of Soft Clay," *ASCE J. Soil Mech. Found. Eng. Div.*, vol. 99, no. SM 12, Dec. 1973.

51. Palmer, D. J., and J. G. Stuart, "Some Observations on the Standard Penetration Test and a Correlation of the Test with a New Penetrometer," *4th Int. Conf. Soil Mech. Found. Eng.*, vol. 1, 1957.

52. Proctor, D. C., "Requirements of Soil Sampling for Laboratory Testing," *Offshore Soil Mechanics*, Cambridge University and Lloyd's Register of Shipping, 1976.

53. Raymond, G. P., D. L. Townsend, and M. J. Lojkasch, "The Effect of Sampling on the Undrained Soil Properties of a Leda Clay," *Can. Geotech. J.*, vol. 8, no. 4, 1971.

54. Rutledge, P. C., "Relation of Undisturbed Sampling to Laboratory Testing," *Trans. ASCE*, vol. 109, 1944.

55. Sandegren, E., "Sample Transportation Problems," *Swed. Geotech. Inst.*, no. 19, 1961.

56. Sanglerat, G., *The Penetrometer and Soil Exploration*, Elsevier, Amsterdam, 1972.

57. Schjetne, K., "The Measurement of Pore Pressure during Sampling," *4th Asian Conf. Int. Soc. Soil Mech. Found. Eng.*, Speciality Session 1, Quality in Soil Sampling, 1971.

58. Schmertmann, J., discussion of V. F. B. deMello, "The Standard Penetration Test," *4th PanAm. Conf. Soil Mech. Found. Eng.*, vol. III, 1971.

59. Schmertmann, J., "Use the SPT to Measure Dynamic Soil Properties?—Yes, But . . . !," *Dynamic Geotechnical Testing, ASTM Spec. Tech. Publ. 654*, 1978.

60. Schmertmann, J. H., "Statics of SPT," *ASCE J. Geotech. Eng. Div.*, vol. 105, May 1979.

61. Schmertmann, J. H., and A. Palacios, "Energy Dynamics of SPT," *ASCE J. Geotech. Eng. Div.*, Aug. 1979.

62. Serota, S., and R. A. Jennings, "Undisturbed Sampling Techniques for Sands and Very Soft Clays," *4th Int. Conf. Soil Mech. Found. Eng.*, vol. 1, 1957.

63. Skempton, A. W., and V. A. Sowa, "The Behavior of Saturated Clays during Sampling and Testing," *Geotechnique*, vol. 13, no. 4, 1963.

64. Sone, S., H. Tsuchiya, and Y. Saito, "The Deformation of a Soil Sample during Extrusion from a Sample Tube," *4th Asian Conf. Int. Soc. Soil Mech. Found. Eng.*, Specialty Session Quality in Soil Sampling, 1971.

65. Sowers, G. F., "Modern Procedures for Underground Investigations," ASCE, separate no. 435, 1954.

66. Swedish Committee on Piston Sampling, "Standard Piston Sampling," *Swed. Geotech. Inst.*, no. 19, 1961.

67. Thornburn, T. H., and W. R. Larsen, "A Statistical Study of Soil Sampling," *ASCE J. Geotech. Eng. Div.*, vol. 85, Oct. 1959.

68. Underwood, L. B., "Classification and Identification of Shales," *ASCE J. Soil Mech. Found. Eng. Div., vol. 93*, no. SM 6, Nov. 1967.

69. van Bruggen, J. P., "Sampling and Testing Undisturbed Sands from Boreholes," *1st Int. Conf. Soil Mech. Found. Eng.*, vol. 1, 1936.

70. Ward, W. H., "Some Field Techniques for Improving Site Investigation and Engineering Design," *Roscoe Memorial Symp. Stress-Strain Behavior of Soils*, Cambridge, 1971.

71. Wineland, J. D., "Borehole Shear Device," *6th Pan Am. Conf. Soil Mech. Found. Eng.*, vol. 1, 1975.

72. Yoshimi, Y., M. Hatanaka, and H. Oh-Oka, "A Simple Method for Undisturbed Sand Sampling by Freezing," *9th Int. Conf. Soil Mech. Found. Eng.*, papers presented at Specialty Session 2, Soil Sampling, vol. 1, 1977.

4

Physical and Index Properties of Soils

4-1 INTRODUCTION

In evaluating the characteristics of a building site, the geotechnical engineer essentially seeks an answer to the basic question: How well does a soil stratum serve a designated function? To answer this question, the engineer may be looking for answers to more specific questions:

1. How well will the strata serve under in situ conditions? For example, does the stratum possess sufficient bearing capacity (discussed in Chapter 14) to support a given load; or would it permit excessive seepage (discussed in Chapter 7) if it were part of a dam design; or will it undergo excessive settlement (discussed in Chapter 9) under certain loads, etc.?
2. Is the soil subject to significant alterations from imposed conditions? For example, will a large sustained load consolidate a clay layer; or will dynamic loads transform a loose sand stratum into a dense one; or will fluctuation of the water table affect the shear strength of a clay, etc.?
3. By what means can one improve on the undesirable characteristics of a given formation? For example, a loose sand and gravel formation may be compacted into a more dense stratum by various means; or a stratum of relatively lower shear strength may be improved by an injection of cement grout, chemical and other stabilizing agents, etc.

The answers to some of these problems are normally derived via a combination of the study of the *physical* and *index properties* of the soil, sound judgment, and relevant experience on the part of the soils engineer.

While it may be generally safe to assume that sand and gravel soils will behave differently than clays, one cannot say that all sands will behave alike or that all clays will behave alike. For example, a loose sand will behave differently than the same sand in a dense state; a saturated clay will behave differently than the same clay in a dry state. Hence the engineer proceeds to evaluate the properties of a given soil in terms of its grain characteristics (e.g., size, shape, specific gravity) as well as the arrangement of the particles within the mass (e.g., relative density for cohesionless soils, consistency for cohesive soils) (3). It is the purpose of this chapter to establish a rational basis for classifying soils according to their physical and index properties.

4-2 VOID RATIO AND POROSITY

Virtually all strata consists of a combination of solids and voids. Furthermore, the void volume encompasses the volume of water and the volume of air. Hence a total soil mass will be assumed to be composed of solid soil particles (although there may exist some negligible voids within the particles themselves), water, and air. For convenience, the mass is separated into these three basic components, as illustrated by Fig. 4-1. One notes that the volumetric relationship takes into account all three quantities. On the other hand the weight of the air is negligible relative to that of the water and solids and is, therefore, neglected in the overall weight consideration.

The *void ratio e* of the mass is defined as the ratio of the volume of voids V_v to the volume of solids V_s, given by Eq. (4-1):

$$e = \frac{V_v}{V_s} \tag{4-1}$$

The void ratio is expressed as a number and falls in the range of

$$0 < e < \infty$$

The *porosity n* of the soil mass is defined as the ratio of the volume of the voids V_v to the total volume of the mass V. It is given by Eq. (4-2):

$$n = \frac{V_v}{V} \times 100 \tag{4-2}$$

Porosity is expressed as a percentage and falls in the range of

$$0 < n < 100$$

The relationship between porosity and void ratio is given by Eq. (4-3):

$$n = \frac{V_v}{V} = \frac{V_v}{V_s + V_v} = \frac{V_v/V_s}{(V_s + V_v)/V_s} = \frac{e}{1 + e} \tag{4-3}$$

where V_v = volume of voids
V_s = volume of solids
V = total volume of soil mass

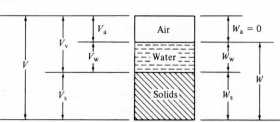

FIGURE 4-1
Phase diagram illustration of volume V and weight W as parts of a unit mass of soil.

It is common to assume that the volume of the solids V_s in a given mass remains a constant, regardless of any imposed conditions. The value of V_v, however, is altered by a change in the volumes of air and/or water. Hence the void ratio varies in direct proportion to the change in V_v. On the other hand, in the expression for n [Eq. (4-3)] one notes that both the numerator and the denominator are a function of the "variable" V_v. Thus of the two, the void ratio is the more explicit and, therefore, the more widely used expression for the volumetric relationship between voids and solids.

The porosity for natural sands depends to a great degree on the shape of the particle and the uniformity of the particle's size, as well as on circumstances related to sedimentation and deposition. The porosity of most sand masses falls in the range of 25 to 50 percent. As mentioned above, the porosity cannot exceed 100 percent.

Although the void ratio can theoretically range from zero to infinity, the common range is between 0.5 and 0.9 for sand and gravel soils and between 0.7 and 1.5 for clays. It may, however, reach higher values, perhaps exceeding 3 or 4, for some colloidal-type clays.

4-3 WATER CONTENT AND DEGREE OF SATURATION

The *water* or *moisture content w* is defined as the ratio of the weight of water to the weight of the solid particles. This is expressed by Eq. (4-4):

$$w = \frac{W_w}{W_s} \times 100 \qquad (4\text{-}4)$$

where W_w = weight of water
W_s = weight of solids

The water content is expressed as a percentage and falls in the range of $0 < w < \infty$.

The *degree of saturation S*, expressed in percent, is defined as the ratio of the volume of water to the volume of voids. It may be written as

$$S = \frac{V_w}{V_v} \times 100 \qquad (4\text{-}5)$$

where V_w = volume of water
V_v = volume of voids

One may view the degree of saturation as the volume of void spaces occupied by water relative to that which could be occupied if all the pores were full of water.

From Eq. (4-5) one notes that the degree of saturation varies from $S = 0$ for a completely dry soil to $S = 100$ percent for a totally saturated state. From a practical point of view, however, the two extremes are seldom approached and never actually reached for soils in their natural states. For example, even for submerged soils, a certain percentage of air exists within a soil mass, and therefore the volume of water may approach but never quite equal the volume of the voids; that is, $S < 100$ percent even for submerged conditions.

The water content can exceed 100 percent, as indicated by Eq. (4-4). In sands it generally varies between 10 and 30 percent, while in clay it may vary from perhaps 10 to over 300 percent—a value associated with very-fine-grain loosely deposited clays. The degree of saturation and the water content in a soil mass may have a significant effect on some of the characteristics and behavior of that soil. This is particularly true for a fine-grain soil. For example, a high water content may greatly reduce the shear strength of a clay stratum and/or its bearing capacity, or the amount and rate of consolidation is appreciably influenced by the degree of saturation.

4-4 SPECIFIC GRAVITY

The general definition for *specific gravity* is the ratio between the unit weight of a substance and the unit weight of pure water at 4° C. This may be represented by Eq. (4-6):

$$G_{\text{subs}} = \frac{\gamma_{\text{subs}}}{\gamma_{\text{w}}} \qquad (4\text{-}6)$$

where G_{subs} = specific gravity of a substance

γ_{subs} = unit weight of substance

γ_{w} = unit weight of water at 4° C (1 g/cm^3 = 9.807 kN/m^3)

The specific gravities of different minerals vary rather widely, with that of the majority of soil particles ranging between 2.6 and 2.8. Soils with high organic content will have lower values. The value commonly used for the specific gravity of soil particles is 2.7, and unless specified otherwise, this value would be assumed as a reasonable average in our example problems and discussions. Table 4-1 gives the specific gravities for a selected group of minerals.

The unit weight of water varies, but rather insignificantly, with temperature. For example, it varies from 1.0 g/cm^3 = 9.807 kN/m^3 at 4° C to approximately 0.996 g/cm^3 = 9.768 kN/m^3 at 25° C (77° F). For our purpose of calculations, and for the degree of accuracy generally expected in soil mechanics, a unit weight of

TABLE 4-1
Specific Gravity of Some Selected Minerals

Mineral	Specific Gravity
Gypsum volcanic ash	2.32
Orthoclase	2.56
Kaolinite	2.61
Quartz	2.67
Calcite	2.72
Dolomite	2.87
Magnetite	5.17

1 g/cm^3 = 9.807 kN/m^3 is a value deemed acceptable. Similarly, the corresponding unit weight of water in the *foot-pound-second* (fps) system is 62.42 lb/ft^3.

EXAMPLE 4-1

Given   The following are useful relationships:

(a)   Unit weight of mass $\gamma = \dfrac{G + Se}{1 + e}\gamma_w$

(b)   Unit weight of mass $\gamma = \dfrac{1 + w}{1 + e}G\gamma_w$

(c)   Dry unit weight $\gamma_d = \dfrac{G}{1 + e}\gamma_w$ $\gamma_d = \dfrac{G}{1 + wG/S}\gamma_w$

(d)   Submerged or *buoyant* unit weight $\gamma_b = \dfrac{G - 1}{1 + e}\gamma_w$

(e) $V_v = \dfrac{e}{1 + e}V$ (f) $V_s = \dfrac{1}{1 + e}V$

(g) $V_w = \dfrac{Se}{1 + e}V$ (h) $W_w = \dfrac{Se}{1 + e}V\gamma_w$

(i) $W_s = \dfrac{1}{1 + e}VG_m\gamma_w$ (j) $W = \dfrac{G + Se}{1 + e}V\gamma_w$

where, γ = unit weight of mass
γ_b = submerged unit weight
γ_d = dry unit weight of soil
γ_s = unit weight of solids
W = total weight of soil mass
V = total volume of soil mass
G = specific gravity of *solids*

The other terms have already been defined.

Find   Derive all of these expressions from basic definitions.

Procedure   (a) $\gamma = \dfrac{W}{V} = \dfrac{W_s + W_w}{V_s + V_v} = \dfrac{(W_s + W_w)/V_s}{(V_s + V_v)/V_s} = \dfrac{G + V_w/V_s}{1 + e}\gamma_w$; but $V_w = SV_v$. Thus,

$\gamma = \dfrac{G + SV_v/V_s}{1 + e}\gamma_w$

Answer

$$\gamma = \dfrac{G + Se}{1 + e}\gamma_w$$

(b) $\gamma = \dfrac{W_s + W_w}{V_s + V_v} = \dfrac{(W_s + W_w)/W_s}{(V_s + V_v)/W_s} = \dfrac{1 + w}{(V_s + V_v)/\gamma_s V_s}$

$\gamma = \dfrac{1 + w}{1 + e}\, \gamma_s\, ; \gamma_s = G\gamma_w$

Answer

$$\gamma = \frac{1 + w}{1 + e}\, G\gamma_w$$

(c) From (a) above, $S = 0$,

$$\gamma_d = \frac{G + 0}{1 + e}\, \gamma_w$$

Answer

$$\gamma_d = \frac{G}{1 + e}\, \gamma_w$$

(c') $\gamma_d = \dfrac{W}{V} = \dfrac{W_s + W_w}{V_s + V_v}$

If dry, $W_w = 0$. Hence,

$\gamma_d = \dfrac{W_s}{V_s + V_v} = \dfrac{W_s/V_s}{(V_s + V_v)/V_s} = \dfrac{G\gamma_w}{1 + V_v/V_s}$

But $V_v = V_w/S;\ V_w = W_w/\gamma_w;\ V_s = W_s/G\gamma_w$ or

$$\gamma_d = \frac{G\gamma_w}{1 + V_w/V_s S} = \frac{G\gamma_w}{1 + W_w G/W_s S}$$

Answer

$$\gamma_d = \frac{G\gamma_w}{1 + wG/S}$$

(d) Using (a) above and $S = 1$, and an upward lift $\approx \gamma_w$,

$$\gamma_b = \gamma - \gamma_w = \frac{G + Se}{1 + e}\, \gamma_w - \gamma_w = \left(\frac{G + e}{1 + e} - 1\right)\gamma_w$$

$$\gamma_b = \left(\frac{G + e - 1 - e}{1 + e}\right)\gamma_w$$

Answer

$$\gamma_b = \frac{G - 1}{1 + e}\, \gamma_w$$

(e) $V_v = eV_s = e(V - V_v) = eV - eV_v$ and $V_v + eV_v = eV$. Thus,

Answer

$$V_v = \frac{eV}{1 + e}$$

(f) $V_s = V - V_v = V - eV_s$ and $V_s(1 + e) = V$. Thus,

Answer

$$V_s = \frac{V}{1 + e}$$

(g) $V_w = SV_v$; but $V_v = eV/(1 + e)$ from (e) above. Thus,

$$V_w = S\left(\frac{eV}{1 + e}\right)$$

Answer

$$V_w = \frac{SeV}{1 + e}$$

(h) $W_w = W - W_s = V\gamma - V_s G\gamma_w$. From (a),

$$\gamma = \frac{G + Se}{1 + e}\gamma_w$$

and from (f),

$$V_s = \frac{1}{1 + e}V$$

Thus,

$$W_w = V\left(\frac{G + Se}{1 + e}\right)\gamma_w - \left(\frac{1}{1 + e}V\right)G\gamma_w$$

Answer

$$W_w = \frac{Se}{1 + e}V\gamma_w$$

(i) $W_s = W - W_w = V\gamma_m - V_w\gamma_w = V\left(\frac{G_m + Se}{1 + e}\right)\gamma_w - \left(\frac{Se}{1 + e}\right)V\gamma_w$

Then,

Answer

$$W_s = \frac{1}{1 + e}G_m V\gamma_w$$

(j) $\quad W = W_w + W_s = \dfrac{Se}{1 + e} V\gamma_w + \dfrac{1}{1 + e} G_m V\gamma_w$

Thus,

Answer

$$W = \dfrac{Se + G_m}{1 + e} V\gamma_w$$

EXAMPLE 4-2

Given A moist soil sample weighs 346 g. After drying at 105° C its weight is 284 g. The specific gravity of the mass and of the solids is 1.86 and 2.70, respectively.

Determine (a) The water content. (b) The void ratio. (c) The degree of saturation. (d) The porosity.

Procedure The air is assumed to be weightless; $\gamma_w = 1$ gm/cm^3.

(a) $\quad w = \dfrac{W_w}{W_s} = \left(\dfrac{346 - 284}{284}\right) \times 100$

Answer

$$w = 21.83\%$$

(b) $\quad e = V_v/V_s$ and

$$V_s = \dfrac{W_s}{G\gamma_w} = \dfrac{284}{(2.7)(1)} = 105.18 \text{ cm}^3$$

The volume of mass V is:

$$V = \dfrac{W_m}{G_m\gamma_w} = \dfrac{346}{(1.86)(1)} = 186.02 \text{ cm}^3$$

and

$$V_v = V - V_s = 186.02 - 105.18 = 80.84 \text{ cm}^3$$

Thus,

$$e = \dfrac{V_v}{V_s} = \dfrac{80.84}{105.18}$$

Answer

$$e = 0.77$$

(c) $\quad S = \dfrac{V_w}{V_v} = \left(\dfrac{346 - 284}{80.84}\right) \times 100$

Answer

$$S = 76.69\%$$

(d) $n = \dfrac{V_v}{V} = \left(\dfrac{80.84}{186.02}\right) \times 100$

Answer

$$n = 43.46\%$$

EXAMPLE 4-3

Given

A soil deposit is being considered as a fill for a building site. In its original state in the borrow pit the void ratio is 0.95. Based on laboratory tests, the desired void ratio in its compacted state at the building site is to be no greater than 0.65.

Find

The percentage decrease (or loss) of volume of the deposit from its original state.

Procedure

Let V_i = initial total volume and V_f = final total volume. Then

$$e_i = \left(\frac{V_v}{V_s}\right)_i = 0.95$$

$$e_f = \left(\frac{V_v}{V_s}\right)_f = 0.65$$

Assuming no change in V_s (typical assumption),

$$V_{si} = V_{sf}$$

or

$$\frac{V_{vi}}{0.95} = \frac{V_{vf}}{0.65}$$

But

$$V_i = V_{vi} + V_s = 0.95V_s + V_s = 1.95V_s$$

$$V_f = V_{vf} + V_s = 0.65V_s + V_s = 1.65V_s$$

The volume loss due to compaction is then

$$\frac{(1.95 - 1.65)V_s}{1.95V_s} \times 100$$

Answer

$$\text{Volume loss due to compaction} = 15.38\%$$

EXAMPLE 4-4

Given

A Shelby-tube sampler is cut such that the volume of the soil in the cut piece is determined to be 413 cm³. (From the constant cross section or area and the average length of the specimen, one can estimate the specimen's volume expediently and reasonably accurately.) The weight of the mass was 727 g. After drying, the sample's weight is 607 g. Assume $G = 2.65$; $\gamma_w = 1$ gm/cm³ = 9.807 kN/m³.

Find   (a) The water content. (b) The void ratio. (c) The porosity. (d) The degree of saturation. (e) The specific gravity of the mass.

Procedure

(a) $w = \dfrac{W_w}{W_s} = \left(\dfrac{727 - 607}{607}\right) \times 100$

Answer

$$w = 19.77\%$$

(b) $V_s = \dfrac{W_s}{\gamma_w G} = \dfrac{607}{1(2.65)} = 229.06 \text{ cm}^3$

$V_v = V - V_s = 413 - 229.06 = 183.94 \text{ cm}^3$

$e = \dfrac{V_v}{V_s} = \dfrac{183.94}{229.06}$

Answer

$$e = 0.80$$

(c) $n = \dfrac{V_v}{V} = \left(\dfrac{183.94}{413}\right) \times 100$

Answer

$$n = 44.54\%$$

(d) $S = \dfrac{V_w}{V_v} = \left(\dfrac{727 - 607}{183.94}\right) \times 100$

Answer

$$S = 65.24\%$$

(e) $G_m = \dfrac{W}{V} = \dfrac{727}{413}$

Answer

$$G_m = 1{:}76$$

EXAMPLE 4-5

Given   A soil sample has a water content of 8 percent. $G_m = 1.9$ and $G = 2.66$.

Find (a) The void ratio of the sample. (b) The degree of saturation. (c) The porosity. (d) How much water (in kilograms) would have to be added to 1 m³ of this soil in order to bring the water content to 13 percent, assuming that the void ratio remains constant.

Procedure (a) Assume $W_s = 100$ g; $W_w = 8$ g; and $\gamma_w = 1$ gm/cm³ $= 9.807$ kN/m³.

$$V_s = \frac{W_s}{G\gamma_w} = \frac{100}{(2.66)(1)} = 37.60 \text{ cm}^3$$

$$V_w = 8 \text{ cm}^3$$

$$V = \frac{W}{\gamma_m} = \frac{W}{G_m\gamma_w} = \frac{W_s + W_w}{(1.9)(1)} = \frac{108}{1.90} = 56.84 \text{ cm}^3$$

Therefore,

$$V_v = V - V_s = 56.84 - 37.60 = 19.20 \text{ cm}^3$$

and

$$e = \frac{19.20}{37.60}$$

Answer

$$e = 0.51$$

(b) $S = \dfrac{V_w}{V_v} = \left(\dfrac{8}{19.24}\right) \times 100$

Answer

$$S = 42\%$$

(c) $n = \dfrac{V_v}{V} = \left(\dfrac{19.24}{56.84}\right) \times 100$

Answer

$$n = 34\%$$

(d) The volume added per 56.84 cm³ is $(13 - 8)/56.84 = 5$ g/56.84 cm³. By proportion, since 1 m = 100 cm,

$$\frac{5 \text{ g}}{56.84 \text{ cm}^3} = \frac{x}{(100)^3}$$

$$x = \frac{5 \times 10^6}{56.84} = 8.80 \times 10^4 \text{ g} = 88 \text{ kg}$$

Answer

Thus the total weight of water per cubic meter is 88 kg

4-5 RELATIVE DENSITY OF GRANULAR SOILS

The state of compactness of a natural granular soil is commonly expressed by its *relative density* D_r. It is defined as

$$D_r = \left(\frac{e_{max} - e}{e_{max} - e_{min}}\right) \times 100 \tag{4-7}$$

or

$$D_r = \frac{\gamma_{d\,max}}{\gamma_d} \times \frac{\gamma_d - \gamma_{d\,min}}{\gamma_{d\,max} - \gamma_{d\,min}} \tag{4-7a}$$

where e_{max} = void ratio of soil in loosest state

e_{min} = void ratio of soil in densest state

e = void ratio of soil deposit (in situ state)

$\gamma_{d\,max}$ = dry unit weight of soil in densest state

$\gamma_{d\,min}$ = dry unit weight of soil in loosest state

γ_d = dry unit weight of soil deposit (in situ state)

In determining the void ratios one is confronted with the problem of measuring solid volumes. Since it is much easier to measure unit weights, the second expression [Eq. (4-7a)] is more appealing (see Example 4-6). The procedure for determining the unit weights is detailed in various ASTM standards (ASTM D-2049-69 for $\gamma_{d\,max}$ and and $\gamma_{d\,min}$; ASTM D-2167-66, 1977; D-1556-66, 1974; D-2922-70, etc., for γ_d). Briefly, $\gamma_{d\,min}$ is determined by pouring, from a fixed height, dry sand into a mold in the loosest form. $\gamma_{d\,max}$ is determined by vibrating a sample subjected to a surcharge weight. The largest $\gamma_{d\,max}$ obtained from densifying either a dry or a saturated sample is used (dry or wet method). γ_d may be obtained by any of several ASTM approved methods; see also Section 16-2f.

Relative density is commonly used as a measure of density of compacted fills (e.g., part of the specification requirements), or as an indication of the state of compactness of in situ soils. Indirectly it also reflects on the stability of a stratum. For example, a loose (small D_r) granular soil is rather unstable, especially if subjected to

TABLE 4-2
Commonly Used Designations Associated with Relative Density for Granular Soils

Designation	D_r (%)
Very loose	0–15
Loose	15–35
Medium dense	35–70
Dense	70–85
Very dense	85–100

shock or vibrating loads; vibratory loads would "compress" it into perhaps more dense and more stable formation (6–9) (see Example 4-7 for perfect spheres).

The state of compactness and the relative density are related in a rather empirical way (10). For example, for very loose sand, D_r is very small; for very dense sand, D_r is very large. A commonly used range of values for D_r and the associated descriptions for the state of compactness are given in Table 4-2. During subsurface exploration the Standard Penetration Test (SPT) (described in Section 3-7) is a commonly used method to characterize the density of a natural soil (see Fig. 3-5).

EXAMPLE 4-6

Given   A test of the density of soil in place was performed (explained in more detail in Section 16-2) by digging a small hole in the soil, weighing the extracted soil, and measuring the volume of the hole. The soil (moist) weighed 895 g; the volume of the hole was 426 cm^3. After drying, the sample weighed 779 g. Of the dried soil, 400 g was poured into a vessel in a very loose state. Its volume was subsequently determined to be 276 cm^3. That same 400 g was then vibrated and tamped to a volume of 212 cm^3. $G = 2.71$; $\gamma_w = 1$ gm/cm^3.

Find   Find D_r. (a) Via Eq. (4-7) and (b) via Eq. (4-7a).

Procedure

(a) $V_s = \dfrac{W_s}{G\gamma_w} = \dfrac{779}{(2.71)(1)} = 287.5$ cm^3

$V_v = 426 - 287.5 = 138.5$ cm^3

$e = \dfrac{V_r}{V_s} = \dfrac{138.5}{287.5} = 0.48$

$e_{max} = \dfrac{276 - 400/2.71}{400/2.71} = 0.88$

$e_{min} = \dfrac{212 - 400/2.71}{400/2.71} = 0.44$

Then

$$D_r = \frac{e_{max} - e}{e_{max} - e_{min}} = \left(\frac{0.88 - 0.48}{0.88 - 0.44}\right) \times 100$$

Answer

$$D_r = 91\%$$

(b) $\gamma_{d\,max} = \dfrac{400}{212}\,\gamma_w = 1.89$

$\gamma_{d\,min} = \dfrac{400}{276}\,\gamma_w = 1.45$

$\gamma_d = \dfrac{779}{426}\,\gamma_w = 1.83$

where units are relative. Then

$$D_r = \frac{1.89}{1.83} \times \left(\frac{1.83 - 1.45}{1.89 - 1.45}\right) \times 100$$

Answer

$$D_r = 90\%$$

Note that part (b) was determined without knowledge of the value of G.

EXAMPLE 4-7

Given   A soil sample consists of sand grains uniform in size and spherical in shape.

Find   A general expression for the relative density, assuming this sand to be perfectly cohesionless.

Procedure Ideally the range of void ratios e_{min}/e_{max} for a sand of uniform size and round particles approaches that for perfect spheres. Hence we shall assume the sand to consist of perfect spheres of equal size.

Figure 4-2a is a plan view showing the loosest arrangement of identical spheres. This is known as *simple packing*, with each sphere making contact with its adjacent sphere (four from sides, one from top, and one from bottom). Each of the spheres may be assumed to fit within a cube of side dimension d, as shown in Fig. 4-2c. Thus the volume of the cube V_C is d^3. The volume of the sphere V_S is $\pi d^3/6$. Thus the maximum void ratio (loosest state) is

$$e_{max} = \frac{V_v}{V_S} = \frac{V_C - V_S}{V_S} = \frac{d^3 - \pi d^3/6}{\pi d^3/6} = 0.91$$

Now assume that the total column is permitted to translate horizontally along a 45° line with respect to the x axis such that each sphere falls within the pocket formed by four adjacent spheres in a layer below. This forms a *face-centered cubic*, as shown in Fig. 4-2b. Now let us extract a typical cube as shown in Fig. 4-2d. This is a top view which shows a hemisphere on the top plane and four octants. There is a direct duplication of this arrangement at the bottom and the four vertical sides of the cube. Hence we note a total of six hemispheres and eight octants, or a total of four complete spheres. The volume of the cube, therefore, is equal to $(d\sqrt{2})^3 = 2d^3\sqrt{2}$, and that for the spheres equals $4(\pi d^3/6) = \frac{2}{3}\pi d^3$. Hence the minimum void ratio (densest state) is

$$e_{min} = \frac{2d^3\sqrt{2} - \frac{2}{3}\pi d^3}{\frac{2}{3}\pi d^3} = 0.35$$

Substituting the corresponding values in our basic definition for relative density, we have

Answer

$$D_r = \frac{0.91 - e}{0.56}$$

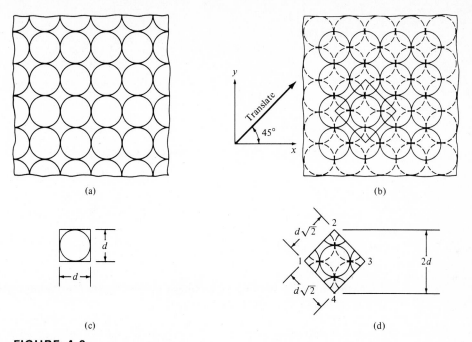

FIGURE 4-2

Soil sample, assuming perfect spheres. (a) Simple cubic packing. (b) Face-centered cubic. (c) Cubic element from simple cubic. (d) Cubic element from face-centered cubic.

4-6 CONSISTENCY OF CLAYS; ATTERBERG LIMITS

In the previous section we have developed some rather broad guidelines for describing the physical state of a granular stratum in terms of its relative density (Table 4-2). This is not an adequate description for the state of a very-fine-grain soil. For clays the usual classification is derived from their engineering properties under varying conditions of moisture. *Consistency* is a term which is frequently used to describe the degree of firmness (e.g., soft, medium, firm, or hard). The *Atterberg limits* are an empirically developed but widely used procedure for establishing and describing the consistency of soil. The limits were named after A. Atterberg who first introduced the concept, although subsequent modifications were made by Terzaghi (13) and later by Casagrande (3, 4) who improved on the test procedures and amplified the relationship these limits hold to various soil types.

The consistency of cohesive soils is greatly affected by the water content of the soil. A gradual increase of the water content, for example, may transform a dry clay from perhaps a *solid* state to a *semisolid* state, to a *plastic* state, and, after further moisture increase, into a *liquid* state. The water contents at the corresponding junction points of these states are known as the *shrinkage limit*, the *plastic limit*, and the *liquid limit*, respectively. This is shown schematically in Fig. 4-3. The detailed test procedures

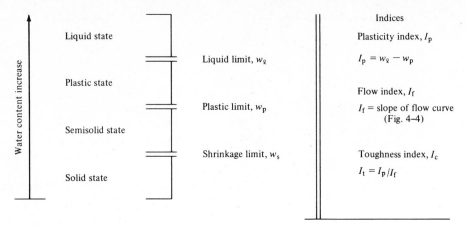

FIGURE 4-3
Atterberg limits and indices.

for determining these limits may be found in most laboratory manuals dealing with soil tests or in ASTM, AASHTO standards, etc.

The *liquid limit* w_1 (LL) is the water content at the point of transition of the clay sample from a liquid state to the plastic state, whereby it acquires a certain shearing strength (ASTM D-423). Briefly, the liquid limit of a clay is the water content, in percent, at which a grooved sample in a standard apparatus, and which was cut by a standard tool, closes along the groove for approximately 10 mm when subjected to 25 drops in a liquid-limit apparatus. By trying a number of moisture contents, a series of points could be plotted on a semilogarithmic scale as shown in Fig. 4-4. For the sample shown, the liquid limit is 47.5.

The *plastic limit* w_p (PL) is the smallest water content at which the soil begins to crumble when rolled out into thin threads, approximately 3 mm in diameter (ASTM D-424). Briefly, the samples are rolled slowly at decreasing water contents until the water content is reached at which a thread of approximately 3-mm diameter begins to crumble. This may be done by hand on a glass or some other smooth surface.

The *shrinkage limit* w_s (SL) is the smallest water content below which a soil sample will not reduce its volume any further, that is, it will not shrink any further with further drying (ASTM D-427). Briefly, the test is conducted by measuring the volume at the various water contents in the process of drying.

As mentioned before, the Atterberg limits are of a somewhat empirical nature, useful mostly for soil identification and classification. For example, the liquid limit appears to be directly proportional to the compressibility of the clay. Referring to the plasticity chart of Fig. 4-7, one notes that the inorganic clays lie above the *A* line, the organic ones below. Thus by plotting the plasticity index and the liquid limit of the clay sample, we are able to approximate the soil classification as organic or inorganic. It is possible to have an inorganic clay and an organic clay fall within a "common" region. However, the strength of the dry organic soil is significantly greater

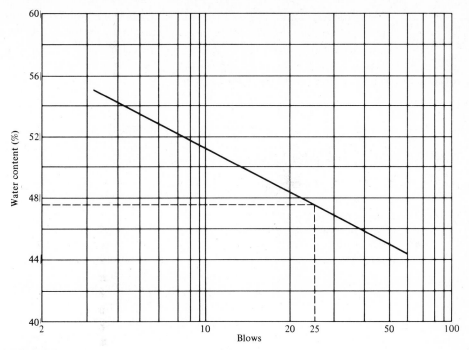

FIGURE 4-4
Liquid limit chart.

than that of the inorganic. Furthermore, organic soils have a dark-gray or black color, and possess certain odors of organic matter.

As was discussed in Section 2-4, clay absorbs water to its surface. As a guide, therefore, the water absorbed by a soil provides some estimate of the amount of clay present in that soil. Skempton (12) proposed a relationship between the plasticity index and the percentage (by weight) of clay sizes finer than 2 μm. He called it the *activity of clay A*, which is expressed by

$$A = \frac{I_p}{\% \text{ finer than 2 } \mu\text{m}}$$

Information regarding "activity" may provide one with an inclination as to the type of clay present, and subsequently the behavioral nature of the soil. For example, the relative level of activity expected is low for kaolinite, medium for illite, and high for montmorillonite (e.g., certain results reported by Skempton indicated that the level of activity for kaolinite was less than 0.5; for illite, it was approximately 1; and for montmorillonite, greater than 7).

Because of the high degree of disturbance during the extraction, handling, and testing phases, a typical soil specimen may display properties that are appreciably different from those of an in situ soil. This observation may be particularly applicable to the Atterberg limits. That is, except for the shrinkage limit, appreciable remolding and structural disturbances of the clay specimen are inherent in the test procedures, not to mention the disturbance from sampling and handling. Further evidence of the effects of disturbance is manifested clearly in the results from unconfined compression tests (ASTM D-2166-72). The strength of an undisturbed clay is several times that of the remolded clay. The ratio of the compressive strength of an undisturbed sample to that of a remolded sample is referred to as the degree of *sensitivity*. Hence the results of tests from disturbed samples, including the Atterberg limits tests, should be viewed within the context of such limitations, and should be regarded perhaps only as a part of or a supplement to a more detailed evaluation program.

4-7 SOIL CLASSIFICATION

Up to now we have categorized different soils mostly by such general terms as coarse grained or fine grained, cohesive or cohesionless. We associated sands or gravels with the coarse-grain particles and regarded them as cohesionless; silts and clays were viewed as fine grained and cohesive. Such classifications are too general; they provide neither a reasonable delineation of these categories nor an acceptable description of a *mixture* of different size grains. More systematic and uniform identification means are needed for grouping soils for engineering purposes and to establish a more unified and rational basis for communicating their properties from one user to another.

The classification systems are empirical in nature. Most were developed to serve a specific need related to a particular type of engineering work. For example, the AASHTO (American Association of State Highway and Transportation Officials, formerly the Bureau of Public Roads) system provided a systematic grouping of various soils in accordance with their suitability for use in highway subgrades and embankment construction. The Unified Classification System evolved mostly in connection with Casagrande's work (4) on military airfields. The Corps of Engineers has developed a classification for soils that display similar frost behavior.

The early classification systems were generally based on grain size. Although such classifications are widely used and may prove useful in many instances, they are generally inadequate. For example, it is not good practice to predict the permeability of two soils of like grain sizes; the permeability is affected to a significant degree by the grain shape. Similarly it may be folly to compare the compressibility of two clays of identical particle sizes, using size as the relevant characteristic, and ignore the mineral content, environmental factors, and behavioral nature of the clays (see Section 2-4). Hence many proposals have been made to expand the classification system to include properties beyond those based on grain size.

Of the number of classification systems proposed over the past few decades, the Unified Classification System and the AASHTO system appear to be the most widely used by current-day practitioners. However, most adopt the grain-size characteristics as a basis for separating the ingredients into gravel, sand, silt, or clay. Also, the

Atterberg limits are normally used as an additional criterion for identifying consistency and plasticity characteristics of the fine-grained particles.

Classification Based on Grain Size

Figure 4-5 shows the delineation between various different grain-size fractions (e.g., gravel, sand, silt, or clay) for some of a number of classification systems. The textural composition of coarse-grained soils is usually determined by screening the soil through a series of sieves of various sizes, weighing the material retained by each sieve. This is usually referred to as *sieve* or *mechanical analysis*. Material finer than the openings of a No. 200 mesh sieve (openings of 0.075 mm) is generally analyzed by a method of *sedimentation*. The most common test, the *hydrometer test*, is based on the principle that grains of different sizes fall through a liquid at different velocities. The essence of this concept is that a sphere falling through a liquid will reach a terminal velocity expressed by Stoke's law: $v = (\gamma_s - \gamma_w)D/18\mu$, where γ_s and γ_w are unit weights of sphere and liquid, respectively, μ is the viscosity of the liquid, and D is the diameter of the sphere. The details for performing these tests may be found in laboratory manuals dealing with soil testing or ASTM D-442-63.

The results of a grain-size analysis are commonly represented in the form of a graph as shown in Fig. 4-6. The aggregate weight, as a percentage of the total weight, of all grains smaller than any given diameter is plotted on the ordinate using an arithmetic scale; the size of a soil particle, in millimeters, is plotted on the abscissa, which uses a logarithmic scale. Reference is made to Fig. 4-6, showing the typical results of a mechanical analysis test plotted in this case using the MIT system. One notes the range of particle sizes for the three basic designations of sand, silt, and clay,

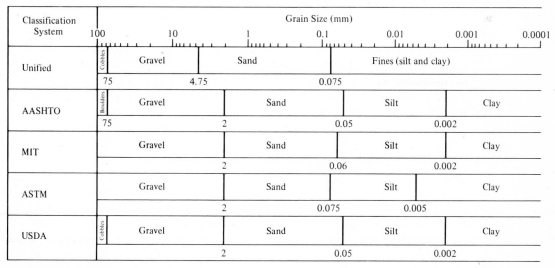

FIGURE 4-5
Soil classification based on grain size.

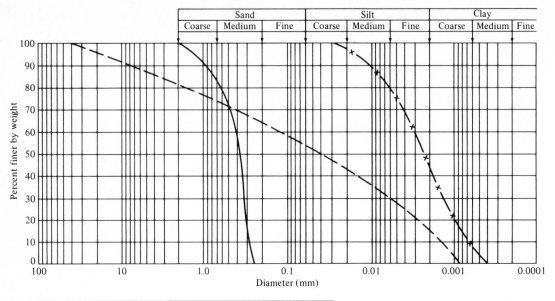

Soil	Legend	w_l	w_p	I_p
A	——————	NA	NA	—
B	— — —	36	22	14
C	— x —— x —	42	26	16

FIGURE 4-6
Grain-size distribution curves, MIT classification.

with each further subdivided into categories of coarse, medium, or fine grained. This scale of classification is rather typical of several arbitrarily chosen such scales.

Some rather relevant and useful information may be obtained from a grain-size curve, such as indicated in Fig. 4-6: (1) the total percentage of a given size; (2) the total percentage larger or finer than a given size; and (3) the uniformity or the range in grain-size distribution. One indication of the gradation is given by the *uniformity coefficient* C_u, attributed to Allen Hazen, and defined as

$$C_u = \frac{D_{60}}{D_{10}} \qquad (4\text{-}8)$$

It expresses the ratio of the diameter of the particle size at 60 percent to the diameter of the particle size at 10 percent finer by weight on the grain-size distribution curve (see Example 4-8). A large coefficient corresponds to a large range in grain sizes, and the soil is regarded as well graded. A coefficient of 1 would represent soil sizes of the same magnitude. Generally soils whose coefficient is less than 4 are considered uniform.

Another is the *coefficient of curvature* (or *coefficient of concavity*) C_z, sometimes used as a measure of the shape of the grain-size distribution curve. It is defined as

$$C_z = \frac{D_{30}^2}{D_{10}D_{60}} \qquad (4\text{-}9)$$

For a C_z value of about 1, the soil is considered well graded. For a C_z much less or much larger than 1, the soil is viewed as poorly graded.

Unified Soil Classification System

Subsequent to the airfield classification system developed by Casagrande (4), the Bureau of Reclamation and the Army Corps of Engineers developed the Unified Classification System in 1952. The soils are divided into two main groups, coarse grained and fine grained, and defined by a set of two letters, a prefix and a suffix. The coarse-grain designation is assigned to soils for which over 50 percent, by weight, of the material is retained by the No. 200 sieve (0.075 mm). Within the coarse-grained group, the prefix G is assigned to the soil if more than 50 percent of the particles are retained by the No. 4 sieve (4.76 mm), and S if more than 50 percent pass through the No. 4 sieve. G or S is followed by a suffix that describes the gradation: W—well graded; P—poorly graded; M—containing silt; C—containing clay. For example, a well-graded gravel would be represented by GW; a poorly graded sand by SP. This is shown in Fig. 4-7.

The fine-grained designation represents the soils for which more than 50 percent pass the No. 200 (0.075 mm) sieve. These are divided into silts (M), clays (C), and organic silts or clays. The suffix following one of these designations is L—low plasticity, or H—high plasticity (L for a liquid limit < 50 percent; H for a liquid limit > 75 percent). The fine-grained soils are classified according to their plasticity index and liquid limit via the plasticity chart of Fig. 4-7. The *A* line separates the inorganic clays from the silts and organic soils. The general classification criteria are given in Fig. 4-7.

AASHTO Classification System

As mentioned previously, the AASHTO classification system is a widely used method for classifying soils for earthwork structures, particularly subgrades, bases, subbases, and embankments. Its present form represents the culmination of several revisions and alterations since its development by the U.S. Bureau of Public Roads about 1929. It classifies soils into seven groups (A-1 through A-7) based on particle-size distribution, liquid limit, and plasticity index. This is shown in Fig. 4-8, including subgroups for A-1, A-2, and A-7. A brief description of the various groups is given in Fig. 4-9 (1). The system separates soils into granular and silt-clay groups. A-1 through A-3 are granular, with 35 percent or less passing the No. 200 sieve (0.075 mm). If more than 35 percent

Major Divisions			Group Symbols	Typical Names
Coarse-Grained Soils More than 50% retained on No. 200 sieve*	Gravels 50% or more of coarse fraction retained on No. 4 sieve	Clean gravels	GW	Well-graded gravels and gravel–sand mixtures, little or no fines
			GP	Poorly graded gravels and gravel–sand mixtures, little or no fines
		Gravels with fines	GM	Silty gravels, gravel–sand–silt mixtures
			GC	Clayey gravels, gravel–sand–clay mixtures
	Sands More than 50% of coarse fraction passes No. 4 sieve	Clean sands	SW	Well-graded sands and gravelly sands, little or no fines
			SP	Poorly graded sands and gravelly sands, little or no fines
		Sands with fines	SM	Silty sands, sand–silt mixtures
			SC	Clayey sands, sand–clay mixtures
Fine-Grained Soils 50% or more passes No. 200 sieve*	Silts and Clays Liquid limit 50% or less		ML	Inorganic silts, very fine sands, rock flour, silty or clayey fine sands
			CL	Inorganic clays of low to medium plasticity, gravelly clays, sandy clays, silty clays, lean clays
			OL	Organic silts and organic silty clays of low plasticity
	Silts and Clays Liquid limit greater than 50%		MH	Inorganic silts, micaceous or diatomaceous fine sands or silts, elastic silts
			CH	Inorganic clays of high plasticity, fat clays
			OH	Organic clays of medium to high plasticity
Highly Organic Soils			PT	Peat, muck, and other highly organic soils

FIGURE 4-7
Unified Classification System, ASTM D-2487-69.

$$C_u = D_{60}/D_{10} > 4 \qquad C_z = \frac{D_{30}^2}{D_{10}D_{60}} \text{ between 1 and 3}$$

Not meeting both criteria for GW

Atterberg limits plot below A line or plasticity index less than 4	Atterberg limits plotting in hatched area are borderline classifications requiring use of dual symbols
Atterberg limits plot above A line and plasticity index greater than 7	

Classification on Basis of Percentage of Fines

Less than 5% pass. No. 200 sieve	GW, GP, SW, SP GM, GC, SM, SC
More than 12% pass No. 200 sieve	Borderline classification requiring use of dual symbols
5–12% pass No. 200 sieve	

$$C_u = D_{60}/D_{10} > 6 \qquad C_z = \frac{D_{30}^2}{D_{10}D_{60}} \text{ between 1 and 3}$$

Not meeting both criteria for SW

Atterberg limits plot below A line or plasticity index less than 4	Atterberg limits plotting in hatched area are borderline classifications requiring use of dual symbols
Atterberg limits plot above A line and plasticity index greater than 7	

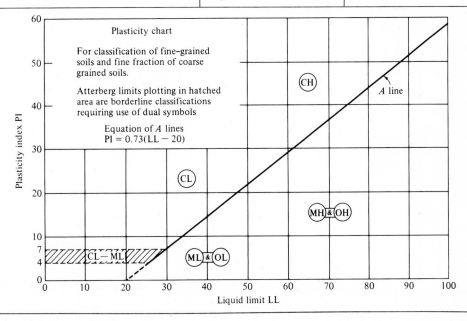

Plasticity chart

For classification of fine–grained soils and fine fraction of coarse grained soils.

Atterberg limits plotting in hatched area are borderline classifications requiring use of dual symbols

Equation of A lines
PI = 0.73(LL − 20)

Plasticity index PI

Liquid limit LL

Visual–manual identification, see ASTM D-2488.

*0.075 mm sieve.

General Classification	Granular Materials (35% or less passing 0.075 mm)							Silt-Clay Materials (More than 35% passing 0.075 mm)			
	A-1		A-3	A-2				A-4	A-5	A-6	A-7-5 A-7-6
Group Classification	A-1-a	A-1-b		A-2-4	A-2-5	A-2-6	A-2-7				
Sieve analysis, percent passing:											
2.00 mm (No. 10)	50 max	—	—								
0.425 mm (No. 40)	30 max	50 max	51 min								
0.075 mm (No. 200)	15 max	25 max	10 max	35 max	35 max	35 max	35 max	36 min	36 min	36 min	36 min
Characteristics of fraction passing 0.425 mm (No. 40)											
Liquid limit				40 max	41 min	40 max	41 min	40 max	41 min	40 max	41 min
Plasticity index*	6 max		NP	10 max	10 max	11 min	11 min	10 max	10 max	11 min	11 min
Usual types of significant constituent materials	Stone fragments, gravel, and sand		Fine sand	Silty or clayey gravel and sand				Silty soils		Clayey soils	
General rating as subgrade	Excellent to good							Fair to poor			

* Plasticity index of A-7-5 subgroups is equal to or less than LL − 30, Plasticity index of A-7-6 subgroup is greater than LL −.

FIGURE 4-8
The AASHTO Classification of soils and soil-aggregate mixtures.

Description of Classification Groups	
Subgroup A-1a	Includes those materials consisting predominantly of stone fragments or gravel
Subgroup A-1b	Includes those materials consisting predominantly of coarse sand, either with or without a well-graded soil binder
Subgroup A-3	Fine beach sand or fine desert loess sand without silty or clay fines or with a very small amount of nonplastic silt
Subgroups A-2-4 and A-2-5	Include various granular materials containing 35% or less passing the 0.075-mm sieve and with a minus 0.425 mm in having the characteristics of the A-4 and A-5 groups
Subgroups A-2-6 and A-2-7	Include material similar to that described under subgrades A-2-4 and A-2-5, except that a fine portion contains plastic clay having the characteristics of the A-6 or A-7 group
Subgroup A-4	The typical materials of this group are the nonplastic or moderately plastic silty soils
Subgroup A-5	Similar to that described under group 2-4, except that it is usually of diatomaceous or micaceous character
Subgroup A-6	Usually a plastic clay having 75% or more passing the 0.075-mm sieve
Subgroup A-7-5	Includes materials with moderate plasticity indexes in relation to liquid limit
Subgroup A-7-6	Includes materials with high plasticity indexes in relation to liquid limit

FIGURE 4-9
Description of AASHTO groups.

passes through a No. 200 sieve, the material falls in the silt-clay group. These groups are evaluated via a formula referred to as a *group index* (GI):

$$GI = (F - 35)[0.2 + 0.005(LL - 40)] + 0.01(F - 15)(PI - 10) \quad (4\text{-}10)$$

where F = percentage passing No. 200 sieve, expressed as a whole number. This percentage is based only on the material passing the No. 200 sieve.

LL = liquid limit

PI = plasticity index

The higher the value of GI, the less the suitability (e.g., a material of GI = 15 is less suitable than one whose GI = 1). Likewise, A-1 is more suitable than A-4, as indicated in Fig. 4-8.

EXAMPLE 4-8

Given The grain-size distribution shown in Fig. 4-6 and the corresponding values for w_l and w_p shown.

Find The coefficient of uniformity for soil *A*.

Procedure For soil A, $D_{60} = 0.38$ mm, $D_{10} = 0.28$ mm, and $C_u = D_{60}/D_{10} = 1.36$. The soil is considered uniform in size.

EXAMPLE 4-9

Given The data shown in Fig. 4-6.

Find Classification for the three soils based on the unified soil classification system.

Procedure **FOR SOIL A.** From Fig. 4-6 we see that more than 50 percent (virtually 100 percent) is retained by the No. 200 sieve (0.075 mm). Thus the soil is coarse grained. Also, since more than 50 percent passes through the No. 4 sieve (4.75-mm size), the soil is sand, S. Furthermore, since less than 12 percent passes through the No. 200 sieve, the soil is poorly graded, P. Hence the material is a poorly graded sand, SP.

FOR SOIL B. From Fig. 4-6 we note that more than 50 percent (about 55 percent) is retained by the No. 200 sieve. Thus the soil is coarse grained. Furthermore, more than 50 percent passes through the No. 4 sieve and more than 12 percent (about 30 percent) passes through the No. 200 sieve. We determine the classification from the plasticity chart, Fig. 4-7. For $w_l = 36$ and $I_p = 14$, the material plots just above the A line. Thus the material may be classified as SC.

FOR SOIL C. Since more than 50 percent (about 85 percent) passes the No. 200 sieve, the soil is fine grained. For $w_l = 42$ and $I_p = 16$, the soil plots almost on the A line. Hence an approximate classification is CL.

EXAMPLE 4-10

Given The data of Fig. 4-6.

Find The AASHTO classification for the three soils.

Procedure **FOR SOIL A.** The amount of soil passing the No. 10, 40, and 200 sieves is:

No. 10 (2-mm) sieve	100%
No. 40 (0.425-mm) sieve	70%
No. 200 (0.075-mm) sieve	0%

Hence the classification is A-3 (excellent to good category).

FOR SOIL B. The amount passing, for corresponding sieves, is:

No. 10 sieve	82%
No. 40 sieve	72%
No. 200 sieve	57%

For $w_l = 36$ and $I_p = 22$, the soil is A-6, clay.

FOR SOIL C. The amount passing, for the respective sieves, is:

$$\begin{array}{ll} \text{No. 10 sieve} & 100\% \\ \text{No. 40 sieve} & 100\% \\ \text{No. 200 sieve} & 100\% \end{array}$$

For $w_1 = 42$ and $I_p = 16$, the soil is A-7, clay.

The group index is

$$\begin{aligned} \text{GI} &= (F - 35)[0.2 + 0.005(w_1 - 40)] + 0.01(F - 15)(I_p - 10) \\ &= (100 - 35)[0.2 + 0.005(42 - 40)] + 0.01(100 - 15)(16 - 10) \\ &= 65(0.2 + 0.010) + 0.01(85)(6) \\ &= 13.65 + 5.10 \end{aligned}$$

Answer

$$\text{GI} = 18.75\text{—clay}$$

PROBLEMS

For the following problems assume $\gamma_w = 1 \text{ gm/cm}^3 = 9.807 \text{ kN/m}^3$ and, unless specified otherwise, $G = 2.7$. $G_m = G_{\text{mass}}$.

4-1. Why is the dry weight (weight of solids) rather than the total weight used in defining the water content? Can the water content exceed 100 percent? Explain.

4-2. A soil sample was determined to possess the following characteristics: $G = 2.74$, $e = 0.69$, and $w = 14$ percent. Determine:
(a) The degree of saturation.
(b) The porosity.
(c) The unit dry weight of the sample.

4-3. A moist soil sample was found to have a volume of 22.3 cm³ and to weigh 29.7 g. The dry weight of the sample was determined to be 23 g. Determine:
(a) The void ratio.
(b) The water content.
(c) The porosity.
(d) The degree of saturation of the sample.

4-4. Laboratory tests on a soil sample yielded the following information: $G = 2.71$, $G_m = 1.92$, $w = 13$ percent. Calculate:
(a) The void ratio.
(b) The degree of saturation.
(c) The porosity.

4-5. A soil sample was determined to have a water content of 8 percent and a degree of saturation of 42 percent. After adding some water, the degree of saturation was altered to 53 percent. Assuming no change in the volume of the voids, determine:
(a) The void ratio.
(b) The water content.
(c) The specific gravity of the mass of the sample in the altered state.

4-6.   A soil sample taken from a borrow pit has an in situ void ratio of 1.15. The soil is to be used for a compaction project where a total of 100,000 m³ is needed in a compacted state with the void ratio predetermined to be 0.73. Determine how much volume is to be excavated from the borrow pit.

4-7.   Laboratory tests determined the water content in a certain soil to be 14 percent at a degree of saturation of 62 percent. Determine:
(a)   The specific gravity.
(b)   The porosity.
(c)   The void ratio of the mass.

4-8.   A soil in its natural state was found to have a void ratio of 0.88 and a degree of saturation of 72 percent. Determine:
(a)   The specific gravity.
(b)   The water content.
(c)   The porosity of the soil mass.

4-9.   Material for an earth fill was available from three different borrow sites. In the compacted state the fill measured 100,000 m³ at a void ratio of 0.70. The corresponding in situ void ratio and cost (material and transportation) of the material for the three sites is as follows:

Borrow Site	Void Ratio	Total Cost per Cubic Meter
1	0.8	$6.40
2	1.7	$6.00
3	1.2	$5.15

Determine the most economical site.

4-10. A saturated soil sample weighs 1015 g in its moist, natural state and 704 g after drying. $G = 2.68$. Determine:
(a)   The water content.
(b)   The void ratio.
(c)   The porosity.

4-11. The mass specific gravity of an undisturbed soil sample was determined to be 1.96 at a water content of 14 percent. The void ratios in the loosest and densest states were determined to be 0.81 and 0.48, respectively. Determine the relative density of the mass.

4-12. The relative density of a soil sample was found to be 0.81. The void ratio in its natural state was 0.53, and in its densest state it was 0.48. Determine the void ratio of the sample in its loosest state.

4-13. An undisturbed soil mass has a weight of 788 g and a volume of 406 cm³. The weight of the dry sample was 670 g, $G = 2.65$. The void ratios in the loosest and densest states were found to be 0.77 and 0.52, respectively. Determine the relative density of the mass.

4-14. During compaction, a 1-m-thick stratum was consolidated a total of 3 cm via a vibrating roller. Initially the void ratio was determined to be 0.94, the water content 16 percent, and $G = 2.67$. Determine:
(a)   The void ratio.
(b)   The porosity.
(c)   The specific gravity of the mass in the compacted state.

Use the following data for Problems 4-15 through 4-20:

Percent Passing

Sieve		Soil Sample					
No.	*Opening (mm)*	*A*	*B*	*C*	*D*	*E*	*F*
4	4.76	100	90	100	100	94	100
8	2.38	97	64	100	90	84	100
10	2.00	92	54	96	77	72	98
20	0.85	87	34	92	59	66	92
40	0.425	53	22	81	51	58	84
60	0.25	42	17	72	42	50	79
100	0.15	26	9	49	35	44	70
200	0.075	17	5	32	33	38	63
Characteristics of −40 Fraction							
LL—ω_1		35	—	48	46	44	47
PL—ω_p		20	—	26	29	23	24

4-15. Draw the grain-size distribution curves for samples *A*, *C*, and *E*.

4-16. Draw the grain-size distribution curves for samples *B*, *D*, and *F*.

4-17. Classify soil samples *A*, *C*, and *E* by the unified classification system.

4-18. Classify soil samples *B*, *D*, and *F* by the unified classification system.

4-19. Classify soil samples *A*, *C*, and *E* by the AASHTO classification.

4-20. Classify soil samples *B*, *D*, and *F* by the AASHTO classification.

BIBLIOGRAPHY

1. American Association of State Highway and Transportation Officials, pt 1, Specifications (M 145–73, 1974.

2. Atterberg, A., "Über die Physikalishe Bodenuntersuchung und über die Plastizität der Tone," *Int. Mitt. Boden.*, vol. 1, 1911.

3. Bishop, A. W., "Soil Properties and Their Measurements," *5th Int. Conf. Soil Mech. Found. Eng.*, vol. 3, 1961.

4. Casagrande, A., "Research on the Atterberg Limits of Soils," *Public Roads*, Oct. 1932.

5. Casagrande, A., "Classification and Identification of Soils," *ASCE J. Geotech. Eng. Div.*, June 1947.

6. D'Appolonia, E., "Loose Sands—Their Compaction by Vibroflotation," *ASTM Spec. Tech. Publ.* no. 156, *Soil Mech. Found. Eng.*, Am. Soc. Test. Mater., Philadelphia, Pa., 1967.

7. D'Appolonia, D. J., and E. D'Appolonia, "Determination of the Maximum Density of Cohesionless Soils," *3rd Asian Conf. Soil Mech. Found. Eng.*, 1967.

8. D'Appolonia, D. J., R. V. Whitman, and E. D'Appolonia, "Sand Compaction with Vibratory Rollers," *ASCE Speciality Conf. Placement and Improvement of Soil to Support Structures*, 1968.

9.   Forssblad, L., "Investigation of Soil Compaction by Vibroflotation," *Acta Polytech. Scand.*, no. C134, Stockholm, Sweden, 1965.

10. Holtz, W. G., "The Relative Density Approach—Uses, Testing Requirements, Reliability, and Short-Comings," *ASTM Spec. Tech. Publ.* no. 523, Am. Soc. Test Mater., Philadelphia, Pa., 1973.

11. Moorhouse, D. C., and G. L. Baker, "Sand Densification by Heavy Vibratory Compactor," *3rd Pan Am. Conf. Soil Mech. Found. Eng.*, 1968.

12. Skempton, A. W., "The Colloidal Activity of Clays," *Proc. 3rd Int. Conf. Soil Mech. Found. Eng.*, vol. 1, 1953.

13. Terzaghi, K., "Principles of Soil Mechanics II—Compressive Strength of Clay," *Eng. News Rec.*, vol. 94, 1925.

14. Wagner, A. A., "The Use of the Unified Soil Classification System by the Bureau of Reclamation," *4th Int. Conf. Soil Mech. Found. Eng.*, London, England, vol. I, 1957.

5

Permeability of Soils

5-1 INTRODUCTION

As used in geotechnical engineering, *permeability* is that property of the soil which permits water to flow through it—through its voids (8). Thus soils with large voids are more permeable than those whose voids are small. Furthermore, since most soils with large voids usually have large void ratios, we may deduce that, other factors not withstanding, permeability increases with increasing void ratios.

Among the engineering problems where soil permeability may play quite a prominent role are:

1. Quantity of leakage through and under dams (discussed in Chapter 7)
2. Rate of consolidation and related settlements (discussed in Chapter 9)
3. Infiltration into the dewatering of deep excavations
4. Stability of slopes, embankments, and hydrostatic uplift evaluations
5. Seepage velocity through the soil which may create erosion (piping effects) via transportation of fine-grained particles

The resistance to flow is greatly affected by the size and geometry of the voids (9). In turn, the void characteristics are related to the size, shape, and the degree of packing of the soil grains. Generally a coarse-grain soil is more permeable than a fine-grain soil such as clay. A greater resistance to flow is offered by the much smaller pores associated with the small-grain-size soils (1, 4, 5, 7).

5-2 FLOW THROUGH CIRCULAR CAPILLARY TUBES

The voids in a soil are interconnected with each other, resembling a series of irregularly sized and shaped, winding, and twisted conduits through which the water flows. This is apparent even in fine-grained soils (3). Obviously, therefore, the flow through soils is not in a straight line, as may be expected in an ordinary tube, does not occur at a constant velocity throughout, and, generally, does not comply with the typical assumptions made in the fluid mechanics courses. Nevertheless experience has shown that a reasonable tie exists between flow through soils and flow through tubes when appropriate "adjustments" are made.

Pipe flow may be *laminar* or *turbulent*. In laminar flow the particles of water follow a designated path such that the path of one particle never intersects the path of any other; the flow is irregular and unsteady for turbulent flow. In the fluid mechanics course where paths of flow are well defined, Reynolds' number may be used with reasonable accuracy to determine the zones of laminar and turbulent flow. In soil mechanics this number is not directly relevant. The concept of steady and laminar flow, however, is of appreciable value in our effort to analyze the flow in soils and in explaining the phenomenon of groundwater movement. On the other hand, while turbulent flow is common in pipes, in most soils it is rather unlikely; it may be found at times in coarse-grain soils. Hence our analysis will assume laminar flow.

For laminar flow in tubes the velocity varies from zero at the wall to a maximum at the center, as shown in Fig. 5-1. Over a distance L the change in head is $(h_1 - h_2)$. The pressure difference, therefore, is $\gamma_w(h_1 - h_2)$. The corresponding difference in end thrusts at points 1 and 2 is $\pi r^2 \gamma_w(h_1 - h_2)$. This must be balanced by the shearing stresses on the sides of the annular ring. The unit shear force at distance r is proportional to the coefficient of viscosity μ and the velocity gradient $(-dv/dr)$. The minus sign indicates that the velocity decreases with an increase in radius. The total shear force acting on the surface of the annular cylinder is $(2\pi r L)\mu(-dv/dr)$. Summing forces and recognizing that the shear force is opposite to the direction of flow, we obtain

$$\pi r^2 \gamma_w(h_1 - h_2) = 2\pi r L \mu\left(-\frac{dv}{dr}\right) \tag{a}$$

or, separating variables,

$$\frac{\gamma_w}{2\mu}\left(\frac{h_1 - h_2}{L}\right)r\, dr = -dv \tag{b}$$

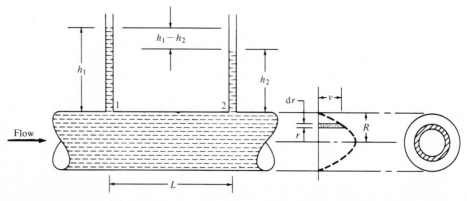

FIGURE 5-1
Laminar flow through pipes.

The value of $(h_1 - h_2)/L$ is known as the *hydraulic gradient i*. Hence integrating equation (b), we get

$$-v = \frac{\gamma_w}{4\mu} i r^2 + C \qquad (c)$$

where C is the constant of integration. At the tube's walls $r = R$ and $v = 0$. Hence from equation (c), we get

$$v = \frac{\gamma_w}{4\mu}(R^2 - r^2)i \qquad (d)$$

The rate of flow through the annular ring is $v\, dA$, or

$$Q = \int_0^R v \cdot 2\pi r\, dr = \frac{\pi \gamma_w}{2\mu} i \int_0^R (R^2 - r^2)r\, dr \qquad (e)$$

Integrating and substituting limits, we get

$$Q = \frac{\pi \gamma_w}{8\mu} i R^4 \qquad (5\text{-}1)$$

In terms of tube's area, $A = \pi R^2$, Eq. (5-1) becomes

$$Q = \frac{\gamma_w}{8\mu} R^2 i A \qquad (5\text{-}1a)$$

The average velocity of flow is Q/A, or

$$v_a = \frac{\gamma_w}{8\mu} R^2 i \qquad (5\text{-}1b)$$

5-3 DARCY'S LAW

In 1856 the French hydraulic engineer Henri Darcy demonstrated experimentally that the rate of flow of water through a soil is proportional to the hydraulic gradient. Referring to Fig. 5-2 and assuming laminar flow, Darcy's law may be written as

$$Q = kiA \qquad (5\text{-}2)$$

or

$$Q = k\left(\frac{\Delta h}{L}\right)A \qquad (5\text{-}2a)$$

where Q = rate of flow
$\quad k$ = coefficient of permeability, $\approx (\gamma_w/8\mu)R^2$ from Eq. (5-1a)
$\quad i$ = gradient or head loss between two given points, $=(h_1 - h_2)/L$
$\quad A$ = total cross-sectional area of tube
$\quad \Delta h$ = difference in heads at the two ends of soil sample
$\quad L$ = length of sample

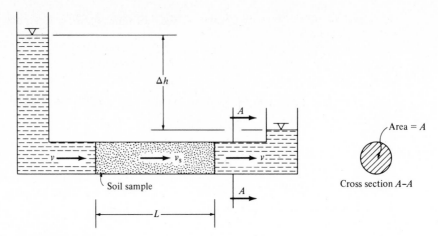

FIGURE 5-2
Uniform gravitational flow through soils.

In Fig. 5-2 one notes that the "tube" velocity v of the water must necessarily be different from the velocity of the water through the soil voids, or *seepage velocity* v_s, since the tube area A is much larger than the cross-sectional area of the voids A_v. For continuity of flow, the quantity of flow Q must be the same throughout the system. Hence,

$$Q = Av = A_v v_s \qquad \text{(a)}$$

from which

$$v_s = \left(\frac{A}{A_v}\right)v = \left(\frac{AL}{A_v L}\right)v = \frac{V}{V_v} v \qquad \text{(b)}$$

or

$$v_s = \frac{1}{n} v \qquad \text{(5-3)}$$

or

$$v = n v_s \qquad \text{(5-3a)}$$

where V = total volume of mass
$\quad V_v$ = volume of voids
$\quad n$ = porosity ratio

It is rather obvious then that the velocity v is a *superficial* velocity; the use of v instead of v_s, however, is a convenient and widely used notation.

From Eq. (5-2), $Q/A = v = ki$. Hence, substituting in Eq. (5-3a),

$$v_s = \frac{ki}{n} \qquad \text{(5-4)}$$

The common units for k are those of velocity, usually centimeters per minute or centimeters per second.

From Eq. (5-4) we note that the coefficient of permeability k of a soil is the constant of proportionality between the tube velocity v and the hydraulic gradient i. This is not to imply a constant value for a given soil. Indeed, as we shall see in subsequent discussions, the value of k may vary quite greatly for a given soil with the direction of flow.

5-4 LABORATORY PERMEABILITY TESTS

There are two basic designs of apparatus used in laboratories for estimating the coefficient of permeability k of soils: the *constant-head* and the *falling-head* permeameters. Permeability is sometimes also estimated from the rate of consolidation, but the values are generally unreliable. There are other factors influencing the consolidation rate which are difficult to account for in a relevant manner. A more advisable procedure entails subjecting the specimen to a constant-head permeability or falling-head permeability test, as discussed in the following. Only the constant-head and variable-head permeability tests will be presented in this section.

Constant-Head Permeameter

Figure 5-3 is a schematic illustration of a constant-head permeameter. The hydrostatic head remains constant, with the quantity of water flowing through the soil sample for any period of time measured by means of a graduate. Besides the quantity Q, one measures the length of the soil sample L, the gross cross-sectional area A, the value for h, and the time over which Q is collected in the graduate. From Darcy's law,

$$k = \frac{QL}{hAt} \tag{5-5}$$

L and A are shown in Fig. 5-3; Q is the total volume of water collected during time t.

The constant-head permeability test is most reliable and accurate for relatively permeable soils, such as sand, where the quantity of discharge is rather large. For rather impervious soil the accuracy of this test decreases appreciably. Hence the falling-head permeameter is preferable in that case.

Falling-Head Permeameter

For most fine-grain soils the falling-head permeameter, shown in Fig. 5-4, gives results which are better than those of the constant-head type. As in the case of the constant-head test, the length of the specimen is denoted by L and the cross-sectional area by A. The area of the standpipe is given by a. The discharge Q is measured by means of a graduate. If we let h_0, h, and h_f represent the head at the beginning, at any intermediate time, and at the end of the test, respectively, we may derive an expression for the coefficient of permeability as follows: The discharge dQ through the soil may

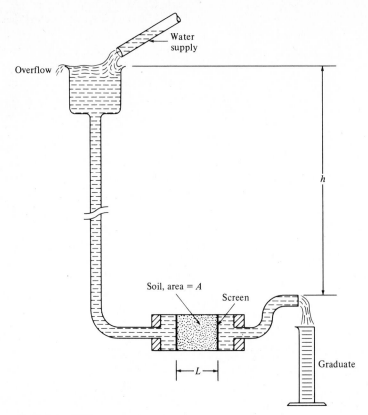

FIGURE 5-3
Schematic of constant-head permeameter test setup.

be expressed as the area of the standpipe times the velocity of fall. Since the head decreases with time, the velocity will be negative. Hence $dQ = -a(dh/dt)$. From Darcy's law we can write $dQ = kiA$. Equating the two expressions, we get

$$-a\,\frac{dh}{dt} = kiA = k\left(\frac{h}{L}\right)A$$

or separating variables, we have

$$-a\left(\frac{dh}{h}\right) = \frac{k}{L}\,A\,dt$$

The only variables are the head and time. Thus,

$$-a\int_{h_0}^{h_f}\frac{dh}{h} = \frac{k}{L}\,A\int_{t_0=0}^{t_f}dt$$

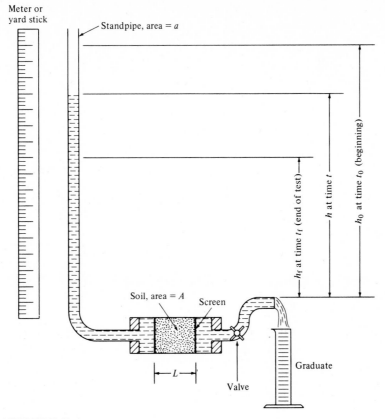

FIGURE 5-4
Schematic of falling-head permeability test setup.

Integrating and solving for k, we get

$$k = \frac{aL}{At_f} \ln \frac{h_0}{h_f} \tag{5-6}$$

or

$$k = 2.3 \frac{aL}{At_f} \log_{10} \frac{h_0}{h_f} \tag{5-6a}$$

EXAMPLE 5-1

Given   A soil sample in a constant-head permeameter is 7.30 cm (2.875 in) in diameter and 14 cm (5.5 in) long. 320 cm³ of water pass through the soil in 3 hr 22 min. The difference in head between points of intake and discharge (value of h in Fig. 5-3) is 24 cm.

Find   The coefficient of permeability of the soil.

Procedure   Using Eq. (5-5), we have

$$k = \frac{QL}{hAt}$$

$$= \frac{320 \times 14}{(24)[\pi(7.3)^2/4] \times (180 + 22)}$$

$$= 2.21 \times 10^{-2} \text{ cm/min} = 3.68 \times 10^{-4} \text{ cm/s}$$

Hence from Eq. (5-1),

$$k = \frac{Q}{iA} = \frac{1.58}{1.71 \times 41.35} = 2.21 \times 10^{-2} \text{ cm/min}$$

Answer

$$k = 3.68 \times 10^{-4} \text{ cm/s} \ (1.21 \times 10^{-5} \text{ ft/s})$$

EXAMPLE 5-2

Given   A soil sample in a variable-head permeameter 7.30 cm (2.875 in) in diameter and 14 cm (5.5 in) long. The standpipe has a diameter of 1 cm (0.4 in). The head on the sample dropped from 130 cm (51.2 in) to 72 cm (28.3 in) in 1 hr 18 min.

Find   The coefficient of permeability.

Procedure   In Eq. (5-6a)

$$a = \frac{\pi}{4}(1)^2 = 0.785 \text{ cm}^2$$

$$A = \frac{\pi}{4}(7.3)^2 = 41.85 \text{ cm}^2$$

$$t_f - t_0 = 78 \text{ min}$$

$$h_0 = 130 \text{ cm}; h_f = 72 \text{ cm}$$

$$\log_{10} = \tfrac{130}{72} = 0.2566$$

Hence,

$$k = 2.3 \frac{0.785 \times 14}{41.85 \times 78}(0.2566) = 1.99 + 10^{-3} \text{ cm/min} = 3.32 \times 10^{-5} \text{ cm/s}$$

Answer

$$k = 3.32 \times 10^{-5} \text{ cm/s} \ (1.09 \times 10^{-6} \text{ ft/s})$$

5-5 AUTHENTICITY OF k VALUES FROM LABORATORY TESTS

The coefficient of permeability determined via laboratory tests may be far different from that of the in situ soil. Among some of the factors that affect such deviations are:

1. *Environmental differences.* It is readily apparent that a soil extracted from some depth undergoes changes from its in situ state:
   a. Effective as well as pore pressure changes may be substantial.
   b. Degree of saturation may be altered; under laboratory conditions, near saturation is typical and usually higher than for in situ state.
   c. Density is likely to be less for test samples, especially for the more granular soils.
2. *Disturbance.* The process of carving into the soil formation, extracting, handling, and testing results in at least some, perhaps large, sample disturbance:
   a. The void ratio is likely to be altered in the process.
   b. The particle arrangement (grain structure) may experience some relevant changes.
3. *Test conditions.* We may deduce that in situ conditions are virtually impossible to duplicate in the laboratory. In fact, in most instances we do not know the degree and quantity of the conditions we should duplicate.
   a. The hydraulic gradient of the test sample is usually, for expediency, much higher than that which the in situ soil experiences.
   b. It is virtually impossible to measure such elements as air content, degree of saturation, pore pressure, and particle orientation, and to subsequently duplicate them in the laboratory.
   c. It is difficult to account for the boundary effects associated with a small test sample.
4. *Representative sampling.* It is difficult to ascertain the reliability of test results from rather minute samples and subsequently to correlate them with and predict the behavior of a relatively large formation of the in situ soil.
5. *Direction of flow.* The orientation of the "test" flow is usually transverse to the usual stratification since, generally, the typical sample extracted is a "vertical" one, usually via a Shelby tube through conventional drilling methods. Yet the coefficient of permeability in the horizontal direction is appreciably larger than that in the vertical direction, perhaps as much as 1000 times or more.

From the above it is apparent that the laboratory-determined value of k is in question. Using "undisturbed" samples, to the extent that this is practical, is a step in the right direction. The author has designed and built such an apparatus whereby a thin-wall Shelby tube anywhere from 3 to 50 cm in length could be tested; longer samples may also be used by merely using longer clamping rods. A rubber ring seal is used at the pressure end of the tube, with a screen at the discharge end. The cross-sectional area A is calculated by using the inside diameter of the tube. Beside being expedient, this setup provides for some additional distinct advantages: (1) it minimizes the disturbance of the sample; (2) it permits a permeability reading over a greater length; (3) it permits the testing of a granular sample which may otherwise fall apart during the extraction and setup process.

5-6 FIELD TESTS FOR PERMEABILITY

As mentioned in the preceding section, tests for permeability are not fully reliable if the test sample has been disturbed. Yet some disturbance, particularly apparent in sandy samples, is a virtual certainty to varying degrees, depending on the type of soil and the method and diligence associated with the testing effort. Furthermore, the relatively small samples may not even be representative of the whole stratum when one considers the lack of homogeneous and isotropic characteristics associated with the typical soil stratum. This emerges as particularly conspicuous when one considers the differences in the values for k in the vertical and horizontal directions. For this reason various attempts have been made to determine the coefficient of permeability of the soil in its natural state. One such method is represented in the following discussion.

Figure 5-5 represents the basic schematic of one of several methods used in determining the coefficient of permeability in the field. The method involves the use of three wells (although two wells may suffice) and a pump. A perforated casing is

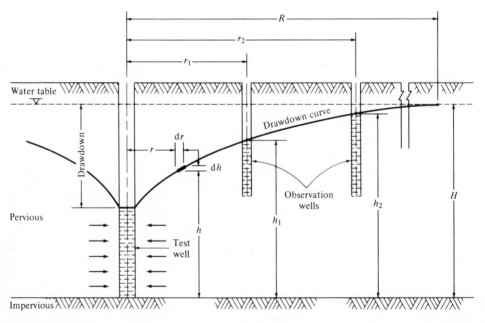

FIGURE 5-5
Diagram of a typical setup for the field permeability test.

sunk through the pervious stratum into the impervious statum, if one does exist, or to a considerable depth below the water table if an impervious stratum does not exist. This is to be used as the *test well*. Two additional perforated casings are sunk at some distance from the test well (perhaps 30- to 60-m spacing) to a depth well below the anticipated draw-down curve shown in Fig. 5.5. These are *observation wells*.

The water in the test well is then pumped out until a steady-state flow into the well is apparent. This could be determined by observing that the level of water in the test well remains at a relatively fixed elevation with continuous pumping. When this occurs, the level of the water in the two observation wells is recorded; also recorded is the distance of each observation well from the test well.

Now let us assume that the water flows into the well in a horizontal, radial direction through the walls of the casing. The surface area of a cylindrical section of radius r and height h is $2\pi rh$. Hence using Eq. (5-1), where $i = dh/dr$, and assuming that the total discharge Q equals the flow into the well (steady-state flow),

$$Q = kiA = 2\pi k \int rh\left(\frac{dh}{dr}\right)$$

Separating variables,

$$Q \int_{r_1}^{r_2} \frac{dr}{r} = 2\pi k \int_{h_1}^{h_2} h\, dh$$

$$Q \ln \frac{r_2}{r_1} = \pi k(h_2^2 - h_1^2)$$

from which

$$k = \frac{Q \ln (r_2/r_1)}{\pi(h_2^2 - h_1^2)} \tag{5-7}$$

EXAMPLE 5-3

Given   Refer to Fig. 5-5. The steady-state discharge is 189.6×10^{-3} m³/min, $h_1 = 4.3$ m (14.1 ft), and $h_2 = 4.5$ m (14.7 ft). The radial distances r_1 and r_2 are: $r_1 = 22$ m (72.2 ft), $r_2 = 40$ m (131.2 ft).

Find   The coefficient of permeability.

Procedure   From Eq. (5-7),

$$k = \frac{Q \ln (r_2/r_1)}{\pi(h_2^2 - h_1^2)} = \frac{189.6 \ln (40/22)}{\pi(4.5^2 - 4.3^2)}$$

$$k = 20.5 \times 10^{-3} \text{ m/min}$$

Answer

$$k = 3.42 \times 10^{-2} \text{ cm/s } (1.12 \times 10^{-3} \text{ ft/s})$$

5-7 FACTORS THAT INFLUENCE PERMEABILITY

As was mentioned throughout this chapter, permeability varies (1) with the constituents within a given stratum and (2) with stratification.

Constituents

The late D. W. Taylor (10) derived an equation which relates the coefficient of permeability k to a number of factors,

$$k = D_s^2 \frac{\gamma_w}{\mu} \cdot \frac{e^3}{1+e} C \qquad (5\text{-}8)$$

where D_s = representative or effective grain diameter of the sphere of equivalent volume-to-area ratio

μ = coefficient of viscosity (or viscosity) of water

e = void ratio of soil mass

C = shape factor, which depends on shape and arrangement of pores

Stratification

It is sometimes desirable to express the composite effects on permeability of a number of layers by an equivalent coefficient of permeability k_e for all the strata. For example, the question may be posed as to what is the equivalent horizontal and/or vertical coefficient of permeability for a stratified arrangement as shown in Fig. 5-6. To evaluate this condition, the following are assumed:

1. k_x and k_y for given layers are average constants for the thicknesses of the respective layers; $k_x \neq k_y$
2. One-dimensional flow, either horizontal or vertical
3. Continuity of flow

Horizontal Flow

For a unit thickness (normal to the page) and for a uniform gradient i the quantity of flow in each layer may be expressed by (from $q = kiA$)

$$q_1 = k_{x1} i H_1; \qquad q_2 = k_{x2} i H_2; \qquad q_n = k_{xn} i H_n$$

Also, the total flow is $q = k_{xe} i H_{tot} = q_1 + q_2 + \cdots + q_n$, and

$$k_{xe} i \cdot \sum H_n = i \cdot \sum (k_x H)_n$$

Thus the equivalent coefficient of permeability k_{xe} for the strata is

$$k_{xe} = \frac{\sum (k_x H)_n}{\sum H_n} \qquad (5\text{-}9)$$

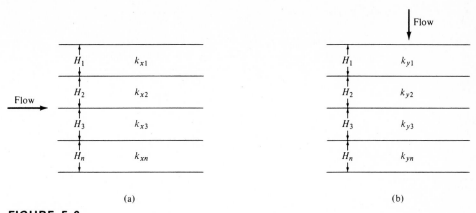

FIGURE 5-6

Undirectional flows through a stratified formation. (a) Horizontal flow. (b) Vertical flow.

Vertical Flow

Assuming flow normal to a uniform cross section A and a hydrostatic head for each layer of h_1, h_2, etc., we have

$$q_1 = k_{y1}\left(\frac{h_1}{H_1}\right)A; \qquad q_2 = k_{yz}\left(\frac{h_2}{H_2}\right)A; \qquad q_n = k_{yn}\left(\frac{h_n}{H_n}\right)H$$

From continuity of flow, $q = q_1 = q_2 = \cdots = q_n$. Thus,

$$k_{ye}iA = k_{y1}\left(\frac{h_1}{H_1}\right)A = k_{y2}\left(\frac{h_2}{H_2}\right)A = \cdots = k_{yn}\left(\frac{h_n}{H_n}\right)A$$

Hence,

$$\frac{h_1}{k_{ye}i} = \frac{H_1}{k_{y1}}; \qquad \frac{h_2}{k_{ye}i} = \frac{H_2}{k_{y2}}; \qquad \frac{h_n}{k_{ye}i} = \frac{H_n}{k_{yn}}$$

$$\frac{1}{k_{ye}i} \cdot \sum (h)_n = \sum \left(\frac{H}{k_y}\right)_n$$

But $i = \sum (h)_n / \sum (H)_n$. Thus,

$$\frac{\sum (H)_n}{k_{ye}} = \sum \left(\frac{H}{k_y}\right)_n$$

or

$$k_{ye} = \frac{\sum (H)_n}{\sum (H/k_y)_n} \qquad (5\text{-}10)$$

Sometimes an estimate of the coefficient of permeability is all that is needed. Table 5-1 provides a rough approximation of the range of permeability values for different soils.

TABLE 5-1
Approximate Range of Values of Coefficient of Permeability k

Particle Size	Coefficient of Permeability k ($\times 10^{-4}$ cm/s)
Sand	
Coarse	3000–5000
Medium	1000
Fine	50–150
Silt	
Coarse, sandy	1–20
Medium	0.1–1
Fine, clayey	0.01–0.1
Clay	
Coarse, silty	0.001–0.01
Medium	0.0001–0.001
Fine, colloidal	0.00001–0.0001

The values for permeability given in Table 5-1 should be used with caution. They are merely rough estimates for various types of soil. Nevertheless for someone familiar with soil and who is able to classify the soils (particularly granular soils) via visual examinations, these values may be useful for comparative purposes.

EXAMPLE 5-4

Given

1 m	$k_x = 4 \times 10^{-3}$ cm/s;	$k_y = 2 \times 10^{-4}$ cm/s
2 m	$k_x = 3 \times 10^{-4}$ cm/s;	$k_y = 5 \times 10^{-5}$ cm/s
3 m	$k_x = 1 \times 10^{-4}$ cm/s;	$k_y = 3 \times 10^{-5}$ cm/s

Find   The equivalent coefficients of permeability in the x and y directions, k_{xe} and k_{ye}.

Procedure   From Eq. (5-9),

$$k_{xe} = \frac{(4 \times 10^{-3} \text{ cm/s})(1 \text{ m}) + (3 \times 10^{-4} \text{ cm/s})(2 \text{ m}) + (1 \times 10^{-4} \text{ cm/s})(3 \text{ m})}{(1 + 2 + 3) \text{ m}}$$

$$k_{xe} = \frac{49 \times 10^{-4}}{6}$$

Answer

$$k_{xe} = 8.17 \times 10^{-4} \text{ cm/s}$$

From Eq. (5-10),

$$k_{ye} = \frac{(1 + 2 + 3)\,\text{m}}{(1\,\text{m}/2 \times 10^{-4}\,\text{cm/s}) + (2\,\text{m}/5 \times 10^{-5}\,\text{cm/s}) + (3\,\text{m}/3 \times 10^{-5}\,\text{cm/s})}$$

Answer

$$k_{ye} = 4.14 \times 10^{-5}\,\text{cm/s}$$

PROBLEMS

5-1. A coarse-sand sample, 12 cm long and 7.3 cm in diameter, was tested in a constant-head permeameter under a head of 100 cm for 1 min 12 s. The quantity discharged was exactly 5 liters. Determine the coefficient of permeability.

5-2. How much water, per hour, would pass through a mass of soil described in Problem 5-1, if the mass is 60 cm long and 100 cm^2 in cross section, under a constant head of 2 m?

5-3. A constant-head permeability test is run on a soil sample 9.6 cm in diameter and 20 cm long. The total head at one end of the sample is 100 cm; at the other end the head is 26 cm. Under these conditions the quantity of flow is determined to be 20 cm^3/min. Determine the coefficient of permeability.

5-4. How much water, per minute, would flow through the mass of soil in Problem 5-3 if the length was changed to 30 cm and the distance between the head-water and tail-water surfaces was 174 cm?

5-5. A constant-head permeability test was run on a soil sample 7.3 cm in diameter and 14 cm long, $Q = 10$ cm^3/min, $h = 100$ cm. The soil (grains) has a dry weight of 985 g, a specific gravity of 2.7, and a void ratio of 0.610. Determine:
    (a) The coefficient of permeability.
    (b) The seepage velocity v_s.
    (c) The superficial velocity v.

5-6. Determine the quantity of flow per minute through a soil sample 7.3 cm in diameter and 28 cm long under a constant head of 100 cm if the coefficient of permeability is 5×10^{-5} cm/s.

5-7. A falling-head permeability test was run on a soil sample 7.3 cm in diameter and 18 cm long. The diameter of the standpipe was 1 cm. The water level in the standpipe dropped from 65 cm to 50 cm in 3 h 13 min. Determine the coefficient of permeability.

5-8. How much would the water level in Problem 5-7 be expected to drop in 6 h?

5-9. A falling-head permeability test was run on a soil sample 9.6 cm in diameter and 10 cm long. The head at the start of the test was 90 cm. Determine how much head was lost during the first 30 min if the coefficient of permeability of the soil was found to be 5×10^{-6} cm/s. The diameter of the standpipe was 1 cm.

5-10. How much water would flow through the soil sample of Problem 5-9 in 30 min under a constant head of 90 cm?

5-11. The quantity of discharge from a constant-head permeability test of a sandy silt averaged 23 cm^3/min during the first few minutes and 26 cm^3/min later. Assuming the test setup to be as represented in Fig. P5-11, determine:
    (a) The coefficient of permeability that you feel is most reliable for this sample if $h = 100$ cm.

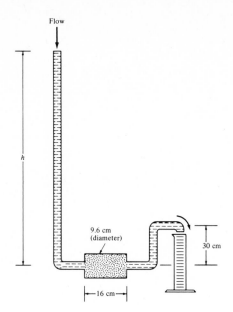

Flow

9.6 cm
(diameter)

30 cm

|←16 cm→|

FIGURE P5-11

(b) Explain the probable reasons for the difference in discharge quantities given above.

(c) Which discharge of the two is probably more representative of the permeability characteristics of the soil.

5-12. The coefficient of permeability of a soil sample to be tested under a falling-head test setup schematically shown in Fig. P5-12 is 4.3×10^{-6} cm/s. How long would it take for h to be reduced from 200 to 160 cm?

5-13. A 16-cm-diameter perforated casing is sunk in a sandy stratum a depth of 13 m below the water table to an impervious stratum. Two observation wells were installed at 20 m and 40 m distance from the test well. During steady-state pumping of 200 liters/min the water level in the two wells was lower by 145 and 125 cm, respectively. Determine the coefficient of permeability.

5-14. Four soil layers have the following characteristics:

H (m)	k_x	k_y
		k (cm/s)
1.5	8×10^{-3}	4×10^{-4}
2	2×10^{-3}	1.5×10^{-4}
2	4×10^{-4}	3×10^{-5}
3	3×10^{-4}	2×10^{-5}

Determine the equivalent coefficients of permeability in the horizontal and vertical directions for the strata.

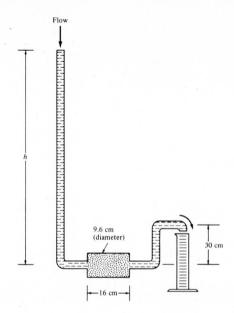

FIGURE P5-12

BIBLIOGRAPHY

1. Bjerrum, L., and J. Huder, "Measurement of the Permeability of Compacted Clays," *4th Int. Conf. Soil Mech. Found. Eng.*, vol. 1, 1957.
2. Childs, E. C., and N. Collins-George, "The Permeability of Porous Materials," *Ro. Soc.*, London, England, ser. A, vol. 201, 1950.
3. Garcia-Bengochea, I., C. W. Lovel, and A. G. Altschaeffl, "Pore Distribution and Permeability of Silty Clays," *ASCE J. Geotech. Eng. Div.*, July 1979.
4. Lambe, T. W., "The Permeability of Fine-Grained Soils," *Permeability of Soils, ASTM Spec. Tech. Publ.* no. 163, Am. Soc. Test. Mater., Philadelphia, Pa., 1954.
5. Lambe, T. W., "The Permeability of Compacted Fine Grained Soils," *ASTM Spec. Tech. Publ.* no. 163, Am. Soc. Test. Mater., Philadelphia, Pa., 1955.
6. Lohnes, R. A., E. R. Tuncer, and T. Demirel, "Pore Structure of Selected Hawaiian Soils," *Trans. Res. Rec.*, no. 612, 1976.
7. Mitchell, J. K., D. R. Hooper, and R. G. Campanella, "Permeability of Compacted Clay," *ASCE J. Soil Mech. Found. Div.*, vol. 21, no. SM4, Proc. Paper 4392, July 1965.
8. Scheidegger, A. E., *The Physics of Flow Through Porous Media*, 3rd ed., University of Toronto Press, Toronto, Canada, 1974.
9. Sridharan, A., A. G. Altschaeffl, and S. Diamond, "Pore Size Distribution Studies," *ASCE J. Soil Mech. Found. Div.*, vol. 97, no. SM5, Proc. Paper 8151, May 1971.
10. Taylor, D. W., *Fundamentals of Soil Mechanics*, Wiley, New York, 1948.
11. U.S. Waterways Experiment Station, "Field and Laboratory Investigation of Design Criteria for Drainage Wells," Tech. Memo. no. 195–1, 1942.

6

Capillarity

6-1 INTRODUCTION

We might expect water to seep down through the soil and regard this phenomenon as a natural occurrence related to forces of gravity. We may not be quite so ready to admit, however, that water moves in the upward direction as well, thereby defying gravitational forces. Yet if we were to immerse the end of a dry wick, the edge of a piece of paper, cloth, etc., into water, the observation would be made that these items will gradually become moist for some height above the water level. Obviously water has been sucked up or absorbed by these articles. Similarly, if one were to immerse the end of a very-small-diameter tube into water, one would notice a certain rise of the water into the tube for some height above the free-water surface. Perhaps in a less explicit manner, a similar occurrence would emerge if a fine-grain soil mass came in partial contact with water. The forces that pull up water into the tube, a soil mass, etc., above the free-water surface are known as *capillary forces*. The height of the water column thus drawn up or retained is called *capillary head*. The phenomenon which explains this rise is known as *capillarity*.

The basis of soil capillarity and related forces stems from the interaction of water and soil particles. The water rise above the free surface is attributed to a combination of *surface tension* (to be discussed in the next section) and the tendency of the water to "wet" the soil particles. That is, *cohesion* (molecular attraction of like particles) is responsible for the development of a state of tension of the surface water molecules, while *adhesion* (molecular attraction of unlike particles) results in the "wetting" of the soil particles.

Capillarity is at least in part responsible for some soil moisture above the water table. In other words, except for some entrapped moisture and/or some surface infiltration, soils above the groundwater table would eventually dry up were it not for capillary forces. Capillarity makes it possible for a dry fine-grained soil to draw up water to elevations well above the water table or to retain moisture above this table. Depending upon a given set of circumstances, soil engineers may view such moisture as beneficial or detrimental. For example, capillary forces may develop an increase in intergranular pressure and, thereby, improve the stability and shear strength of some fine-grained soils. On the other hand, capillary moisture near the surface may cause pavement heave during frost through the formation and subsequent growth of ice crystals or ice lenses in colder regions.

6-2 SURFACE TENSION

Virtually all liquids display a molecular attraction such that surfaces of liquids appear to possess filmlike characteristics. This interaction of surface molecules creates a condition where water is in a state of tension analogous to a thin flexible membrane subjected to tension.

The phenomenon of surface tension manifests itself via numerous examples: an insect could travel on a water surface without breaking through; a drop of mercury assumes an almost spherical shape on a clean glass plate; the surface of water in a glass is not perfectly level—it is somewhat curved where the liquid comes in contact with the walls of the container; the water rises in a small-diameter tube; a razor blade or a pin will float if carefully placed on the surface of still water.

If the pin or razor blade in the above example were to be released under the surface of the water, it would sink quite readily. Perhaps this is not unexpected since the unit weight of these metals is approximately 8 times that of water. This serves to demonstrate that (1) the interaction of the molecules below the surface of the liquid is different from that of the surface; and (2) the molecular attraction at the surface develops sufficient surface tension to provide for the support of the razor blade, pin, etc.

TABLE 6-1
Typical Surface Tension Values for Some Common Liquids at 20° C (68° F)

Liquid	Surface Tension (dyn/cm)
Water	72.8
Mercury	
In air	514.6
In vacuum	486.8
Carbon tetrachloride	26.8
Alcohol, ethyl	22.3
Oil, lubricating	36.6

Although the explanation of the molecular interaction has not been completely unified, experimental findings have established numerical values of acceptable accuracy for a number of liquids, including some given in Table 6-1. Generally these values decrease somewhat with an increase in temperature. For water, the value decrease is approximately 0.2 dyn/cm/° C.

6-3 PRESSURE, CURVATURE, AND SURFACE-TENSION RELATIONSHIP

It is rather obvious that a balloon would not stay inflated if pressure differences did not exist between the inside and outside of the balloon. When inflated, the difference

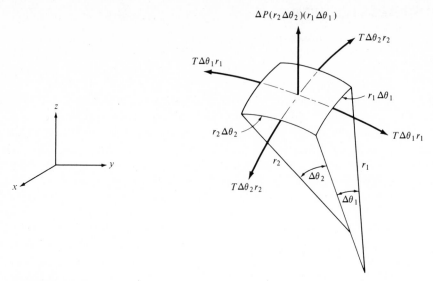

FIGURE 6-1
Element from a flexible membrane.

in pressure is resisted by tensile forces within the skin of the balloon. As we may remember from engineering mechanics in connection with thin-wall pressure vessels, a basic relationship could be developed between pressure differences, container curvature, and skin stresses. The analysis of a thin flexible membrane under tension provides a suitable analogy relating surface tension, pressure differences, and curvature associated with capillary moisture.

Figure 6-1 represents an element from a curved surface with two radii of curvature. For example, this may be a segment from a thin-wall pressure vessel with more than one radius of curvature, such as a football or an elliptical balloon.

Let T represent a surface tensile force per unit length and $\Delta P = P_i - P_o$ the difference in pressure between the inside and outside of the membrane. The total force on the projected surface of the element must be resisted by the surface tension components. From $\sum F_z = 0$, we have

$$(r_2 \Delta\theta_2)(r_1 \Delta\theta_1)\Delta P = 2(T\Delta\theta_2 r_2) \sin\left(\frac{\Delta\theta_1}{2}\right) + 2(T\Delta\theta_1 r_1) \sin\left(\frac{\Delta\theta_2}{2}\right)$$

For small angles the values of $\sin(\Delta\theta/2) = \Delta\theta/2$. Thus,

$$r_1 r_2 \Delta P = T r_2 + T r_1$$

or

$$\Delta P = T\left(\frac{1}{r_1} + \frac{1}{r_2}\right) \tag{6-1}$$

When the radii are the same (e.g., spherical shape), Eq. (6-1) becomes

$$\Delta P = \frac{2T}{r} = \frac{4T}{d} \tag{6-2}$$

A value for $T = 75$ dyn/cm ($= 75 \times 10^{-8}$ kN/cm) is frequently assumed for water temperatures below $20°$ C. Thus for water as a special case, Eq. (6-2) becomes

$$\Delta P = \frac{2T}{r} = \frac{4T}{d} = 4\frac{75 \times 10^{-8} \text{ kN/cm}}{d \text{ (cm)}} = \frac{300 \times 10^{-8}}{d} \text{ kN/cm}^2$$

or

$$\Delta P = \frac{0.03}{d \text{(cm)}} \text{ kN/m}^2 \tag{6-2a}$$

where r is the "single" radius of curvature, such as in a straight pipe or sphere, and $d = 2r$.

6-4 CAPILLARY RISE IN TUBES

We mentioned previously that if we were to insert the end of a small-diameter tube (*capillary* tube) into water, the water would rise into the tube, as indicated in Fig. 6-2. Surface tension combines with the attraction between the glass and water molecules (adhesion) to pull the water up into the tube to a height h_c, known as *height of capillary rise*

The surface of the liquid meets the tube at a definite angle, α, known as the *contact angle*. Pure water meets a clean glass at a contact angle $\alpha = 0°$; the contact angle for mercury is approximately $140°$. The surface of the water is curved rather than flat, forming what is known as a *meniscus*. When $\alpha = 0$, and for small-diameter tubes, the curvature of the meniscus is about equal to the radius of the tube.

In Fig. 6-2 point 1 is under atmospheric pressure; point 2 is under equal pressure since it is at the same level as point 1. However, point 3 (just under the meniscus) is higher than point 2 by an amount equal to h_c. Hence it experiences a pressure less than the atmospheric pressure by an amount of $\Delta P = h_c \gamma_w$. If the liquid is water, Eq. (6-2) can be written as

$$\Delta P = h_c \gamma_w = \frac{2T}{r} = \frac{4T}{d}$$

or

$$h_c = \frac{4T}{\gamma_w d} \tag{6-3}$$

For the special case where $T \doteq 75$ dyn/cm and for $\gamma_w = 9.807$ kN/m^3, Eq. (6-3) becomes

$$h_c = \frac{0.306}{d \text{(cm)}} \text{ cm} \tag{6-3a}$$

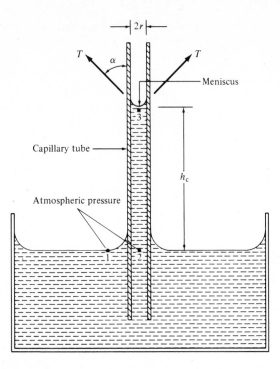

FIGURE 6-2
Capillary rise of water in a
small-diameter tube.

where h_c = height of capillary rise

T = surface tension of water

r = radius of curvature of meniscus $\approx$ radius of tube

$d = 2r$

γ_w = unit weight of water

    Equation (6-3) could also have been derived by assuming the column of water in the tube to be supported by the surface tension over the circumference of the tube. The weight of the water is $h_c(\pi d^2/4)\gamma_w$. The surface tension force is $\pi dT \cos \alpha$. As mentioned previously, for a clean tube $\alpha = 0°$. Therefore $\cos \alpha = \cos 0° = 1$. Thus equating the weight of the water to the surface tension forces, one obtains again Eq. (6-3).

EXAMPLE 6-1

Given   Two capillary tubes A and B of diameters $d_A = 0.10$ mm $= 0.01$ cm and $d_B = 0.01$ mm $= 0.001$ cm.

Find   (a)   The capillary rise in each tube h_{cA} and h_{cB}, respectively.
        (b)   The change in and values of capillary pressures in tubes A and B at a point just under the meniscus.

Procedure   Assume $T = 75$ dyn/cm and $\gamma_w = 9.807$ kN/m³. Equations (6-2a) and (6-3a) apply. From Eq. (6-3a), d is in centimeters,

$$h_{cA} = \frac{0.306}{d} = \frac{0.306}{0.01} = 30.6 \text{ cm (12.05 in)}$$

$$h_{cB} = \frac{0.306}{0.001} = 306 \text{ cm (120.5 in)}$$

Answer

> $h_{cA} = 30.6$ cm (12.05 in)
>
> $h_{cB} = 306$ cm (120.5 in)

From Eq. (6-2a),

$$\Delta P_{cA} = \frac{0.03}{d} = \frac{0.03}{0.01} = 3 \text{ kN/m}^2$$

$$\Delta P_{cB} = \frac{0.03}{d} = 30 \text{ kN/m}^2$$

Answer

> $\Delta P_{cA} = 3$ kN/m²
>
> $\Delta P_{cB} = 30$ kN/m²

1 atmosphere $= 1.013 \times 10^6$ dyn/cm² $= 101.3$ kN/m² (absolute)

$$P_{cA} = 101.3 - 3 = 98.3 \text{ kN/m}^2 \text{ (14.27 psi)}$$
$$P_{cB} = 101.3 - 30 = 71.3 \text{ kN/m}^2 \text{ (10.47 psi)}$$

Answer

> $P_{cA} = 98.3$ kN/m² (14.27 psi)
>
> $P_{cB} = 71.3$ kN/m² (10.47 psi)

EXAMPLE 6-2

Given   The end of a clean glass tube, $d = 0.1$ cm, is to be immersed in mercury held in a large container.

Find   The level of the mercury in the tube relative to that in the container.

Procedure $\alpha = 140°$ and $\gamma_{Hg} = 13.6 \times 9.807 = 133.4$ kN/m³. From Table 6-1, $T = (514.6$ dyn/cm) · $(10^{-8}$ kN/dyn$) = 514.6 \times 10^{-8}$ kN/cm. From Eq. (6-3), for corresponding terms,

$$h = \frac{4(514.6 \times 10^{-8} \text{ kN/cm}) \cos 140°}{(133.4 \text{ kN/m}^3)(0.1 \text{ cm})}$$

$$h = -118.2 \times 10^{-4} \text{ m} = -1.18 \text{ cm} (-0.465 \text{ in})$$

Answer

$$h = -1.18 \text{ cm} (-0.465 \text{ in}) \quad \text{(below)}$$

6-5 CAPILLARY TUBES OF VARIABLE RADIUS

Figure 6-3 will be used to illustrate the effects of the tube diameter on the capillary rise of water into the tube. Assume the tube shown in Fig. 6-3a–d to be of small diameter d_1, except for the one section which bulges out to an appreciably larger diameter d_2. If the tube is lowered into the water as shown in Fig. 6-3a, the water level will rise into the tube to a height h_c commensurate with the terms given by Eq. (6-3). If the tube is further lowered, as indicated in Fig. 6-3b, the water rise into the large section of the tube will be less than for case 6-3a. Again the capillary rise must correspond to that calculated on the basis of Eq. (6-3) with the corresponding diameter of d_2. If the tube is pushed further into the water, as shown in Fig. 6-3c, the rise h_{c1} will reach a definite value as dictated by Eq. (6-3) for a value of d_1. However, if the tube is raised as shown in Fig. 6-3d, the capillary rise of h_{c1} will be maintained. The same would occur in the case of Fig. 6-3a and b if the water table were to rise above the level of h_{c1}, then subsequently lowered. The water would theoretically remain in the tube at the height of h_{c1} when the water level outside the tube was lowered to its original

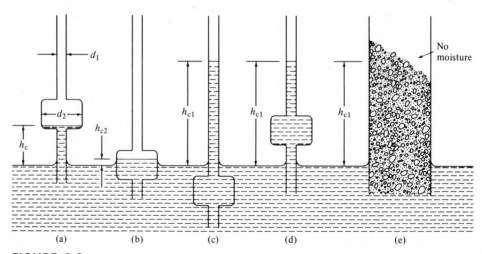

FIGURE 6-3
Capillary rise in tubes of variable cross sections.

level. In the same manner, if a water supply from above fills the tube, the water level in the tube would remain at the height h_{c1}, once equilibrium was reached.

Figure 6-3e depicts a tube of large diameter filled with soil particles. For the sake of discussion it is assumed that the soil pores or voids form continuous (although not necessarily straight) capillary tubes of a size corresponding to a diameter d_1. Theoretically the water would rise to a height h_{c1} within the tube where the soil is present. Note that in the part of the tube where no soil is present no capillary rise takes place in view of the assumed very large diameter.

6-6 CAPILLARY PHENOMENON IN SOILS

Unlike capillary tubes, the voids in soil are quite irregular in shape and size and not necessarily continuous (interconnected) or vertical. Indeed, the conditions regarding surface tension, capillary rise, and pressure variations in soils are much more complex than for the capillary tube discussed in the previous section, and perhaps rather impossible to investigate with any degree of accuracy. The concept presented in connection with capillary tubes, however, does serve as a good basis for describing the capillary phenomenon in soils.

One of the difficulties in applying the theory developed for capillary tubes to soils lies in determining the value for r or d. The voids within the typical soil mass may vary greatly in size and shape. Even for uniformly sized grains the voids are likely to vary in both size and shape. But for nonuniform grain sizes this variation in void characteristics is even more pronounced. Also, variations in soil stratification further add to this problem. Hence our evaluation of capillarity in soils is perhaps more qualitative than quantitative in nature.

It is convenient to relate the void sizes or pathways to the sizes of the soil particles. Generally, as the grain size decreases, the size of the voids or pores also decreases. Conversely, larger grain sizes would imply larger voids between soil particles. Thus the height of capillary rise in soils is tied, in part, to the size of the soil particles.

Figure 6-4a depicts a column of fine-grained soil, uniform in size, for which the amount of capillary rise is denoted by h_c. In Fig. 6-4b the soil column is composed of fine-grain particles separated by a layer of much coarser particles and presumably correspondingly larger voids. Since the capillary rise is inversely proportional to the size of the voids, the column of water cannot rise much past the fine-grain sizes (h_c') if the voids of the coarse material are very large. On the other hand, if that same layered mass is lowered into the water and raised again, or if the water table is raised and then lowered, the water within the soil mass would remain at a height equal to h_c; that is, $h_c'' = h_c$, as shown in Fig. 6-4. This is analogous to the conditions discussed in the previous section.

The rise of water in soils through capillary action does not necessarily mean total saturation. Far to the contrary, even at small heights above the free surface the degree of saturation for many soils is well below 100 percent, with even lower saturation percentage at higher elevations. Furthermore, not all moisture in a soil mass is necessarily related to the capillary phenomenon. Surface infiltration is a common

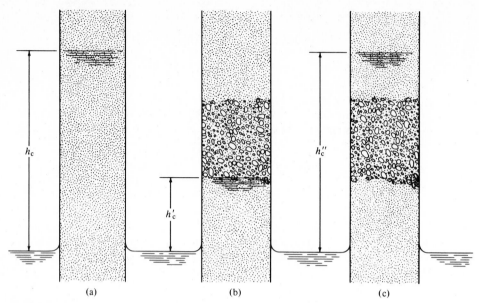

FIGURE 6-4
Capillary rise in columns of soil of varying grain sizes.

source for soil moisture. Fluctuation in the groundwater table may be still another source. Also moisture may get trapped and collect in minute voids within the soil mass and may be retained indefinitely. Such moisture is commonly referred to as *contact moisture*. It may disappear through evaporation if near the surface, but it is unlikely to be drawn down by gravity.

Because of the complex nature of the soil voids, a theoretical prediction of capillary rise in soils is not only of dubious accuracy, but is likely to be even misleading. Perhaps the most reliable approach is by direct observation of this behavior, preferably in situ if possible. Allen Hazen suggested a formula for approximating the capillary rise. This is given by Eq. (6-4):

$$h_c = \frac{C}{eD_{10}} \tag{6-4}$$

where D_{10} = Hazen's effective size (e.g., 10 percent size of grain-size analysis curve), in centimeters

C = empirical constant, which usually varies from 0.1 to 0.5 cm^2

e = void ratio

Another approximation, perhaps equally empirical, is the use of Eq. (6-3a), where D is approximately $D_{10}/5$, given in millimeters. The numerator of Eq. (6-3a) is adjusted accordingly to account for the change in units:

$$h_c = \frac{0.0306}{\frac{1}{5} D_{10} \text{ (mm)}} \quad \text{m} \tag{6-5}$$

6-7 CAPILLARY FORCES

For convenience Eq. (6-2) will be presented again at this point:

$$\Delta P = \frac{4T}{d} \qquad (6\text{-}2)$$

The quantity ΔP represents the *capillary pressure*. In Section 6-2 we have shown ΔP to be equal to the differences in pressure between the inside and the outside faces of the thin membrane. Another way to view the above equation is that for a given capillary tube diameter, the larger the difference ΔP, the larger the surface tension in the liquid.

Figure 6-5a shows the surface tension T at the contact angle α relative to the walls of the tube. The vertical component $T \cdot \cos \alpha$ results in axial compressive stresses within the wall of the tube. Component $T \cdot \sin \alpha$ results in compressive stresses in the wall of the tube in a circumferential direction. In an analogous manner, *contact moisture*, present in appreciable amounts in fine-grained soils, will form menisci which develop surface tension forces. In turn, these forces will pull adjacent soil grains together and in the process develop *intergranular pressure*. Figure 6-5b shows contact capillary moisture around portions of adjacent soil grains. Figure 6-5c shows an enlarged condition of one such contact point. The water around this point forms a meniscus, and thereby surface tensions are developed, analogous to that shown in Fig. 6-5a. The vertical component of surface tension around the point of contact develops the normal forces N, similar to the compressive forces developed in the walls of the tube in the axial direction. Quite apparently the normal forces N translate into intergranular pressures which are directly related to the magnitude of pull of the menisci on the soil grains.

Such increase in intergranular pressure may have appreciable effect upon the increase in the shearing strength in a given soil mass. This phenomenon may be demonstrated by reflecting upon the shear strength of a fine sand under three moisture conditions: (1) perfectly dry; (2) completely saturated; and (3) barely damp. For either a totally saturated or a totally dry sand no capillary forces exist; hence the shear strength of the sand is relatively small. On the other hand, in a damp condition the capillary forces increase the shear strength of the sand, thereby giving it more stability and increased resistance to deformation. This is readily apparent to one walking on a sandy beach when the soil is either totally dry or totally saturated; in both instances the walking is more difficult than for the damp condition. The damp state provides a relatively rigid surface and relatively minimal penetration compared to the totally saturated or dry state. This could be further demonstrated by being able to cut a steep cut in a moist sand versus a much shallower slope in the case of a totally dry or totally saturated sand. The capillary effects give the sand added shear strength, or *apparent cohesion*, which is nonexistent when the sand is either totally dry or totally saturated.

The total normal stress on a particular soil particle is equal to the sum of effective and neutral stresses (or intergranular pressure and pore pressure, respectively) discussed in Section 8.2. The combined stresses may be expressed by Eq. (6-6):

$$\sigma_t = \bar{\sigma} + u_w \qquad (6\text{-}6)$$

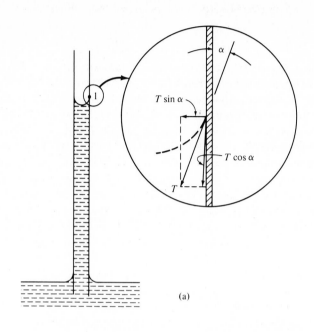

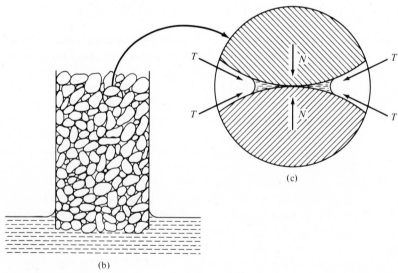

FIGURE 6-5
Capillary stresses.

where σ_t = total (combined) pressure from weight and capillary pressures

$\bar{\sigma}$ = normal component of intergranular pressure

u_w = pore pressure

The total normal stress across any unit area within a soil mass remains about constant. However, the stress in the water (pore stress) may vary appreciably. For example, at a height h above the free-water level, the hydrostatic pressure in the water $u_w = -h\gamma_w$. Thus from Eq. (6-6) we see that when the stress in the water is negative, the intergranular pressure must be greater than the combined pressure by an amount equal to the pore-water pressure u_w.

Significant reversal in pore-water stress may be introduced with fluctuations of the water table or alterations in the general state of the sample. For example, say a clay sample at some depth below the surface is subjected to an intergranular pressure plus a neutral pressure of $\bar{\sigma}$ and u, respectively. Obviously the combined pressure is $\sigma_t = \bar{\sigma} + u$. If the sample were extracted from the ground, the combined pressure would become 0. Hence it has a tendency to expand. However, it is likely to be restricted against expansion by the capillary forces formed in the process. That is, the menisci are formed at the surface of the soil grains, and thereby capillary forces will be likewise developed. Thus the pore-water pressure changes from a pressure of magnitude u to a tension of magnitude $\bar{\sigma}$ in the extracted state. (i.e., $u = \bar{\sigma}$, provided the "tensile" strength of water is not exceeded).

6-8 SHRINKAGE, SWELLING, AND FROST HEAVE

A large reduction of the moisture content in wet clayey soils introduces intergranular pressures, as described above and similar to those caused by overburden pressures. In the process significant reduction in volume, or *shrinkage*, will develop. This could be easily demonstrated by running a shrinkage test as described in Section 4-6. Conversely, if a dry clay is exposed to moisture, the clay will experience *swelling*. As discussed in Section 2-4, this may be attributed to a number of the clay properties: (1) the clay minerals' affinity for water; (2) the base exchange behavior and electrical repulsion feature; (3) the expansion of entrapped air; and (4) the elastic rebound of the soil particles. If a clay whose water content is below the shrinkage limit is immersed in water, capillary attraction forces the water into the soil voids. In the process, air which entered the voids during drying becomes compressed by the entering water. In turn, this increases the interparticle tension until an explosion of the voids occurs, with subsequent disintegration of the soil. This is known as *slaking.*

In cold regions a common problem related to capillarity is *frost* and *frost heave.* If outside cold temperatures persist for some time, the soil near the surface may freeze. However, the soil below a certain depth, the *frost line*, will be above 0° C (32° F). Hence if sufficient saturation exists, some of the water in the voids will freeze, thereby creating some moisture deficiency and increased capillary tension in areas adjoining the ice crystals. Thus additional water is sucked up from below the frost line, thereby adding to the size or enlarging the ice crystals. This process results in the formation of *ice lenses* which increase in size with prolonged low temperatures, causing the familiar frost heave so common under pavements.

Prevention or minimizing of frost damages may be accomplished by removing the frost-susceptible soil to near the frost line and replacing it with clean (e.g., free of silt or clay) sands and gravels. When the water source is via capillarity from the water table, one possible remedy may be to lower the water table by means of drains or ditches if the topography permits natural gravitational flow. Impervious blankets such as plastics or asphalt cements may be an effective but perhaps more costly approach. On the other hand, the removal of water from surface infiltration is equally important if frost-susceptible surface material is present.

PROBLEMS

6-1.   Determine the capillary rise of water in clean glass tubes of: (a) $d = 0.04$ mm. (b) $d = 0.08$ mm. (c) $d = 0.15$ mm.
Assume the water temperature at 20°C.

6-2.   Determine the rise of water in Problem 6-1 if the contact angle is 30°.

6-3.   The rise in a capillary tube is 52 cm above the free-water surface. Determine the surface tension if the radius of the tube is 0.03 mm.

6-4.   What is the diameter of a clean glass tube if water rises to a total of 76 cm at approximately 20°C temperature?

6-5.   Determine the water pressure in the capillary tube of Problem 6-3.
   (a)   At 30 cm above the free surface.
   (b)   At 40 cm above the free surface.
   (c)   Just below the meniscus.
   (d)   Just 1 cm above the free-water surface.

6-6.   Determine the water pressure in the capillary tube of Problem 6-4.
   (a)   At 30 cm above the free surface.
   (b)   At 50 cm above the free surface.
   (c)   Just below the meniscus.
   (d)   Just 1 mm above the water surface.

6-7.   The effective size of a sandy silt is 0.04 mm and its void ratio is 0.7. Determine the approximate capillary rise for this soil. Assume $C = 0.32$.

6-8.   Rework Problem 6-7 using Eq. (6-5). How do the two results compare? Give plausible reasons for the difference.

6-9.   Rework Problem 6-7 for different void ratios: (a) 0.8. (b) 0.9. (c) 0.95. (d) 1.0. How does the void ratio appear to affect the capillary rise? Explain.

6-10.   Rework Problem 6-7 for different effective sizes: (a) 0.05 mm. (b) 0.06 mm. (c) 0.07 mm. (d) 0.08 mm. Plot the values of h_c vs. grain sizes. What appears to be the relationship between the two?

7

Seepage

7-1 INTRODUCTION

It may be quite apparent that if one were to pour water on a sandy or gravelly surface, the water would disappear into the ground. It may be equally apparent that one may not be successful in constructing an efficient dam from a sandy or gravelly soil; the water would *seep* out through the dam quite easily. On the other hand, the water flow through a fine-grain soil, such as silt or clay, would take place with more difficulty. In short, the quantity of flow, other conditions being equal (e.g., hydrostatic heads, stratum thickness, time, etc.), would be much greater in the granular soil than in the silty or clayey soil. The process of water flow through soil is commonly referred to as *seepage.*

Problems associated with the seepage phenomenon are likely to fall into one of the three groups dealing with: (1) flow into pits or out of reservoirs; (2) seepage pressures and related effects which they may have on the stability of slopes, cuts, foundations, etc.; (3) drainage from fine-grain soils subjected to load increase.

Like so many problems in soil mechanics, seepage analysis is frequently not much more than a reasonable estimate. The reason for this may very likely lie in the many assumptions that are made and which are most difficult to verify with any degree of accuracy. For example, it is not unusual to assume that: (1) the stratum characteristics (e.g., permeability, thickness, stratification) are homogeneous, yet it is far more probable than not that this is not so; (2) the laws of hydraulics (e.g., hydraulic gradient, flow pattern, continuity) are suitable tools and applicable in the flow analysis; (3) any apparent variables (e.g., permeability differences in the vertical and horizontal directions, boundary conditions, steady-state flow) could be anticipated and accounted for with reasonable accuracy. Yet accuracy of these assumptions is perhaps more a matter of conjecture than of substantiated proof. Nevertheless methodologies do exist which give acceptable results. It is the purpose of this chapter to present and evaluate some of the techniques for analyzing seepage flow. For the sake of simplicity, only a two-dimensional analysis will be presented here.

7-2 SEEPAGE FORCES

Figure 7-1 will be used as the basis of our discussion in connection with *seepage stresses* and *seepage forces.*

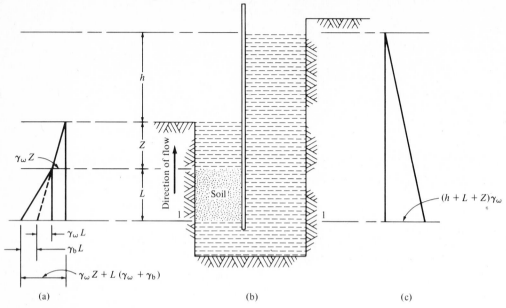

FIGURE 7-1
Water pressures on soil sample of cross section A.

In Fig. 7-1b we show a soil sample totally submerged. For convenience we shall assume total saturation, although, as we have noted in previous sections, total saturation is virtually impossible.

Figure 7-1a depicts the pressures associated with this arrangement. Assuming no water flow in the upward or downward directions, we note that a force from the hydrostatic head acts in all directions on a given particle of soil. Obviously this force would tend to compress the particles, but would not tend to move them relative to each other. Thus this force does not create any shear effects between soil particles. This type of pressure is commonly referred to as *neutral pressure* or *pore water pressure*. On the other hand, the weight of the particles above any given level of soil is supported by the corresponding particles beneath. In the process, forces between particles are induced by this weight. It is rather evident that the lower particles will thus experience greater stress or pressure from such loads than is experienced by the higher level particles; they simply carry more overburden (see also Section 8-2). This is referred to as *intergranular pressure* or *effective pressure*. In Fig. 7-1a $\gamma_w L$ is equal to the pore water pressure, and $\gamma_b L$ is the intergranular pressure.

Now let us reflect on Fig. 7-1c. This is a sketch of the hydrostatic pressures induced by the column of water. Since the pressure will act in all directions, the particles of soil at a given section, say 1–1 in Fig. 7-1b at the bottom of soil column, experience downward as well as upward loads. This is indicated in Fig. 7-2 for an

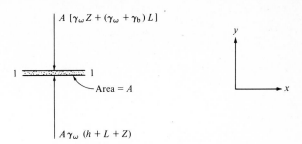

FIGURE 7-2
Forces on soil sample, section 1–1 in Fig. 7-1b.

assumed cross section A. The horizontal pressures on the vertical faces of the sample cancel. Hence by summing forces in the y direction, we get

$$F_{1-1} = A(h + L + Z)\gamma_w - A[\gamma_w Z + L(\gamma_w + \gamma_b)]$$

Simplifying,

$$F_{1-1} = (\gamma_w h - \gamma_b L)A \qquad (7\text{-}1)$$

where $\gamma_w hA$ = seepage force

$$\gamma_b LA = \text{buoyant force} = \gamma_w L\left(\frac{G + Se}{1 + e} - \gamma_w\right)A$$

F_{1-1} = net force at section 1–1

G, S, e are specific gravity of solids, degree of saturation, and void ratio, respectively.

The term $\gamma_w h$ in Eq. (7-1) represents the seepage pressure, while the product $\gamma_w hA$ represents the seepage force. Seepage forces may be visualized as being the result of the drag force by water against the soil particles and the associated reaction by the soil particles to the water. These forces are in the direction of flow.

Equation (7-1) represents the net force on the soil particles at section 1–1. One notes that without the seepage forces given by the expression $\gamma_w hA$ the net force at this section would be the buoyant force. Hence the intergranular pressure would simply be $\gamma_b L$. Thus it is readily apparent that the introduction of the seepage force alters the net force at this particular section. In turn, the seepage force alters the intergranular pressure in the soil mass. An important consequence of this effect will be illustrated in the following section.

7-3 QUICK CONDITION

Reference is made to Eq. (7-1). As mentioned in the previous section, this equation represents the net force at section 1–1 of Fig. 7-1. By increasing the seepage pressures $\gamma_w h$, a point may be reached where the two terms in parentheses would equal. That is, at this point the seepage forces would equal the buoyant forces. This may be viewed as a condition of impending upward movement. Obviously any further increase in the seepage force would result in actual movement. Such a buoyant condition is generally referred to as a *quick* or a *boiling* condition (conditions which result in impending upward movement of soil and water).

At a point at which a quick condition exists, the net force would equal 0. Hence equating Eq. (7-1) to 0 and solving for the ratio of h to L for which boiling occurs, we have

$$0 = (\gamma_w h - \gamma_b L)A; \qquad A \neq 0$$

Thus,

$$0 = \gamma_w h - \gamma_b L = \gamma_w h - \gamma_w L\left(\frac{G + Se}{1 + e} - \gamma_w\right)$$

Assuming total saturation ($S = 1$) and expanding, we get

$$h = L\left(\frac{G - 1}{1 + e}\right)$$

or

$$\frac{h}{L} = \frac{G - 1}{1 + e} \tag{7-2}$$

By definition, the ratio of h to L is referred to as *gradient* whose magnitude is given by $(G - 1)/(1 + e)$. When its value equals 1, it is commonly known as the *critical gradient*—the condition for impending boiling. Hence for some rather common values of $G = 2.7$ and $e = 0.7$ (approximately), h/L is about unity.

As mentioned above, the quick condition was based on a net or effective stress equal to 0. At this point the shear strength of the soil would theoretically appear to be lost. However, the zero effective stress does not necessarily mean boiling in cohesive soils since they display some shear strength even at zero effective stress. On the other hand, a fine-grained cohesionless soil (fine sand, for example) is most likely to be subject to boiling. Furthermore, the probability of boiling is greater for fine sand than for coarse sand or gravel strata. The coarse-grain soils display a greater porosity and permeability. Therefore a larger supply of water would be needed to maintain a gradient of unity—the critical gradient. Hence although it is theoretically possible to have a quick condition in coarse-grained soils, the volume of water necessary for the critical gradient makes such an occurrence unlikely.

Boiling frequency occurs in fine sands when the depth of excavation is a certain distance below the water table. That is, although the sides of the excavation are shored and properly supported laterally, the bottom of the excavation may flow upward when the critical gradient is reached. Sometimes a solution to such a problem may be found by driving sheet piling to depths such that L is large enough to reduce the gradient h/L to a value of 1 or less. Other common examples may include boiling due to artesian pressures and boiling near the downstream part of an earth dam.

7-4 ELEMENTS OF FLOW NET THEORY

Let us consider the element shown in Fig. 7-3 taken from a soil mass through which we assume a steady-state laminar flow. Furthermore assume that the permeability in the x, y, and z directions varies; the coefficients of permeability in the x, y, and z directions will, therefore, be designated as k_x, k_y, and k_z, respectively. If the total head in the element is h, then the gradient in the x direction, over a distance dx, is $i_x = -\partial h/\partial x$, with the minus sign indicating a loss of head with distance in the direction of flow. The change in the gradient i_x per unit distance in the x direction may be written as $\partial i_x/\partial x = -\partial^2 h/\partial x^2$. Thus using Darcy's law, the component of flow *into* the element q_{xi} is

$$q_{xi} = k_x i_x \, dA = k_x \left(-\frac{\partial h}{\partial x} \right) dy \, dz \qquad (a)$$

Similarly, the flow q_{xo} *out of* the element is

$$q_{xo} = \left(k_x + \frac{\partial k_x}{\partial x} dx \right) \left(-\frac{\partial h}{\partial x} - \frac{\partial^2 h}{\partial x^2} dx \right) dy \, dz$$

Expanding, we obtain

$$q_{xo} = k_x \left(-\frac{\partial h}{\partial x} \right) dy \, dx - \left(k_x \frac{\partial^2 h}{\partial x^2} + \frac{\partial k_x}{\partial x} \frac{\partial h}{\partial x} + \frac{\partial k_x}{\partial x} dx \frac{\partial^2 h}{\partial x^2} \right) dx \, dy \, dz \qquad (b)$$

Now let Δq_x represent the difference in water volume entering and exiting. Thus we have

$$\Delta q_x = q_{xi} - q_{xo} = \left(k_x \frac{\partial^2 h}{\partial x^2} + \frac{\partial k_x}{\partial x} \frac{\partial h}{\partial x} + \frac{\partial k_x}{\partial x} dx \frac{\partial^2 h}{\partial x^2} \right) dx \, dy \, dz \qquad (c)$$

If the permeability in the x direction does not change, then $\partial k_x/\partial x = 0$ and Eq. (c) becomes

$$\Delta q_x = k_x \frac{\partial^2 h}{\partial x^2} dx \, dy \, dz \qquad (d)$$

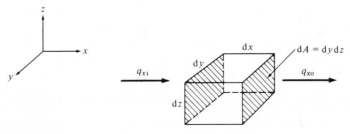

FIGURE 7-3

q_{xi} = quantity of flow into element in the x direction. q_{xo} = quantity of flow out-of-element in the x direction.

Similarly, the corresponding expression for flow in the y and z directions may be written as

$$\Delta q_y = k_y \frac{\partial^2 h}{\partial y^2} \, dx \, dy \, dz \qquad \text{(e)}$$

$$\Delta q_z = k_z \frac{\partial^2 h}{\partial z^2} \, dx \, dy \, dz \qquad \text{(f)}$$

Thus the expression for total flow is

$$\Delta q = \Delta q_x + \Delta q_y + \Delta q_z$$

or

$$\Delta q = \left(k_x \frac{\partial^2 h}{\partial x^2} + k_y \frac{\partial^2 h}{\partial y^2} + k_z \frac{\partial^2 h}{\partial z^2} \right) dx \, dy \, dz \qquad \text{(g)}$$

But since we assumed a steady-state condition, $\Delta q = 0$. Also, for two-dimensional flow (say, in the y–z plane), the flow in the y direction is zero. Hence for two-dimensional flow and steady-state condition, Eq. (g) becomes

$$\left(k_x \frac{\partial^2 h}{\partial x^2} + k_z \frac{\partial^2 h}{\partial z^2} \right) dx \, dy \, dz = 0 \qquad \text{(h)}$$

But since $dx \, dy \, dz \neq 0$, Eq. (h) becomes

$$k_x \frac{\partial^2 h}{\partial x^2} + k_z \frac{\partial^2 h}{\partial z^2} = 0 \qquad \text{(7-3)}$$

For the special case where the permeability in both the x and the z directions is the same, Eq. (7-3) becomes

$$\frac{\partial^2 h}{\partial x^2} + \frac{\partial^2 h}{\partial z^2} = 0 \qquad \text{(7-4)}$$

This is Laplace's equation, which gives the fundamental relationship for steady-state flow in isotropic soils. Simply it states that gradient change in the x direction plus gradient change in the z direction equals zero. It represents the equation of families of two groups of curves which intersect in the x–z plane. One group is referred to as *stream lines* or *flow lines*; the second group consists of lines of constant head, or *equipotential lines*. Furthermore the lines of the groups intersect at right angles. This could be shown as follows. Let S_1 and S_2 in Fig. 7-4 represent two stream lines, and H_1 and H_2 two equipotential lines, as shown. Let v represent the tangential velocity of a fluid particle passing through a point $M(x, z)$; the x and z components of v are v_x and v_z, as shown. Hence $v_x/v_z = dx/dz$. But $v_x = \partial h/\partial x$ and $v_z = \partial h/\partial z$. Therefore,

$$\frac{dx}{dz} = \frac{\partial h/\partial x}{\partial h/\partial z} \qquad \text{(i)}$$

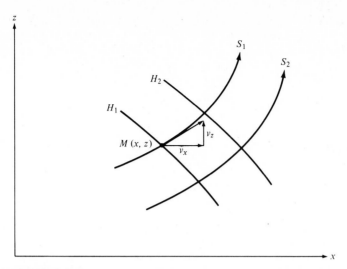

FIGURE 7-4
Stream lines and equipotential lines.

Along an equipotential line (say, H_1) d$h = 0$. Thus,

$$\frac{\partial h}{\partial x}\, \mathrm{d}x + \frac{\partial h}{\partial z}\, \mathrm{d}z = 0$$

or

$$\frac{\mathrm{d}x}{\mathrm{d}z} = -\frac{\partial h/\partial z}{\partial h/\partial x} \qquad\qquad (\mathrm{j})$$

Equation (j) is the negative reciprocal of Eq. (i), thus completing the proof.

Plotting the flow or stream lines and the equipotential lines results in what is commonly referred to as a *flow net*, to be discussed in the next three sections, for two-dimensional flow. The more common condition of flow is for $k_x \neq k_z$.

7-5 FLOW NETS—TWO-DIMENSIONAL FLOW

Seepage losses through the ground or through earth dams and levees and the related flow pattern and rate of energy loss, or dissipation of hydrostatic head, are frequently estimated by means of a graphical technique known as a *flow net* (3).

Figure 7-5 represents an example of a flow net. The path followed by a particle of water as it moves through a saturated soil mass is called a *flow line* or *stream line*. These are shown as solid lines. Assuming laminar flow (typical assumption for such analysis), the flow lines never cross each other. Each of the flow lines in Fig. 7-5 starts at a point where the hydrostatic head is equal to h and ends up on the free surface where the hydrostatic head is equal to 0. The viscous friction in the soil mass has dissipated a hydrostatic head of value h in each of these flow lines along its path of flow.

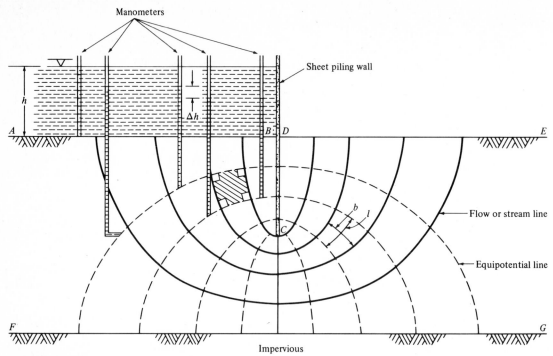

FIGURE 7-5
Flow net for sheet pile cofferdam.

Hence along each flow line there must be a point where the total head or energy is the same for any other line. A line which connects these points of equal head is called an *equipotential line*. In Fig. 7-5 the dashed lines are identified as equipotential lines. It is readily apparent that an infinite number of flow lines and corresponding equipotential lines exists for any given condition.

The hydraulic gradient between two adjacent equipotential lines is the difference in head divided by the distance between these two lines, that is, $i = \Delta h/\Delta L$. The gradient is a maximum along a flow path perpendicular to the equipotential lines. Since ΔL is the shortest distance, i has a maximum value for any given Δh. Hence the flow lines cross the equipotential lines at right angles, since in isotropic soil, flow occurs along paths of greatest or steepest gradient. Therefore the flow lines and the equipotential lines in Fig. 7-5 form a family of mutually orthogonally intersecting curves. This was shown in Section 7-4.

Although an infinite number of flow lines and equipotential lines could be drawn for any given condition, in drawing a flow net it is convenient to limit the number of flow lines and equipotential lines. The number should be greatly influenced by one important consideration: *the geometric figures formed by the equipotential lines and flow lines in the construction of a flow net should approach a square shape as much as possible.* Obviously not all of the blocks in Fig. 7-5 are squares. What is important, however, is to proportion the majority of these figures into approximate squares. If this is adhered to, then (1) the diagonals of the "squares" will have approximately equal lengths; and (2) the intersection of the equipotential and flow lines will be a 90° angle.

Figure 7-5 illustrates a flow net for a two-dimensional case. It assumes that all flow conditions in other parallel planes are similar. Although the flow of water through a soil mass usually poses a three-dimensional situation, the analysis for the three-dimensional case is rather complex and of limited practical value for the purpose of illustrating the fundamentals involved in flow net construction. Hence our discussion will be limited to the two-dimensional case.

7-6 BOUNDARY CONDITIONS

Prior to the construction of a flow net, one needs to evaluate the hydraulic boundary conditions associated with that particular problem and subsequently establish the characteristics of the flow lines and equipotential lines. Sometimes these conditions are well defined and easily established. However, quite frequently the conditions have to be established on the basis of a combination of subsurface exploration and testing, reasonable assumptions, and sound judgment.

For a given set of boundary conditions and steady-state flow through a soil mass, only one flow net could be drawn. Should the boundary conditions be altered, however, adequate time must be provided for the flow to reach steady state. Hence a new flow net could be drawn to conform to the new or altered conditions. For example, let us assume that instead of the sheet piling wall shown in Fig. 7-5 we were to use a heavy concrete wall as shown in Example 7-2. Obviously the pattern of flow, and therefore the flow net, is appreciably different from that for the sheet piling case. Furthermore, should the stratum support a very heavy concrete overburden and be a highly compressible one, additional changes may occur in the flow rate as a result of the changes in the permeability of the soil mass. In fact, some time would pass before a steady flow would be resumed and until a new flow net could be drawn to reflect on the changed conditions.

Let us refer to Fig. 7-5 and focus on the boundary conditions associated with this particular case:

1. On the upstream face (face of the dam) line AB represents an equipotential line of a value equal to h.
2. On the downstream face the surface of exit denoted by line DE is an equipotential line of value equal to 0.
3. The line BCD following on each side of the sheet piling is a flow line.
4. The boundary between the pervious and the impervious strata designated by line FG is a flow line.

The boundary conditions are not always as well defined or easily determined. For example, the delineation of a pervious and an impervious layer is not always as well established, nor is the degree of permeability and the transition from one to the other quite as idealistic as presented. Furthermore, frequently the stratum which is classified as pervious may indeed vary from the upstream to the downstream surface. Also, the hydrostatic head on the upstream side may vary with changes in topography (e.g., sedimentary deposits, etc.) and water elevation or pool level.

The actual head at any time is the combination of pressure head and elevation head; the velocity head is inconsequential. In Fig. 7-5 we see manometers which could be used for direct measurement of the head. For any given equipotential line the reading on the manometers would be the same for the same equipotential line. This method could be used throughout the site and, in fact, is used on occasions to determine the new flow conditions when the boundary conditions appear to be appreciably altered or the flow appears to be different from that anticipated by the flow net determination.

7-7 DRAWING OF FLOW NET FOR TWO-DIMENSIONAL FLOW

Although flow nets for many hydraulic structures may be obtained experimentally via model studies, the graphical or trial-and-error method is generally the most practical, least expensive, quickest, most convenient, and, perhaps, the most common. Other methods such as electrical analogy or mathematical analogies (4, 7, 10) are generally more demanding, requiring special knowledge or elaborate equipment, or both. They are, however, sometimes used as a check on the graphical method. Hence in this section we shall summarize the steps required for the successful determination or construction of a flow net.

The following procedure should prove helpful in the construction of flow nets:

1. Draw the hydraulic structure, the head water elevation (and tail water, if any), and the soil profiles to a convenient scale.
2. Establish the boundary conditions. Usually this means two boundary flow lines and two boundary equipotential lines.
3. Sketch one flow line or one equipotential line adjacent to a boundary flow line or a boundary equipotential line. Keep in mind that they must intersect at right angles.
4. Expand the sketching to more equipotential lines and flow lines, always keeping in mind that roughly square figures should result in the process. In spite of the fact that many of the figures are far from square (in fact, none are perfect squares), this procedure gives remarkably accurate results.
5. Examine the flow net as it is drawn up to this point. It is very unlikely that a completely acceptable flow net will be obtained during the first trial. It is seldom that even the experienced sketcher will draw a totally acceptable or accurate flow net during the first trial. Hence look for apparent flaws in the net result (usually

nonsquare shapes and poor angles of intersection between equipotential and flow lines) and redraw the flow net correcting some of these apparent inaccuracies. Usually two trial approaches are sufficient for most instances.

Five or six flow lines are usually sufficient for most cases. In fact, too many flow lines, especially during the first trial, usually prove cumbersome and rather difficult to maintain in conformity with the stipulated conditions of squares and right angles.

7-8 DETERMINATION OF SEEPAGE QUANTITY FROM FLOW NETS

The typical soil mass is not isotropic. Normally the coefficient of permeability in the horizontal direction is considerably greater than that in the vertical direction. Hence we must transform the section of the hydraulic structure to account for this difference.

Let us refer to Fig. 7-5. In a general way we may think of the quantity of discharge q as the product of the number of flow paths n_f and the quantity of flow per flow path Δq. Thus,

$$q = n_f \cdot \Delta q \tag{a}$$

The total head h is equal to the product of all the equipotential drops n_d and the increment of head loss Δh. Thus we may express the above as

$$h = n_d \cdot \Delta h \tag{b}$$

From Darcy's law the quantity of flow through any "square" is

$$q = kib = k \frac{\Delta h}{l} b \tag{c}$$

where b represents the distance between flow paths, and l is the distance between equipotential lines, as shown in Fig. 7-5. Since we regard these as square figures and therefore $l = b$, Eq. (c) may be written as

$$\Delta q = k \cdot \Delta h \tag{d}$$

Substituting Eqs. (a) and (b) into Eq. (d), we get

$$q = \frac{n_f}{n_d} kh \tag{7-5}$$

One notes that the value of n_f and therefore n_d, may vary for any given situation, but the ratio n_f/n_d should remain a constant if the flow net is properly drawn.

In the case of anisotropic soils (say, $k_x > k_z$) the flow net is drawn for the transformed section. In Fig. 7-6 we see the same section to transformed and natural scales.

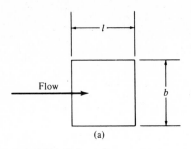

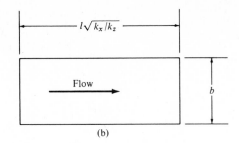

FIGURE 7-6
(a) Transformed scale. (b) Natural scale.

The quantities of flow Δq_T and Δq_N through the two respective sections may be expressed by

$$\Delta q_T = k_e \frac{\Delta h}{l} b = k_e \cdot \Delta h \qquad (a)$$

and

$$\Delta q_N = k_x \frac{\Delta h \cdot b}{l\sqrt{k_x/k_z}} = k_x \frac{\Delta h}{\sqrt{k_x/k_z}} \qquad (b)$$

But $\Delta q_T = \Delta q_N$. Thus we have

$$k_e \cdot \Delta h = k_x \frac{\Delta h}{\sqrt{k_x/k_z}} \qquad (c)$$

Simplifying, we have

$$k_e = \sqrt{k_x k_z} \qquad (7\text{-}6)$$

where k_x = coefficient of permeability in x direction
k_z = coefficient of permeability in z direction
k_e = effective coefficient of permeability

EXAMPLE 7-1

Given $k_x = k_z = 22 \times 10^{-4}$ cm/s for arrangement shown in Fig. 7-7.

Find   (a) The seepage loss per meter of width. (b) The pressure heads (pore pressure) at the 12 points shown in Fig. 7-7.

Procedure   The flow net is drawn as shown in Fig. 7-7.
(a) $n_f = 4, n_d = 10$. Thus,

$$q = \frac{n_f}{n_d} kh = \tfrac{4}{10} (22 \times 10^{-4} \text{ cm/s})(7 \text{ m})(100 \text{ cm/m})(100 \text{ cm/m width})$$

Answer

$$q = 61.60 \text{ cm}^3/\text{s/m}$$

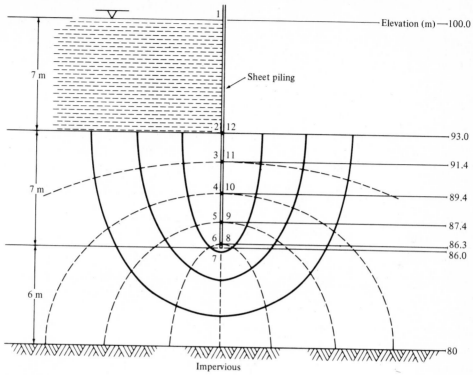

FIGURE 7-7

(b)

Point	Elevation Head (m)	Dissipated Head (m)	Total Head (m)	Pressure Head (m)
1	100.00	0	100.00	0
2	93.00	0	100.00	7.00
3	91.40	0.70	99.30	7.90
4	89.40	1.40	98.60	9.20
5	87.40	2.10	97.90	10.50
6	86.30	2.80	97.20	10.90
7	86.00	3.50	96.50	10.50
8	86.30	4.20	95.80	9.50
9	87.40	4.90	95.10	7.70
10	89.40	5.60	94.40	5.00
11	91.40	6.30	93.70	2.30
12	93.00	7.00	93.00	0

$$\Delta h = \frac{h}{n_d} = \frac{7}{10} = 0.7.$$

Answer

Sample: (pressure head)$_5$ = 97.90 − 87.40 = 10.50 m

EXAMPLE 7-2

Given $k_x = k_z = 30 \times 10^{-4}$ cm/s; other data as shown in Fig. 7-8.

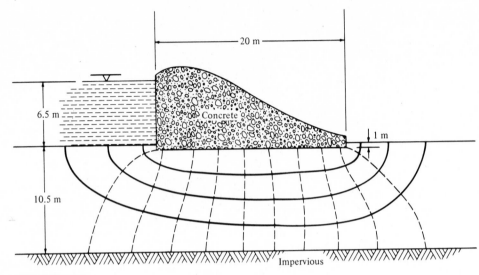

— 20 m —

6.5 m

Concrete

1 m

10.5 m

Impervious

FIGURE 7-8

Find   The quantity of discharge q per meter of width.

Procedure $n_f = 4$, $n_d = 11$. Thus,

$$q = \left(\frac{n_f}{n_d}\right)kh = \left(\frac{4}{11}\right)(30 \times 10^{-4} \text{ cm/s})(6.5 \text{ m})$$

or per 1 m = 100 cm of width,

$$q = (\tfrac{4}{11})(30 \times 10^{-4} \text{ cm/s})(6.5 \text{ m})(100 \text{ cm/m})(100 \text{ cm/m width})$$

Answer $q = 70.91 \text{ cm}^3/\text{s/m} = 7.09 \times 10^{-5} \text{ m}^3/\text{s/m width}$

EXAMPLE 7-3

Given   Refer to Example 7-2. For the given boundary values assume $k_x = 30 \times 10^{-4}$ cm/s, but $k_z = 6 \times 10^{-4}$ cm/s.

Find   The quantity of discharge q per meter of width.

Procedure   Figure 7-9 shows the tranformed section $n_f = 4$, $n_d = 8$. Thus,

$$q = \left(\frac{n_f}{n_d}\right)k_e h = \left(\frac{n_f}{n_d}\right)\sqrt{k_x k_z}\, h$$

$$q = (\tfrac{4}{8})(\sqrt{30 \times 6}) \times 10^{-4} \times 6.5 \text{ m} \times 100 \text{ cm/m} \times 100 \text{ cm/m width}$$

$$q = 43.60 \text{ cm}^3/\text{s/m}$$

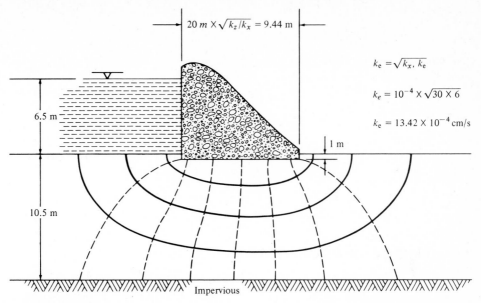

FIGURE 7-9

In the figure:

$20\ m \times \sqrt{k_z/k_x} = 9.44\ m$

$6.5\ m$

$10.5\ m$

$1\ m$

Impervious

$k_e = \sqrt{k_x, k_e}$

$k_e = 10^{-4} \times \sqrt{30 \times 6}$

$k_e = 13.42 \times 10^{-4}\ cm/s$

Answer

$$q = 4.3 \times 10^{-5}\ m^3/s/m\ \text{width}$$

EXAMPLE 7-4

Given $k_x = k_z = 30 \times 10^{-4}$ cm/s (see Example 7-2). There is a 5-m *cutoff wall* (only difference from Example 7-2). See Fig. 7-10.

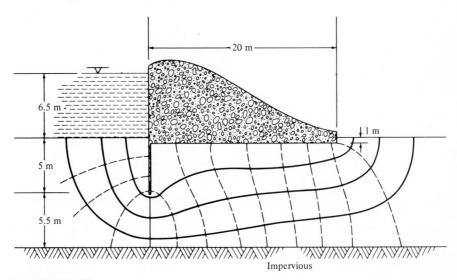

In the figure:

$20\ m$

$6.5\ m$

$5\ m$

$5.5\ m$

$1\ m$

Impervious

FIGURE 7-10

Find   The quantity of discharge q per meter width

Procedure $n_f = 3.5$, $n_d = 13$. Thus,

$$q = \left(\frac{3.5}{13}\right)(30 \times 10^{-4} \text{ cm/s})(6.5 \text{ m})(100 \text{ cm/m})(100 \text{ cm/m width})$$

$$q = 52.50 \text{ cm}^3/\text{s/m}$$

Answer

$$q = 5.25 \times 10^{-5} \text{ m}^3/\text{s/m}$$

Note. In comparison with Example 7-2, the cutoff wall reduced q by 26 percent.

7-9 SEEPAGE THROUGH EARTH DAMS

The same basic idea employed in the construction of flow nets for underdam flow (concrete, cutoff walls, etc.) is applicable to earth dams. However, the boundary conditions for the latter must be adjusted as described below.

The upstream face of the dam, *AD* in Fig. 7-11, is an equipotential line; *DC* is a flow line. Line *AB* represents the top flow line. All the seepage through the dam occurs below this line. Furthermore the pressure head is zero on all points on this line, i.e., the pressure in the soil water is equal to atmospheric pressure, and therefore the total head is equal to the elevation head. This line, commonly referred to as a *phreatic* line, forms the upper flow boundary and is the line of demarcation between the saturated and the relatively dry soil in the earth dam.

Except for the rock-filter toe, the material for the dam in Fig. 7-11 is assumed reasonably homogeneous and isotropic. Hence the flow line *AB* is perpendicular to equipotential line *AD*, and the flow net would be shaped as shown. However, both the upstream and the downstream or discharge points may be different from those shown in Fig. 7-11, thereby resulting in a different flow line character. Figure 7-12 shows some additional forms of dam cross sections and the corresponding upper or

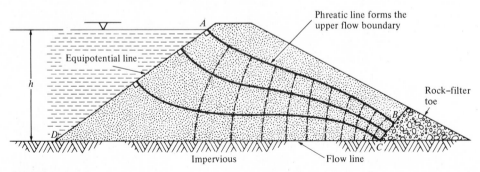

FIGURE 7-11
Flow net for steady-state flow through a homogeneous dam.

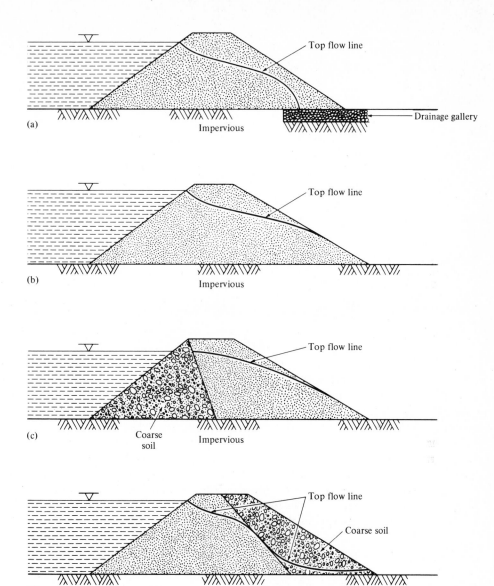

FIGURE 7-12
Top flow lines for various earth dam cross sections.

top flow lines. With the top flow line established and, therefore, the corresponding boundary condition determined, one may proceed to construct the flow net in the manner outlined previously.

The top flow line is close to parabolic in shape for most of its length. It deviates from a parabola (the character of the deviation varies with the cross section of the

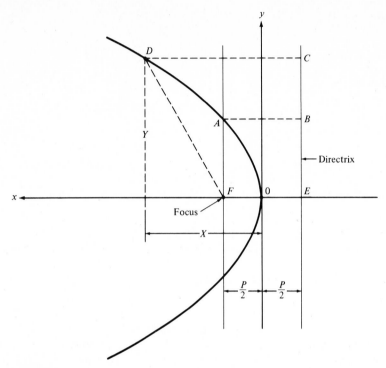

FIGURE 7-13
Parabola.

dam, e.g., Figs. 7-11 and 7-12) only at the upstream and downstream faces. Hence Casagrande (1) suggests that the flow line be shaped as a parabola, with the necessary corrections at the upstream and downstream faces to be made by eye, so as to conform to the basic conditions for flow net sketching.

The *parabola* is the locus of all points which are equidistant from a fixed point, called the *focus*, and a line, called the *directrix*. Figure 7-13 illustrates some basic characteristics of a parabola. Point 0, the *vertex* of the parabola and also a point on the curve itself, is the midpoint of distance FE. By definition, since 0 represents a point on the curve, it must be equidistant between the focus and the directrix. Similarly, distances $FA = AB$ and $FD = DC$, etc.

If vertex 0 is taken as the origin of the rectangular coordinates X, Y, the equation of the parabola can be determined as follows:

$$DC = \frac{P}{2} + X \tag{a}$$

$$(FD)^2 = Y^2 + \left(X - \frac{P}{2}\right)^2 \tag{b}$$

However, $DC = FD$, by definition. Hence we have

$$\left(\frac{P}{2} + X\right)^2 = Y^2 + \left(X - \frac{P}{2}\right)^2 \tag{c}$$

Simplifying Eq. (c), we get the equation of the parabola

$$Y^2 = 2PX \tag{7-7}$$

where $2P$ is the parameter of the parabola.

The following discussion will reflect on the construction of the top flow line by fitting the parabola to the particular physical characteristics of the dams.

CASE A   Horizontal Underdrainage, Fig. 7-14

The case where the discharge is into a drainage gallery is typified by Fig. 7-14. The dam's cross section is drawn to scale, i.e., the slopes, water level, and drainage gallery are represented by a relative proportionate size. Hence we may proceed to construct the top flow line for this case.

Casagrande recommends that $GS = 0.3HS$ (Fig. 7-14). The focus point F may be assumed as the end of the drainage gallery, as shown. Since point G is on the parabola, distance $GF = GI$, by definition; i.e., with GF as a radius and point G as a center, we can draw an arc which intersects the water line extension at point I. The vertical line through point I is the directrix IE. Point 0 is halfway between F and E. We now have points 0 and G on the parabola. As many additional points as desired may be obtained quite readily by keeping in mind that the distance from any point on the parabola to focus F is the same as that to the directrix. For example, distances $FJ = JK$ and $FL = LM$, thereby determining points J and L on the parabola. The additional point (e.g., point L) can be located as follows. Draw a vertical (parallel to the directrix) through an arbitrary point (say, N). With the distance from the directrix to that point (say, EN) as the radius and F as the center, draw an arc. The intersection of that vertical with the arc is a point on the parabola (say, L). Once the parabola is constructed, a perpendicular is drawn to line AB at point S and gradually blended in to the parabola curve. This could be done by eye.

With the top flow line and thus the boundary line established, the rest of the flow lines and equipotential lines could be drawn to complete the flow net. $q = (n_d/n_f)kh$ gives the quantity of seepage, as previously described.

CASE B   Sloping Discharge Faces, Fig. 7-15

The discharge end of the parabola needs to be modified to fit the particular type of discharge face. Figure 7-15 depicts several configurations of downstream portions of dams. In each of these sketches the focus point F is the intersection of the bottom flow line with the discharge face. The angle α, which the discharge face makes with the horizontal, is measured clockwise. The segments designated Δa represent the actual distances from the breakout point R to the parabola. The segments designated a are distances from the breakout point to the focus point F. The relationship between the

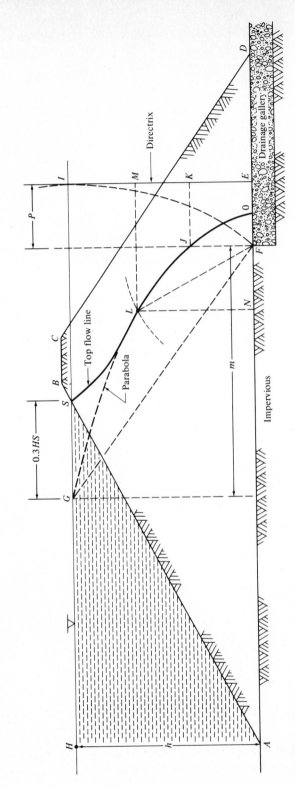

FIGURE 7-14
Development of top flow line from parabola.

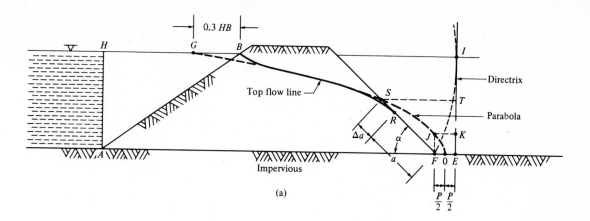

(a)

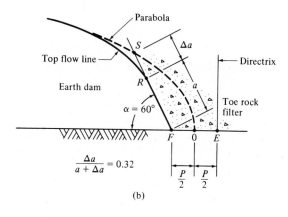

(b)

$$\frac{\Delta a}{a + \Delta a} = 0.32$$

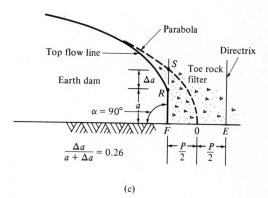

(c)

$$\frac{\Delta a}{a + \Delta a} = 0.26$$

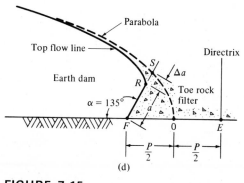

(d)

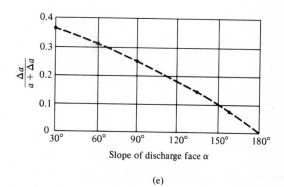

(e)

FIGURE 7-15

Top flow line for various discharge slopes. (a) Top flow line, drainage for general downstream condition. (b) Discharge face $\alpha = 60°$. (c) Discharge face $\alpha = 90°$. (d) Discharge face $\alpha = 135°$. (e) $\Delta a/(a + \Delta a)$ vs. α. (After A. Casagrande.)

slope of the discharge face α and the ratio $\Delta a/(a + \Delta a)$ is given in Fig. 7-15e (after A. Casagrande).

With the focus point F established, the procedure for developing the parabola is much the same as described above and illustrated in Fig. 7-14. More specifically, let us reflect on the cross section shown in Fig. 7-15a. Once the cross section is drawn to scale, point G is determined as shown. Hence we have the distance GF as the radius of the arc FI whose center is G. Subsequently we determine point I, the directrix, and point 0. Any other points, such as J, can be determined and subsequently the parabola drawn as shown. For the slope α the value of $a + \Delta a = FS$ and can be measured directly. Also, from Fig. 7-15e the value of $\Delta a/(a + \Delta a)$ can be obtained for the particular α. Subsequently we determine the breakout point R. The transition section between R and the parabola can be sketched by eye; similarly the segment between F and the parabola can be completed as before to give us the top flow line. Note that Fig. 7-15e is limited to angles α of 30° to 180°. Incidentally, case A (Fig. 7-14) is a particular case for $\alpha = 180°$.

With the top flow line established, the flow net can be completed and the seepage quantity calculated. If the flow net resembles the flow net constructed from confocal parabolas (usually for $\alpha > 30°$), the quantity of seepage can be calculated with acceptable accuracy by assuming a truly parabolic net. Using Darcy's law,

$$q = kiA = k\frac{dY}{dX} A \tag{a}$$

Referring to Fig. 7-13 and assuming unit dam thickness equals 1, the cross section for flow A becomes

$$A = Y(1) = Y$$

From Eq. (7-7), $Y = \sqrt{2PX}$. Thus,

$$A = \sqrt{2PX} \tag{b}$$

and

$$\frac{dY}{dX} = \frac{P}{\sqrt{2PX}} \tag{c}$$

Thus,

$$q = k\left(\frac{P}{\sqrt{2PX}}\right)(\sqrt{2PX}) = kP \tag{7-8}$$

The value of P may be easily related to the characteristics of the parabola shown in Fig. 7-14:

$$FG = GI = m + P \tag{d}$$

Also,

$$(FG)^2 = h^2 + m^2 \tag{e}$$

Substituting,

$$m + P = \sqrt{h^2 + m^2} \tag{f}$$

Solving for P, we get

$$P = \sqrt{h^2 + m^2} - m \tag{g}$$

Thus Eq. (7-8) becomes

$$q = k(\sqrt{h^2 + m^2} - m) \tag{7-9}$$

CASE C   Nonhomogeneous Sections

If the permeability of the soil changes along the flow paths, the flow lines are deflected at the junction point of the two soil masses. Figure 7-16 shows the flow lines entering and leaving the interface of two soil masses of permeability k_1 and k_2. Lines AC and BD are equipotential lines. The drop in head over length AB and CD will be designated Δh. Thus,

$$q = k_1 \left(\frac{\Delta h}{AB}\right) BC = k_2 \left(\frac{\Delta h}{CD}\right) BD \tag{a}$$

But $AB/AC = \tan \alpha_1$ and $CD/BD = \tan \alpha_2$. Hence,

$$\frac{k_1}{k_2} = \frac{\tan \alpha_1}{\tan \alpha_2} \tag{7-10}$$

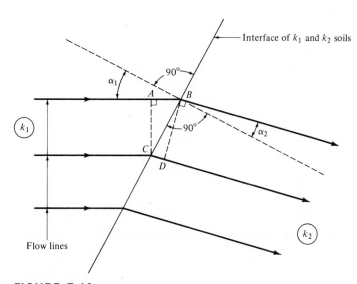

FIGURE 7-16
Change of direction of flow line in soil masses of permeabilities k_1 and k_2.

When the permeability of the second mass is 10 times or more greater than that of the first, the second may be assumed as an open drain (no resistance to flow), and therefore no deflection correction is necessary.

If stratification exists (e.g., $k_x > k_z$) and this is uniform throughout, the procedure consists of transforming, finding the effective permeability coefficient, and subsequently proceeding to evaluate as described in Section 7-8.

EXAMPLE 7-5

Given   See Fig. 7.17.

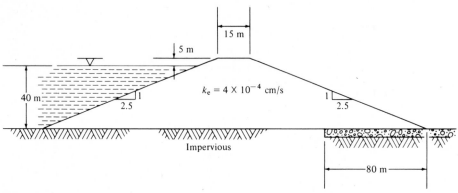

15 m

5 m

40 m

$k_e = 4 \times 10^{-4}$ cm/s

1
2.5

1
2.5

Impervious

80 m

FIGURE 7-17

Find   The quantity of seepage. (a) Via parabolic flow net construction. (b) Via Eqs. (7-8) and (7-9) per meter of width.

Procedure   Figure 7-18 shows the dam drawn to scale (arbitrarily selected). $GS = 0.3HS = 30$ m. Using G as center and GF as radius, an arc drawn as shown determines the directrix. At arbitrary points A, B, C, D, F, etc. draw verticals. With F as center and distances AE, BE, CE, DE, and FE as radii, we determine points 2, 3, 4, 5, and 1, respectively. These are points on the parabola, drawn as shown. Figure 7-18b shows the flow net where the parabola curve from Fig. 7-18a was used in determining the top flow line.

   (a)   From flow net, $n_f = 3$; $n_d = 13$, and $q = (n_f/n_d)kh$. Thus,

$$q = (\tfrac{3}{13})(4 \times 10^{-4} \text{ cm/s})(40 \text{ m})(100 \text{ cm/m})(100 \text{ cm/m width})$$

Answer

$$q = 36.92 \text{ cm}^3/\text{s/m width}$$

   (b)   From Eq. (7-8),

$$q = kP$$

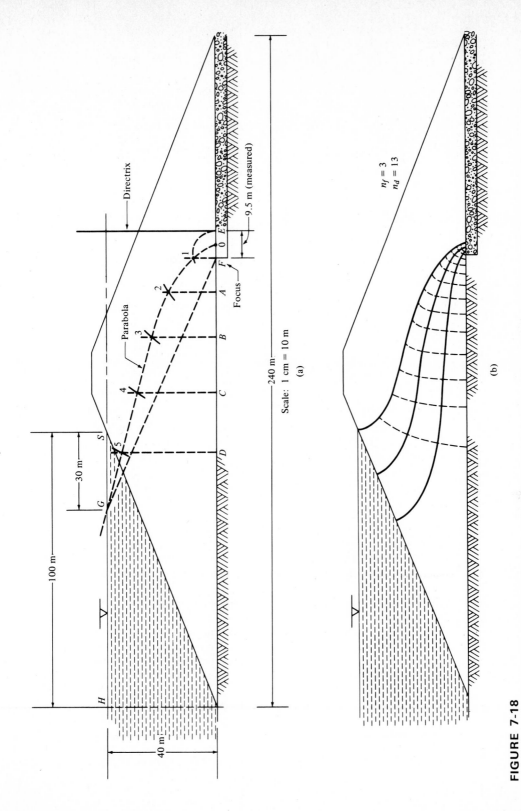

FIGURE 7-18

(a) Dam drawn to scale and construction of parabola curve. (b) Construction of flow net using parabola construction as guide for top flow.

147

where P = distance FE, Fig. 7-18a; also see Fig. 7-14. P = 9.5 m as measured in Fig. 7-18a. Thus,

$$q = (4 \times 10^{-4} \text{ cm/s})(9.5 \text{ m})(100 \text{ cm/m})(100 \text{ cm/m width})$$

Answer

$$q = 38 \text{ cm}^3/\text{s/m width}$$

From Eq. (7-9),

$$q = k(\sqrt{h^2 + m^2} - m)$$

From Fig. 7-18a (also see Fig. 7-14), M = 90 m and h = 40 m. Thus,

$$q = k(40^{\wedge 2} + 90^{-2} - 90) = k(9.48)$$

$$q = (4 \times 10^{-4} \text{ cm/s})(9.48 \text{ m})(100 \text{ cm/m})(100 \text{ cm/m width})$$

Answer

$$q = 37.92 \text{ cm}^3/\text{s/m width}$$

PROBLEMS

7-1. Determine the seepage per lineal meter of wall under the sheet pile wall shown in Fig. P7-1 for h = 3 m, S = 6 m, d = 3 m, and $k_x = k_z = 6 \times 10^{-4}$ cm/s.

7-2. For Problem 7-1 determine the pore pressures at each point where the equipotential lines meet the sheet piling on each side. See Example 7-1.

7-3. Determine the seepage, per lineal meter of wall, under the sheet pile wall shown in Fig. P7-1 if h = 6 m, S = 12 m, d = 6 m, and $k_x = k_z = 6 \times 10^{-4}$ cm/s. Compare this result with that from Problem 7-1 and reflect on the effect these changes have on the seepage quantity. In $q = (n_d/n_f)kh$, would one expect a change in n_d/n_f if the above-mentioned changes were made? Explain.

7-4. For the data in Problem 7-3 determine the pore pressure at each point where the equipotential lines meet the sheet piling, on each side. How do these values compare with those of Problem 7-2?

7-5. (a) Redraw the section in Fig. P7-5 to true scale and construct the flow net for this section.

   (b) Determine the seepage loss in cubic meters per day per meter of dam.

7-6. (a) Redraw the section in Fig. P7-6 to true scale and construct the flow net for this section.

   (b) Determine the seepage loss in cubic meters per day per meter of dam.

7-7. (a) Redraw the section in Fig. P7-7 to true scale and construct the flow net for this section.

   (b) Determine the seepage loss in cubic meters per day per meter of dam.

7-8. (a) Redraw the section in Fig. P7-8 to true scale and construct the flow net for this section.

   (b) Determine the seepage loss in cubic meters per day per meter of dam.

   (c) Compare the seepage loss of this problem to that of Problem 7-7. Does the *location* of the cutoff wall affect the seepage flow here?

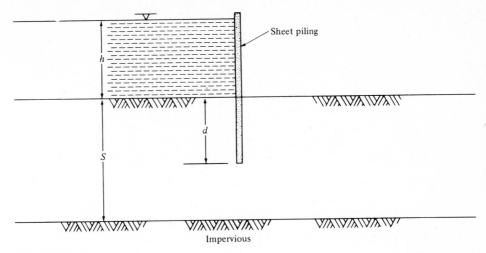

Sheet piling

FIGURE P7-1

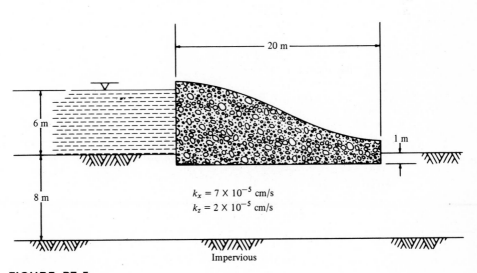

20 m

6 m

1 m

8 m

$k_x = 7 \times 10^{-5}$ cm/s
$k_z = 2 \times 10^{-5}$ cm/s

Impervious

FIGURE P7-5

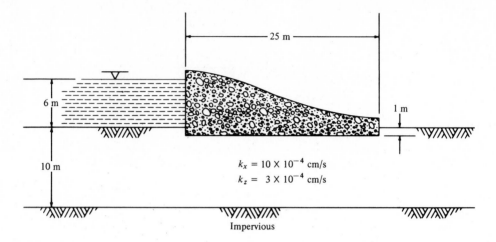

$$k_x = 10 \times 10^{-4} \text{ cm/s}$$
$$k_z = 3 \times 10^{-4} \text{ cm/s}$$

FIGURE P7-6

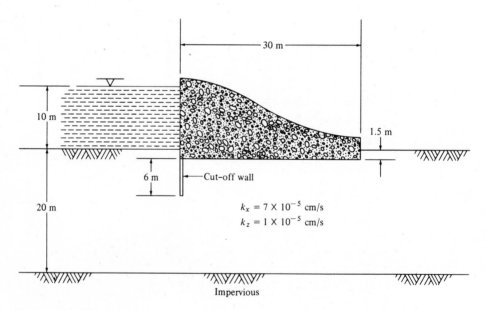

$$k_x = 7 \times 10^{-5} \text{ cm/s}$$
$$k_z = 1 \times 10^{-5} \text{ cm/s}$$

FIGURE P7-7

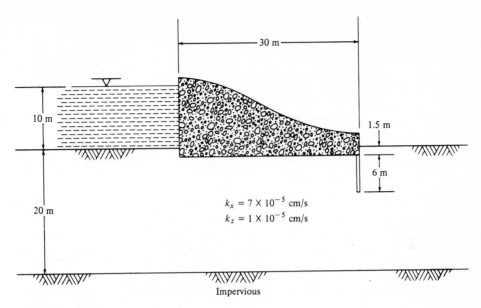

FIGURE P7-8

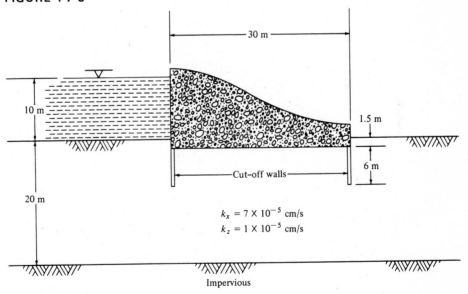

FIGURE P7-9

7-9. (a) Redraw the section in Fig. P7-9 to true scale and construct the flow net for this section.

      (b) Determine the seepage loss in cubic meters per day per meter of dam.

      (c) Compare the seepage loss of this problem to that of Problem 7-8. Is it reasonable to assume that the seepage would be reduced directly as a function of the number of cutoff walls? Explain.

7-10. (a) Redraw the section in Fig. P7-10 to true scale and construct the flow net for this section.

      (b) Determine the seepage loss in cubic meters per day per meter of dam.

7-11. (a) Redraw the section in Fig. P7-11 to true scale and construct the flow net for this section.

      (b) Determine the seepage loss in cubic meters per day per meter of dam.

7-12. (a) Redraw the section in Fig. P7-12 to true scale and construct the flow net for this section.

      (b) Determine the seepage loss in cubic meters per day per meter of dam.

7-13. (a) Sketch the flow net for the example shown in Fig. P7-13.

      (b) Calculate the seepage from the flow net diagram.

      (c) Calculate the seepage by the formula of the parabolic case.

7-14. (a) Sketch the flow net for the example shown in Fig. P7-14.

      (b) Calculate the seepage from the flow net diagram.

      (c) Calculate the seepage by the formula of the parabolic case.

7-15. (a) Sketch the flow net for the example shown in Fig. P7-15.

      (b) Calculate the seepage from the flow net diagram.

      (c) Calculate the seepage by the formula of the parabolic case.

7-16. (a) Sketch the flow net for the example shown in Fig. P7-16.

      (b) Calculate the seepage from the flow net diagram.

      (c) Calculate the seepage by the formula of the parabolic case.

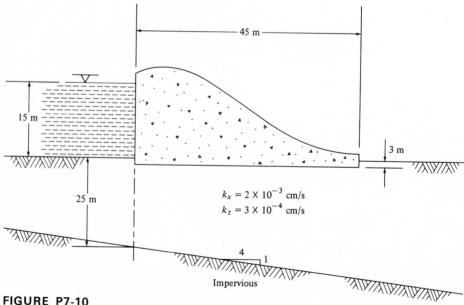

$$k_x = 2 \times 10^{-3} \text{ cm/s}$$
$$k_z = 3 \times 10^{-4} \text{ cm/s}$$

FIGURE P7-10

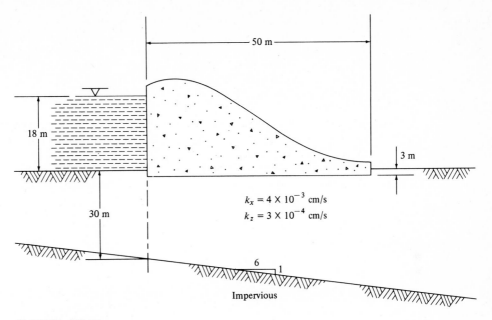

FIGURE P7-11

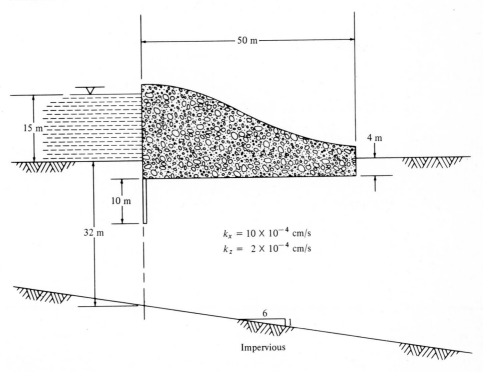

FIGURE P7-12

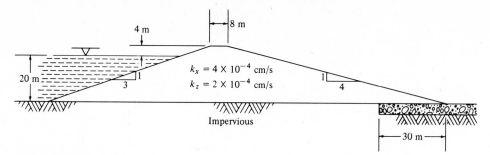

FIGURE P7-13

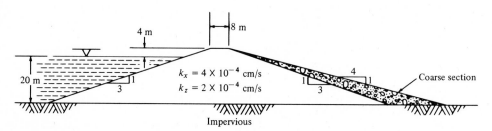

FIGURE P7-14

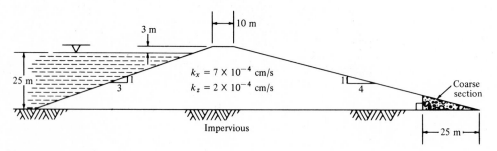

FIGURE P7-15

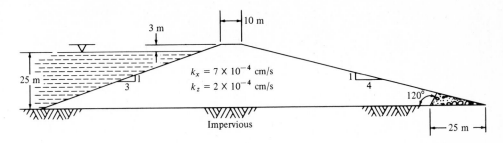

FIGURE 7-16

BIBLIOGRAPHY

1. Casagrande, A., "Seepage through Dams," *Contributions to Soil Mechanics*, ASCE, 1925–1940. (Paper first published in *J. New England Water Works Assoc.*, June 1937.)
2. Cedegren, H. R., "Seepage Requirements of Filters and Pervious Bases," *ASCE J. Soil Mech. Found. Eng. Div.*, vol. 86, no. SM5, Oct. 1960.
3. Cedegren, H. R., *Seepage, Drainage, and Flow Nets*, Wiley, New York, 1967.
4. Christoulas, D. G., "Approximate Solution to Flow Problem under Dams," *ASCE J. Geotech. Eng. Div.*, vol. 97, Nov. 1971.
5. Corps of Engineers, Dept. of Army, "Seepage Control, Soil Mechanics Design," Washington, D.C., 1952.
6. Hall, H. P., "A Historical Review of Investigations of Seepage toward Wells," *J. Boston Soc. Civil Eng.*, vol. 41, 1954.
7. Harr, E., *Groundwater and Seepage*, McGraw-Hill, New York, 1962.
8. Jennings, J. E., "The Heaving of Buildings on Desiccated Clay," *3rd Int. Conf. Soil Mech. Found. Eng.*, Zurich, Switzerland, vol. 1, 1953.
9. Lane, E. W., "Security from Under-Seepage, Masonry Dams on Earth Foundations," *Trans. ASCE*, 1935.
10. Sandhu, R. S., and E. L. Wilson, "Finite Element Analysis of Seepage in Elastic Media," *ASCE J. Eng. Mech. Div.*, vol. 95, no. EM3, Proc. Paper 6615, June 1969.
11. Turnbull, W. J., and C. I. Mansur, "Investigation of Underseepage—Mississippi River Levees," *ASCE J. Geotech. Eng. Div.*, vol. 85, Aug. 1959.

8

Stresses in Soils

8-1 INTRODUCTION

Within the context of geotechnical engineering analysis it is convenient to view the in situ soil stress, at a given depth, in terms of the components of total stress:

1. Stresses induced by the weight of the soil above that level
2. Fluid pressures
3. Stresses introduced by externally applied loads (if any)

Such resolutions facilitate the evaluation and changes in current stress conditions due to different causes. For example, pore pressures may be altered by a fluctuating water table; perhaps this may affect the shear strength of a soil or the stability of a slope. Changes in the overburden loads (e.g., glaciers, buildings, excavation) may provide some reasonable basis for estimating bearing capacity, consolidation predictions, etc. It is within the scope of the soil engineer's responsibility to assess the various stress conditions as they exist, to evaluate the effects of any changes, and to relate these findings to the proposed "project" requirements.

It is difficult to measure with any degree of accuracy actual stresses in a soil mass by experimental means. For example, while piezometers are frequently used to estimate pore-pressure conditions, gauges embedded in a soil mass for the purpose of measuring stress are not really reliable; their very presence would disrupt the stress field which existed prior to their implantation. Hence common approaches for estimating stresses entail a combination of experimental data, analytical evaluations, relevant experience, and engineering judgment. This chapter will be devoted primarily to the presentation of commonly used procedures for estimating stresses due to surface loads.

8-2 EFFECTIVE STRESS

Stress is regarded as intensity of force, generally defined as load per unit area; symbolically $\sigma = P/A$. In soils, stresses are separated into: (1) *intergranular*—stress resulting from particle-to-particle contact; and (2) *pore water*—the stress induced by

156

water pressures. The former is commonly referred to as *effective stress*, while the latter is frequently termed *neutral stress* (or *neutral pressure*). The sum of the effective and neutral stresses is called the *total stress*.

Depicting the various types of stresses might be facilitated by Fig. 8-1. The analysis focuses on a sphere A selected at some arbitrary depth h, as shown. In the analogy, the spheres are considered soil grains of weight W; W_b represents the submerged weight of each sphere. The pore pressure due to the hydrostatic head h is

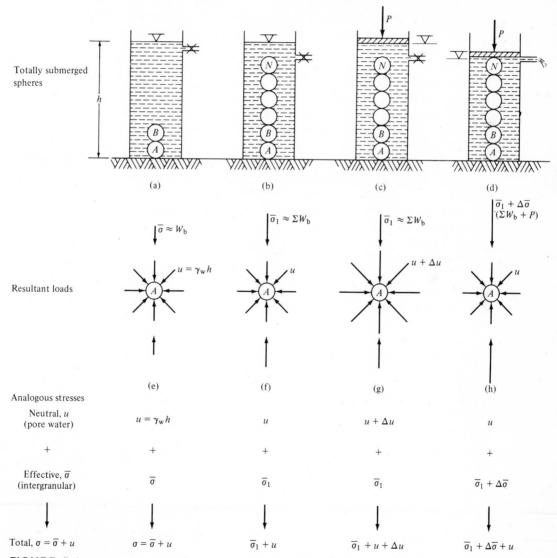

FIGURE 8-1
Analogy for illustrating neutral, effective, and total stresses in soils.

$\gamma_w h$ and acts in all directions perpendicular to the surface of the sphere. If "additional" loads such as shown in Fig. 8-1c are superimposed on the liquid, the pore pressure will increase correspondingly, say by Δu. The common symbol used to denote neutral or pore water stress is u.

The effective stress on sphere A is induced by the contact forces that sphere A has to withstand. It is usually represented by $\bar{\sigma}$. If sphere A supports N spheres of weight W_b, as shown in Fig. 8-1b, the effective stress is $\sum W_b$ divided by an appropriate cross-sectional area. Any additional "surface" loads such as in Fig. 8-1d will correspondingly increase the stress, say by an amount equal to $\Delta\bar{\sigma}$. Note that the additional load in Fig. 8-1c is supported totally by the water and, therefore, does not affect the inter-granular or effective stress.

At the point of contact the stresses are not really meaningful, since the contact area is infinitesimal. It is convenient, therefore, to express the stress as the load divided by a more identifiable area. The gross area of the soil mass is a commonly used value. Correspondingly, any such stress is a nominal rather than an actual or true stress.

The stress caused by the loads above sphere A (not including surcharge load P) might be analogous to the overburden supported by a soil particle. Such soil stresses are commonly referred to as *geostatic stresses.*

Let us apply the above discussion to define corresponding vertical soil stresses in an element within a general state formation represented by Fig. 8-2. The effects from surface loads are not included; their influence will be explained in subsequent sections. The various stresses can be expressed as follows.

Neutral stress:

$$u = \gamma_w h_w \tag{8-1a}$$

Effective stress:

$$\bar{\sigma} = \gamma_d h_1 + \gamma_b h_w \tag{8-1b}$$

Total stress:

$$\sigma = \bar{\sigma} + u \tag{8-1c}$$

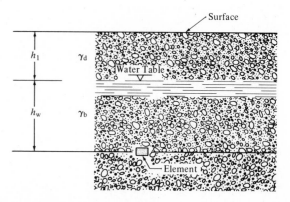

FIGURE 8-2
Stresses on a soil element.

Δu may be added to Eq. (8-1a) to account for any changes in the neutral stress (e.g., fluctuating water table, imposed pressures).

8-3 STRESSES DUE TO SURFACE LOADS; BOUSSINESQ'S EQUATIONS

In 1885 Joseph Valentin Boussinesq advanced theoretical expressions for determining stresses at a point within an "ideal" mass due to surface point loads (3). They are based on the assumption that the mass is an (1) *elastic*, (2) *isotropic*, (3) *homogeneous*, and (4) *semi-infinite medium* which extends infinitely in all directions from a level surface. Boussinesq's equations provide a widely used basis for estimating the stresses within a soil mass caused by a concentrated load applied perpendicularly to the soil surface. In 1938 Westergaard (17) developed a solution for stresses within a soil mass by assuming the material to be reinforced by very rigid horizontal sheets which prevent any horizontal strain. Several load conditions based on Westergaard's development are mentioned in Appendix A. Within this chapter Boussinesq's expressions will receive the primary focus.

Boussinesq's equations may be expressed in terms of either rectangular or polar coordinates. Referring to the elements in Fig. 8-3, the equations are as follows.

In rectangular coordinates:

$$\sigma_z = \frac{3Q}{2\pi} \frac{z^3}{R^5} \tag{8-2a}$$

$$\sigma_x = \frac{3Q}{2\pi} \left\{ \frac{x^2 z}{R^5} + \frac{1-2\mu}{3} \left[\frac{1}{R(R+z)} - \frac{(2R+z)x^2}{R^3(R+z)^2} - \frac{z}{R^3} \right] \right\} \tag{8-2b}$$

$$\sigma_y = \frac{3Q}{2\pi} \left\{ \frac{y^2 z}{R^5} + \frac{1-2\mu}{3} \left[\frac{1}{R(R+z)} - \frac{(2R+z)y^2}{R^3(R+z)^2} - \frac{z}{R^3} \right] \right\} \tag{8-2c}$$

$$\tau_{zx} = -\frac{3Q}{2\pi} \frac{xz^2}{R^5} \tag{8-2d}$$

$$\tau_{xy} = \frac{3Q}{2\pi} \left[\frac{xyz}{R^5} - \frac{1-2\mu}{3} \frac{(2R+z)xy}{R^3(R+z)^2} \right] \tag{8-2e}$$

$$\tau_{yz} = -\frac{3Q}{2\pi} \frac{yz^2}{R^5} \tag{8-2f}$$

In polar coordinates:

$$\sigma_z = \frac{Q}{2\pi} \frac{3z^3}{(r^2 + z^2)^{5/2}} = \frac{Q}{2\pi z^2} (3\cos^5 \theta) \tag{8-3a}$$

$$\sigma_r = \frac{Q}{2\pi} \left[\frac{3r^2 z}{(r^2 + z^2)^{5/2}} - \frac{1 - 2\mu}{r^2 + z^2 + z\sqrt{r^2 + z^2}} \right]$$

$$= \frac{Q}{2\pi z^2} \left[3\sin^2 \theta \cos^3 \theta - \frac{(1 - 2\mu)\cos^2 \theta}{1 + \cos \theta} \right] \tag{8-3b}$$

$$\sigma_t = -\frac{Q}{2\pi} (1 - 2\mu) \left[\frac{z}{(r^2 + z^2)^{3/2}} - \frac{1}{r^2 + z^2 + z\sqrt{r^2 + z^2}} \right]$$

$$= -\frac{Q}{2\pi z^2} (1 - 2\mu) \left[\cos^3 \theta - \frac{\cos^2 \theta}{1 + \cos \theta} \right] \tag{8-3c}$$

$$\tau = \frac{Q}{2\pi} \frac{3rz^2}{(r^2 + z^2)^{5/2}} = \frac{Q}{2\pi z^2} (3 \sin \theta \cos^4 \theta) . \tag{8-3d}$$

In the above equations μ designates Poisson's ratio, which varies between 0 and 0.5. Although Poisson's ratio may be readily obtained from tables for most materials, for soil it cannot. In fact, the experimental results in this regard vary widely and are inconclusive. Because it simplified some of the equations [e.g., Eqs. (8-2c), (8-2e), (8-3b), and (8-3c)], many engineers have used a value of $\mu = 0.5$.

The expression for vertical stress, designated σ_z, is regarded as reasonably accurate and is widely used in problems associated with bearing capacity and settlement analysis. Hence it is this expression that is primarily discussed in this chapter.

Equation (8-3a) is more conveniently expressed in a slightly different form, as shown by Eq. (8-4).

$$\sigma_z = \frac{Q}{z^2} \frac{\frac{3}{2}\pi}{[(r/z)^2 + 1]^{5/2}} \tag{8-4}$$

or

$$\sigma_z = \frac{Q}{z^2} N_B \tag{8-5}$$

where N_B, commonly referred to as the vertical stress coefficient, is given by

$$N_B = \frac{\frac{3}{2}\pi}{[(r/z)^2 + 1]^{5/2}}$$

Table 8-1 gives the values of Boussinesq's vertical stress coefficient for various ratios of r/z.

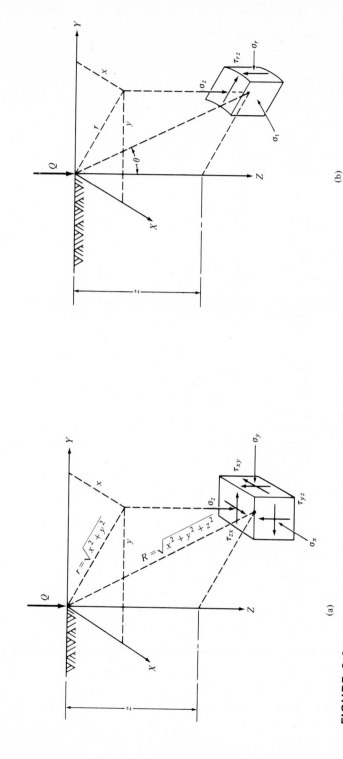

(a)

(b)

FIGURE 8-3
Stresses on elements due to concentrated load Q. (a) Rectangular coordinate notation. (b) Polar coordinate notation.

161

TABLE 8-1 Values of Boussinesq's Vertical Stress Coéfficient N_B

r/z	N_B	r/z	N_B	r/z	N_B	r/z	N_B
0.00	0.47746	2.45	0.00368	4.90	0.00015	7.35	0.00002
0.05	0.47449	2.50	0.00337	4.95	0.00015	7.40	0.00002
0.10	0.46573	2.55	0.00310	5.00	0.00014	7.45	0.00002
0.15	0.45163	2.60	0.00285	5.05	0.00013	7.50	0.00002
0.20	0.43287	2.65	0.00262	5.10	0.00013	7.55	0.00002
0.25	0.41032	2.70	0.00241	5.15	0.00012	7.60	0.00002
0.30	0.38492	2.75	0.00223	5.20	0.00011	7.65	0.00002
0.35	0.35766	2.80	0.00206	5.25	0.00011	7.70	0.00002
0.40	0.32946	2.85	0.00190	5.30	0.00010	7.75	0.00002
0.45	0.30111	2.90	0.00176	5.35	0.00010	7.80	0.00002
0.50	0.27332	2.95	0.00163	5.40	0.00010	7.85	0.00002
0.55	0.24660	3.00	0.00151	5.45	0.00009	7.90	0.00001
0.60	0.22136	3.05	0.00140	5.50	0.00009	7.95	0.00001
0.65	0.19784	3.10	0.00130	5.55	0.00008	8.00	0.00001
0.70	0.17619	3.15	0.00121	5.60	0.00008	8.05	0.00001
0.75	0.15646	3.20	0.00113	5.65	0.00008	8.10	0.00001
0.80	0.13862	3.25	0.00105	5.70	0.00007	8.15	0.00001
0.85	0.12262	3.30	0.00098	5.75	0.00007	8.20	0.00001
0.90	0.10833	3.35	0.00091	5.80	0.00007	8.25	0.00001
0.95	0.09564	3.40	0.00085	5.85	0.00006	8.30	0.00001
1.00	0.08440	3.45	0.00080	5.90	0.00006	8.35	0.00001
1.05	0.07449	3.50	0.00075	5.95	0.00006	8.40	0.00001
1.10	0.06576	3.55	0.00070	6.00	0.00006	8.45	0.00001
1.15	0.05809	3.60	0.00066	6.05	0.00006	8.50	0.00001
1.20	0.05134	3.65	0.00062	6.10	0.00005	8.55	0.00001
1.25	0.04543	3.70	0.00058	6.15	0.00005	8.60	0.00001
1.30	0.04023	3.75	0.00054	6.20	0.00005	8.65	0.00001
1.35	0.03568	3.80	0.00051	6.25	0.00005	8.70	0.00001
1.40	0.03168	3.85	0.00048	6.30	0.00005	8.75	0.00001
1.45	0.02816	3.90	0.00045	6.35	0.00004	8.80	0.00001
1.50	0.02508	3.95	0.00043	6.40	0.00004	8.85	0.00001
1.55	0.02236	4.00	0.00040	6.45	0.00004	8.90	0.00001
1.60	0.01997	4.05	0.00038	6.50	0.00004	8.95	0.00001
1.65	0.01786	4.10	0.00036	6.55	0.00004	9.00	0.00001
1.70	0.01600	4.15	0.00034	6.60	0.00004	9.05	0.00001
1.75	0.01436	4.20	0.00032	6.65	0.00003	9.10	0.00001
1.80	0.01290	4.25	0.00030	6.70	0.00003	9.15	0.00001
1.85	0.01161	4.30	0.00028	6.75	0.00003	9.20	0.00001
1.90	0.01047	4.35	0.00027	6.80	0.00003	9.25	0.00001
1.95	0.00945	4.40	0.00026	6.85	0.00003	9.30	0.00001
2.00	0.00854	4.45	0.00024	6.90	0.00003	9.35	0.00001
2.05	0.00774	4.50	0.00023	6.95	0.00003	9.40	0.00001
2.10	0.00701	4.55	0.00022	7.00	0.00003	9.45	0.00001
2.15	0.00637	4.60	0.00021	7.05	0.00003	9.50	0.00001
2.20	0.00579	4.65	0.00020	7.10	0.00003	9.60	0.00001
2.25	0.00528	4.70	0.00019	7.15	0.00002	9.70	0.00001
2.30	0.00481	4.75	0.00018	7.20	0.00002	9.80	0.00001
2.35	0.00440	4.80	0.00017	7.25	0.00002	9.90	0.00000
2.40	0.00402	4.85	0.00016	7.30	0.00002	10.00	0.00000

EXAMPLE 8-1

Given Q, a concentrated point load, acts vertically at the surface.

Find Vertical stress σ_z distribution along the depth for $r = 2$ m.

Procedure Using Eq. (8-5), we have $\sigma_z = (Q/z^2)N_B$, where N_B may be obtained from Table 8-1. For $r = 2$ m, the values of σ_z at various arbitrarily selected depths, as shown in Fig. 8-4, are given in the following table. The distribution of σ_z with depth is shown in Fig. 8-4.

z_1(m)	r/z	N_B	z^2	Q/z^2	$\sigma_z(Q/m^2)$
0	∞	0	0	∞	Indeterminate
0.4	5.0	0.00014	0.16	$6.250Q$	$0.0009Q$
0.8	2.5	0.00337	0.64	$1.563Q$	$0.0053Q$
1.2	1.67	0.01712	1.44	$0.694Q$	$0.0119Q$
1.6	1.25	0.04543	2.56	$0.391Q$	$0.0178Q$
2.0	1.0	0.08440	4.00	$0.250Q$	$0.0211Q$
2.4	0.83	0.12953	5.76	$0.174Q$	$0.0225Q$
2.8	0.71	0.17243	7.84	$0.128Q$	$0.0221Q$
3.6	0.56	0.24214	12.96	$0.0772Q$	$0.0187Q$
5.0	0.40	0.32946	25.00	$0.0400Q$	$0.0132Q$
10.0	0.2	0.43287	100.00	$0.0100Q$	$0.0043Q$

EXAMPLE 8-2

Given Q, a concentrated point load, acts vertically at the surface.

Find Vertical stress σ_z distribution for depth $z = 2$ m.

Procedure N_B may be found from Table 8-1. σ_z is given in the following table for various values of r, as shown in Fig. 8-5.

r(m)	r/z	N_B	z^2	Q/z^2	$\sigma_z(Q/m^2)$
0	0	0.47746	4.0	$0.25Q$	$0.1194Q$
0.4	0.2	0.43287	4.0	$0.25Q$	$0.1082Q$
0.8	0.4	0.32946	4.0	$0.25Q$	$0.0824Q$
1.2	0.6	0.22136	4.0	$0.25Q$	$0.0553Q$
1.6	0.8	0.13862	4.0	$0.25Q$	$0.0347Q$
2.0	1.0	0.08440	4.0	$0.25Q$	$0.0211Q$
2.4	1.2	0.05134	4.0	$0.25Q$	$0.0129Q$
2.8	1.4	0.03168	4.0	$0.25Q$	$0.0079Q$
3.6	1.8	0.01290	4.0	$0.25Q$	$0.0032Q$
5.0	2.5	0.00337	4.0	$0.25Q$	$0.0008Q$
10.0	5	0.00014	4.0	$0.25Q$	$0.0001Q$

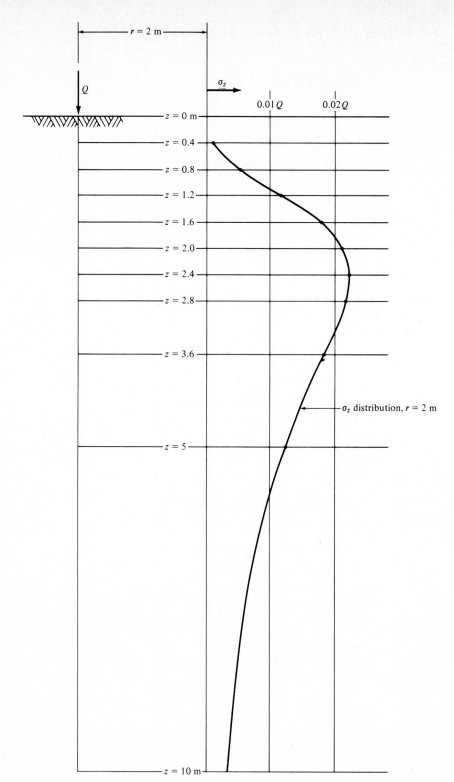

FIGURE 8-4

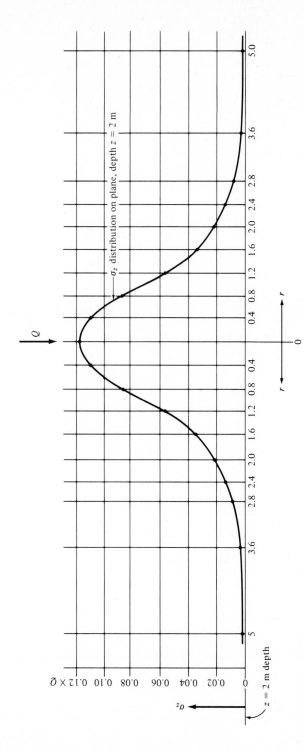

FIGURE 8-5

EXAMPLE 8-3

Given Q, a concentrated point load, acts vertically at the surface.

Find   (a) σ_z for various values of horizontal distances r and at $z = 1, 2, 3$, and 4 m. (b) Plot the σ_z distribution for $z = 1, 2, 3$, and 4 m.

Procedure   Table 8-1 will be used to determine N_B. σ_z for $z = 2$ m was determined in Example 8-2. σ_z for $z = 1, 3, 4$ m is given in the following tables. The σ_z distributions are shown in Fig. 8-6.

z = 1 m

r(m)	r/z	N_B	z^2	Q/z^2	$\sigma_z(Q/m^2)$
0	0	0.47746	1	Q	$0.47746Q$
0.4	0.4	0.32946	1	Q	$0.32946Q$
0.8	0.8	0.13862	1	Q	$0.13862Q$
1.2	1.2	0.05134	1	Q	$0.05134Q$
1.6	1.6	0.01997	1	Q	$0.01997Q$
2.0	2.0	0.00854	1	Q	$0.00854Q$
2.4	2.4	0.00402	1	Q	$0.00402Q$
2.8	2.8	0.00206	1	Q	$0.00206Q$
3.6	3.6	0.00066	1	Q	$0.00066Q$
5.0	5.0	0.00014	1	Q	$0.00014Q$
10	10	0.0000	1	Q	$0.00000Q$

z = 3 m

r(m)	r/z	N_B	z^2	Q/z^2	$\sigma_z(Q/m^2)$
0	0	0.47746	9	$Q/9$	$0.0531Q$
0.4	0.1333	0.45630	9	$Q/9$	$0.0507Q$
0.8	0.2666	0.40200	9	$Q/9$	$0.0447Q$
1.2	0.4000	0.32950	9	$Q/9$	$0.0366Q$
1.6	0.5333	0.25555	9	$Q/9$	$0.0284Q$
2.0	0.6666	0.19060	9	$Q/9$	$0.0212Q$
2.4	0.8000	0.13862	9	$Q/9$	$0.0154Q$
2.8	0.9333	0.09983	9	$Q/9$	$0.0111Q$
3.6	1.2000	0.05134	9	$Q/9$	$0.0057Q$
5.0	1.6666	0.01710	9	$Q/9$	$0.0019Q$
10	3.3333	0.00095	9	$Q/9$	$0.0001Q$

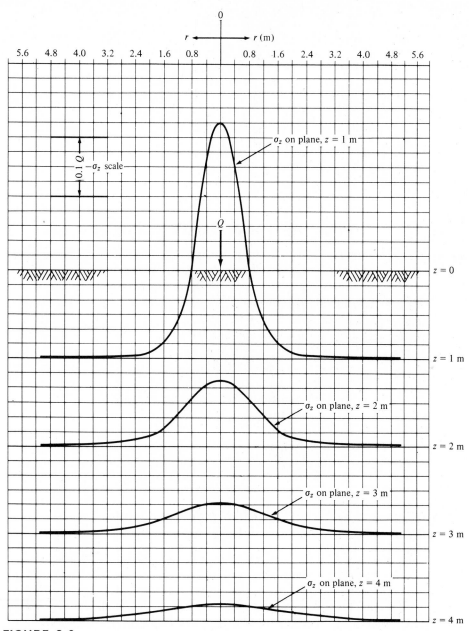

FIGURE 8-6

$$z = 4\,\text{m}$$

r(m)	r/z	N_B	z^2	Q/z^2	$\sigma_z(Q/\text{m}^2)$
0	0	0.47746	16	$Q/16$	$0.02984Q$
0.4	0.1	0.46573	16	$Q/16$	$0.02911Q$
0.8	0.2	0.43287	16	$Q/16$	$0.02705Q$
1.2	0.3	0.38492	16	$Q/16$	$0.02406Q$
1.6	0.4	0.32946	16	$Q/16$	$0.02059Q$
2.0	0.5	0.27332	16	$Q/16$	$0.01708Q$
2.4	0.6	0.22136	16	$Q/16$	$0.01384Q$
2.8	0.7	0,17619	16	$Q/16$	$0.01101Q$
3.6	0.9	0.10833	16	$Q/16$	$0.00677Q$
5.0	1.25	0.04543	16	$Q/16$	$0.00284Q$
10	2.50	0.00337	16	$Q/19$	$0.00021Q$

8-4 DISTRIBUTION OF PRESSURE FROM POINT LOAD

An analysis of Eq. (8-4) reveals that the intensity of vertical stress at a point within a soil mass caused by a given surface point load decreases with an increase in the depth and radial distance from the load to the point within the mass. This is made readily apparent by viewing the plotted results in Example 8-3 (Fig. 8-6). For convenience these results are reintroduced as part of Fig. 8-7.

The intensity of the vertical stress σ_z at various depths and radial distances is plotted to a uniform scale for all four graphs of Fig. 8-6 and is schematically represented in Fig. 8-7 by the arrows under the dashed lines. If one were to connect the points of equal stress for various depths, the result would be a series of *pressure bulbs*, as indicated by the solid lines in Fig. 8-7. That is, the pressure at each point of a particular pressure bulb has the same value. Hence any number of pressure bulbs may be drawn for any given load, with each pressure bulb representing a particular stress magnitude. The value of any given pressure bulb could be obtained by merely reading the intensity of σ_z corresponding to the point where the solid line intersects any of the dashed lines.

8-5 PRESSURE CAUSED BY UNIFORMLY LOADED LINE OF FINITE LENGTH

Boussinesq's expression for the vertical stress σ_z as given by Eq. (8-2a) is not directly applicable for the determination of vertical stresses induced by line loads, perhaps typified by continuous-wall footings. It can be modified, however, to provide us with a tool for estimating the vertical stress or pressure from a line load.

Figure 8-8 shows a line load applied at the surface. For an element selected at an arbitrary fixed point in the soil mass, an expression for σ_z could be derived by integrating Boussinesq's expression for point load as given by Eq. (8-2a). The line load is assumed to be of equal intensity q and applied at the surface. Furthermore, one notes

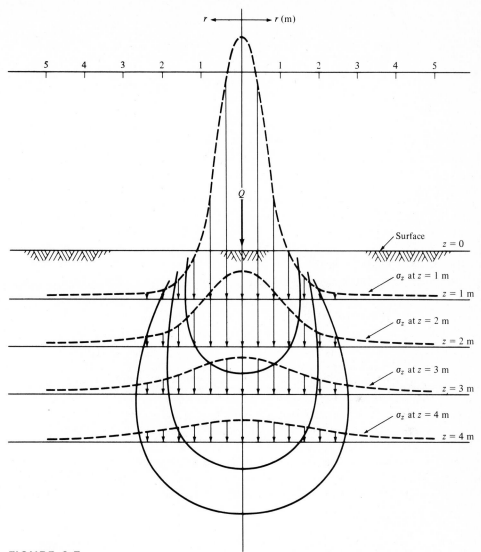

FIGURE 8-7
Distribution of vertical stress σ_z induced by point load Q. Dashed lines represent the σ_z distribution for various values of depth z; solid lines connect points of equal stress.

that the intensity of q is expressed as a force per unit length (perhaps kilonewtons per meter or kips per foot).

With these assumptions established, the expression for σ_z can be determined as follows. From Eq. (8-2a),

$$\sigma_z = \frac{3z^3}{2\pi} \int_0^l \frac{q\,dy}{R^5} \tag{a}$$

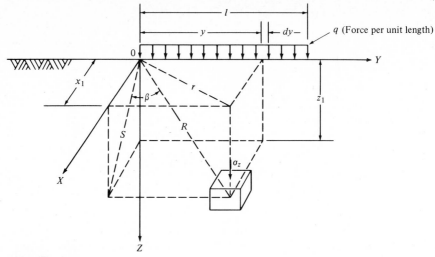

FIGURE 8-8
Vertical stress σ_z induced by line loads.

where $R = \sqrt{x^2 + y^2 + z^2}$. Thus we have

$$\sigma_z = \frac{3z^3}{2\pi} \int_0^l \frac{q \, dy}{(x^2 + y^2 + z^2)^{5/2}} \tag{b}$$

For a specific location of the element, x and z are constants (x_1 and z_1 in Fig. 8-8). Let $x_1^2 + z_1^2 = S^2$, and $y = S \tan \beta$. Then $dy = S \sec^2 \beta \, d\beta$. Equation (b) becomes

$$\sigma_z = \frac{3z^3 q}{2\pi} \int \frac{S \sec^2 \beta \, d\beta}{S^5 (1 + \tan^2 \beta)^{5/2}} = \frac{3z^3 q}{2\pi S^4} \int \frac{\sec^2 \beta \, d\beta}{(1 + \tan^2 \beta)^{5/2}} \tag{c}$$

But $1 + \tan^2 \beta = \sec^2 \beta$. Then Eq. (c) becomes

$$\sigma_z = \frac{3z^3 q}{2\pi S^4} \int \frac{\sec^2 \beta \, d\beta}{(\sec^2 \beta)^{5/2}} = \frac{3z^3 q}{2\pi S^4} \int \frac{d\beta}{\sec^3 \beta} = \frac{3z^3 q}{2\pi S^4} \int \cos^3 \beta \, d\beta$$

$$\sigma_z = \frac{3z^3 q}{2\pi S^4} \left[\sin \beta - \frac{\sin^3 \beta}{3} \right] \tag{d}$$

From Fig. 8-8, $\tan \beta = y/S$ and $\sin \beta = y/R = y/\sqrt{x^2 + y^2 + z^2}$. Thus substituting into Eq. (d), we get

$$\sigma_z = \frac{3z^3 q}{2\pi (x^2 + z^2)^2} \left[\frac{y}{\sqrt{x^2 + y^2 + z^2}} - \left(\frac{y}{\sqrt{x^2 + y^2 + z^2}} \right)^3 \right]$$

Rearranging,

$$\sigma_z = \frac{q/z}{2\pi (x^2/z^2 + 1)^2} \left[\frac{3(y^2/z^2)^{1/2}}{\sqrt{y^2/z^2 + 1 + x^2/z^2}} - \left(\frac{(y^2/z^2)^{1/2}}{\sqrt{y^2/z^2 + 1 + x^2/z^2}} \right)^3 \right] \tag{e}$$

If we let $(x/z) = m$ and $(y/z) = n$, Eq. (e) becomes

$$\sigma_z = \frac{q/z}{2\pi(m^2 + 1)^2} \left[\frac{3n}{\sqrt{n^2 + 1 + m^2}} - \left(\frac{n}{\sqrt{n^2 + 1 + m^2}} \right)^3 \right] \qquad (8\text{-}6)$$

or

$$\sigma_z = \frac{q}{z} P_0 \qquad (8\text{-}6a)$$

where

$$P_0 = \frac{1}{2\pi(m^2 + 1)^2} \left[\frac{3n}{\sqrt{n^2 + 1 + m^2}} - \left(\frac{n}{\sqrt{n^2 + 1 + m^2}} \right)^3 \right]$$

Values for P_0 for various combinations of m and n are given in Table 8-2. In using Table 8-2 one notes that the values for m and n are *not* interchangeable. Furthermore, for values of m and n falling within the range of those given in the table, a straight-line interpolation may be assumed.

The application of this expression and the use of Table 8-2 are illustrated by Example 8-4. If the point at which the stress is desired lies between the two ends of the line, the effects of the load on that point are evaluated separately on each side of the point and subsequently added. On the other hand, if the point lies beyond the end of the line, the value of σ_z is that produced by the full length of the extended line minus the effect of the extension.

EXAMPLE 8-4

Given $q = 100$ kN/m (22.5 kip/m = 6.85 kip/ft).

Find σ_z at points 0, 1, and 2 shown in Fig. 8-9.

Procedure   (a)   Point 0. $m = x/z = 2/4$; $n = y/z = 3.2/4$. From Table 8-2, $P_0 = 0.15775$. Thus,

$$\sigma_{z0} = \frac{100 \text{ kN}}{4} (0.15775)$$

Answer

$$\sigma_{z0} = 3.944 \text{ kN/m}^2 \; (82.4 \text{ lb/ft}^2)$$

(b)   Point 1. $\sigma_{z1} = \sigma_{zL}$ due to load on left + σ_{zR} due to load on right of point 1.
From Fig. 8-10a and Table 8-2, $P_0 = 0.06534$. Thus,

$$\sigma_{zL} = \tfrac{100}{4}(0.06534) = 1.634$$

From Fig. 8-10b and Table 8-2, $P_{0R} = 0.12578$. Thus,

$$\sigma_{zR} = \tfrac{100}{4}(0.12578) = 3.144$$

TABLE 8-2
Influence Values P_0 for Case of Line Load of Finite Length Uniformly Loaded (Boussinesq Solution)

m	0.1	0.2	0.3	0.4	0.5	0.6	0.7	0.8	0.9	1.0	1.2	1.4
0.0	0.04735	0.09244	0.13342	0.16917	0.19929	0.22398	0.24379	0.25947	0.27176	0.28135	0.29464	0.30277
0.1	0.04619	0.09020	0.13023	0.16520	0.19470	0.21892	0.23839	0.25382	0.26593	0.27539	0.28853	0.29659
0.2	0.04294	0.08391	0.12127	0.15403	0.18178	0.20466	0.22315	0.23787	0.24947	0.25857	0.27127	0.27911
0.3	0.03820	0.07472	0.10816	0.13764	0.16279	0.18367	0.20066	0.21429	0.22511	0.23365	0.24566	0.25315
0.4	0.03271	0.06406	0.09293	0.11855	0.14058	0.15905	0.17423	0.18651	0.19634	0.20418	0.21532	0.22235
0.5	0.02715	0.05325	0.07742	0.09904	0.11782	0.13373	0.14694	0.15775	0.16650	0.17354	0.18368	0.19018
0.6	0.02200	0.04322	0.06298	0.08081	0.09646	0.10986	0.12112	0.13045	0.13809	0.14430	0.15339	0.15931
0.7	0.01752	0.03447	0.05035	0.06481	0.07762	0.08872	0.09816	0.10608	0.11265	0.11805	0.12607	0.13140
0.8	0.01379	0.02717	0.03979	0.05136	0.06172	0.07080	0.07862	0.08525	0.09082	0.09546	0.10247	0.10722
0.9	0.01078	0.02128	0.03122	0.04041	0.04872	0.05608	0.06249	0.06800	0.07268	0.07663	0.08268	0.08687
1.0	0.00841	0.01661	0.02441	0.03169	0.03832	0.04425	0.04948	0.05402	0.05793	0.06126	0.06645	0.07012
1.2	0.00512	0.01013	0.01495	0.01949	0.02369	0.02752	0.03097	0.03403	0.03671	0.03905	0.04281	0.04558
1.4	0.00316	0.00626	0.00927	0.01213	0.01481	0.01730	0.01957	0.02162	0.02345	0.02508	0.02777	0.02983
1.6	0.00199	0.00396	0.00587	0.00770	0.00944	0.01107	0.01258	0.01396	0.01522	0.01635	0.01828	0.01979
1.8	0.00129	0.00256	0.00380	0.00500	0.00615	0.00724	0.00825	0.00920	0.01007	0.01086	0.01224	0.01336
2.0	0.00085	0.00170	0.00252	0.00333	0.00410	0.00484	0.00554	0.00619	0.00680	0.00736	0.00836	0.00918
2.5	0.00034	0.00067	0.00100	0.00133	0.00164	0.00194	0.00224	0.00252	0.00278	0.00303	0.00349	0.00389
3.0	0.00015	0.00030	0.00045	0.00060	0.00074	0.00088	0.00102	0.00115	0.00127	0.00140	0.00162	0.00183
4.0	0.00004	0.00008	0.00012	0.00016	0.00020	0.00024	0.00027	0.00031	0.00035	0.00038	0.00045	0.00051
5.0	0.00001	0.00003	0.00004	0.00006	0.00007	0.00008	0.00010	0.00011	0.00012	0.00013	0.00016	0.00018
6.0	0.00001	0.00001	0.00002	0.00002	0.00003	0.00003	0.00004	0.00005	0.00005	0.00006	0.00007	0.00008
8.0	0.00000	0.00000	0.00000	0.00000	0.00001	0.00001	0.00001	0.00001	0.00001	0.00001	0.00001	0.00002
10.0	0.00000	0.00000	0.00000	0.00000	0.00000	0.00000	0.00000	0.00000	0.00000	0.00000	0.00000	0.00001

n

					n					
1.6	1.8	2.0	2.5	3.0	4.0	5.0	6.0	8.0	10.0	∞
0.30784	0.31107	0.31318	0.31593	0.31707	0.31789	0.31813	0.31822	0.31828	0.31830	0.31831
0.30161	0.30482	0.30692	0.30966	0.31080	0.31162	0.31186	0.31195	0.31201	0.31203	0.31204
0.28402	0.28716	0.28923	0.29193	0.29307	0.29388	0.29412	0.29421	0.29427	0.29428	0.29430
0.25788	0.26092	0.26293	0.26558	0.26670	0.26750	0.26774	0.26783	0.26789	0.26790	0.26792
0.22683	0.22975	0.23169	0.23426	0.23535	0.23614	0.23638	0.23647	0.23653	0.23654	0.23656
0.19438	0.19714	0.19899	0.20147	0.20253	0.20331	0.20354	0.20363	0.20369	0.20371	0.20372
0.16320	0.16578	0.16753	0.16990	0.17093	0.17169	0.17192	0.17201	0.17207	0.17208	0.17210
0.13496	0.13735	0.13899	0.14124	0.14224	0.14297	0.14320	0.14329	0.14335	0.14336	0.14338
0.11044	0.11264	0.11416	0.11628	0.11723	0.11795	0.11818	0.11826	0.11832	0.11834	0.11835
0.08977	0.09177	0.09318	0.09517	0.09608	0.09677	0.09699	0.09708	0.09713	0.09715	0.09716
0.07270	0.07452	0.07580	0.07766	0.07852	0.07919	0.07941	0.07949	0.07955	0.07957	0.07958
0.04759	0.04905	0.05012	0.05171	0.05248	0.05310	0.05330	0.05338	0.05344	0.05345	0.05347
0.03137	0.03253	0.03340	0.03474	0.03542	0.03598	0.03617	0.03625	0.03630	0.03632	0.03633
0.02097	0.02188	0.02257	0.02368	0.02427	0.02478	0.02496	0.02504	0.02509	0.02510	0.02512
0.01425	0.01496	0.01551	0.01643	0.01694	0.01739	0.01756	0.01765	0.01768	0.01769	0.01771
0.00986	0.01041	0.01085	0.01160	0.01203	0.01244	0.01259	0.01266	0.01271	0.01272	0.01273
0.00424	0.00453	0.00477	0.00523	0.00551	0.00581	0.00593	0.00599	0.00603	0.00605	0.00606
0.00201	0.00217	0.00231	0.00258	0.00277	0.00298	0.00307	0.00312	0.00316	0.00317	0.00318
0.00057	0.00063	0.00068	0.00078	0.00086	0.00096	0.00102	0.00105	0.00108	0.00109	0.00110
0.00021	0.00023	0.00025	0.00029	0.00033	0.00038	0.00041	0.00043	0.00045	0.00046	0.00047
0.00009	0.00010	0.00011	0.00013	0.00014	0.00017	0.00019	0.00020	0.00022	0.00023	0.00023
0.00002	0.00002	0.00003	0.00003	0.00004	0.00005	0.00005	0.00006	0.00007	0.00007	0.00008
0.00001	0.00001	0.00001	0.00001	0.00001	0.00002	0.00002	0.00002	0.00003	0.00003	0.00004

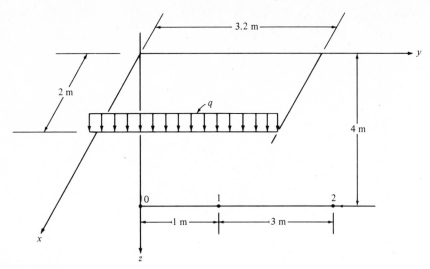

FIGURE 8-9

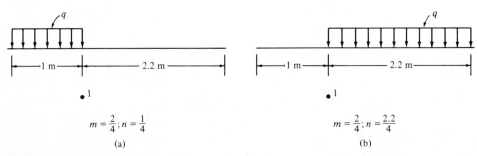

$$m = \frac{2}{4} ; n = \frac{1}{4}$$

(a)

$$m = \frac{2}{4} ; n = \frac{2.2}{4}$$

(b)

FIGURE 8-10
(a) Left load. (b) Right load.

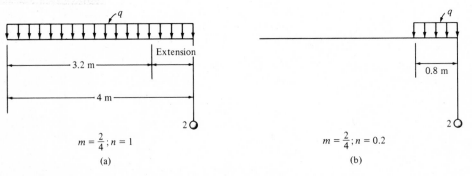

$$m = \frac{2}{4} ; n = 1$$

(a)

$$m = \frac{2}{4} ; n = 0.2$$

(b)

FIGURE 8-11
(a) Actual load plus extra load. (b) Extra load.

Answer

$$\sigma_{z1} = \sigma_{2L} + \sigma_{2R} = 4.778 \text{ kN/m}^2 \ (100 \text{ lb/ft}^2)$$

(c)   Point 2. $\sigma_{z2} = \sigma_z$ due to actual plus extra load $- \sigma_z$ due to extra load.
From Fig. 8-11a and Table 8-2, $P_0 = 0.1735$ and from Fig. 8-11b, $P_0 = 0.05325$. Thus,

$$\sigma_{z2} = \tfrac{100}{4}(0.1735 - 0.0532)$$

Answer

$$\sigma_{z2} = 3.01 \text{ kN/m}^2 \ (63 \text{ lb/ft}^2)$$

8-6 UNIFORMLY LOADED CIRCULAR AREA

The unit vertical stress on any given depth could be determined with acceptable accuracy by extending Boussinesq's equation (8-2a) to a uniformly loaded circular area.

Two separate cases of the vertical stress under circular footings will be considered. Case A considers only the vertical stress under the center of the footing, while case B considers the vertical stress at any point in the soil, including under the center of the footing.

CASE A   Vertical Stress Under the Center of the Footing (Fig. 8-12)

From Boussinesq's equation,

$$\sigma_z = \frac{3qz^3}{2\pi} \int_0^{2\pi} \int_0^r \frac{\rho \, d\beta \, d\rho}{(\rho^2 + z^2)^{5/2}}$$

Integrating with respect to β and substituting limits, we have

$$\sigma_z = \frac{3qz^3}{2\pi} \int_0^r \frac{\rho \, d\rho}{(\rho^2 + z^2)^{5/2}} \left[\beta \right]_0^{2\pi} = \frac{3qz^3}{1} \int_0^r \frac{\rho \, d\rho}{(\rho^2 + z^2)^{5/2}}$$

Integrating,

$$\sigma_z = qz^3 \left[\frac{-1}{(\rho^2 + z^2)^{3/2}} \right]_0^r$$

or

$$\sigma_z = qz^3 \left[\frac{-1}{(r^2 + z^2)^{3/2}} + \frac{1}{(z^2)^{3/2}} \right]$$

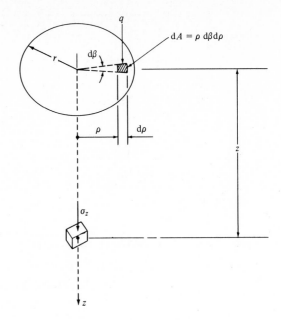

FIGURE 8-12
Vertical stress under center of
loaded circular area.

Hence,

$$\sigma_z = q\left[1 - \frac{1}{(r^2/z^2 + 1)^{3/2}}\right] \tag{8-7}$$

or

$$\sigma_z = qW_0 \tag{8-7a}$$

where W_0, a dimensionless coefficient, is given by

$$W_0 = \left[1 - \frac{1}{(r^2/z^2 + 1)^{3/2}}\right]$$

Values for W_0 for various combinations of r and z are given in Table 8-3.

CASE B   Vertical Stress at Any Point in the Soil (Fig. 8-13)

Equation (8-7) is only valid when σ_z is to be determined under the center of a circular area. Charts and tables are available, however, which provide rather expedient means for estimating σ_z for points lying under as well as outside the center (2, 5, 8). A chart developed by Foster and Ahlvin (5) will be explained in detail. The expression for σ_z takes the form given by Eq. (8-8):

$$\sigma_z = qN_z(m, n) \tag{8-8}$$

where N_z is a shape function of dimensionless variables, $m = z/r$, and $n = q/r$.

TABLE 8-3 Influence Values W_0 for Vertical Stress Under Center of Loaded Circular Area

r/z	W_0	r/z	W_0	r/z	W_0
0.00	0.00000	0.50	0.28446	1.00	0.64645
0.01	0.00015	0.51	0.29304	1.01	0.65171
0.02	0.00060	0.52	0.30162	1.02	0.65689
0.03	0.00135	0.53	0.31019	1.03	0.66200
0.04	0.00239	0.54	0.31875	1.04	0.66703
0.05	0.00374	0.55	0.32728	1.05	0.67198
0.06	0.00537	0.56	0.33579	1.06	0.67686
0.07	0.00731	0.57	0.34427	1.07	0.68166
0.08	0.00952	0.58	0.35271	1.08	0.68639
0.09	0.01203	0.59	0.36112	1.09	0.69104
0.10	0.01481	0.60	0.36949	1.10	0.69562
0.11	0.01788	0.61	0.37781	1.11	0.70013
0.12	0.02122	0.62	0.38608	1.12	0.70457
0.13	0.02482	0.63	0.39431	1.13	0.70894
0.14	0.02870	0.64	0.40247	1.14	0.71384
0.15	0.03282	0.65	0.41085	1.15	0.71747
0.16	0.03721	0.66	0.41863	1.16	0.72163
0.17	0.04183	0.67	0.42662	1.17	0.72572
0.18	0.04670	0.68	0.43454	1.18	0.72975
0.19	0.05180	0.69	0.44239	1.19	0.73373
0.20	0.05713	0.70	0.45018	1.20	0.73763
0.21	0.06268	0.71	0.45789	1.21	0.74147
0.22	0.06844	0.72	0.46553	1.22	0.74525
0.23	0.07441	0.73	0.47310	1.23	0.74896
0.24	0.08057	0.74	0.48059	1.24	0.75262
0.25	0.08692	0.75	0.48800	1.25	0.75622
0.26	0.09346	0.76	0.49533	1.26	0.75076
0.27	0.10017	0.77	0.50259	1.27	0.76323
0.28	0.10704	0.78	0.50976	1.28	0.76666
0.29	0.11408	0.79	0.51685	1.29	0.77003
0.30	0.12126	0.80	0.52386	1.30	0.77334
0.31	0.12859	0.81	0.53079	1.31	0.77660
0.32	0.13604	0.82	0.53763	1.32	0.77981
0.33	0.14363	0.83	0.54439	1.33	0.78296
0.34	0.15133	0.84	0.55106	1.34	0.78606
0.35	0.15915	0.85	0.55765	1.35	0.78911
0.36	0.16706	0.86	0.56416	1.36	0.79211
0.37	0.17507	0.87	0.57058	1.37	0.79507
0.38	0.18317	0.88	0.57692	1.38	0.79797
0.39	0.19134	0.89	0.58317	1.39	0.80083
0.40	0.19959	0.90	0.58934	1.40	0.80364
0.41	0.20790	0.91	0.59542	1.41	0.80640
0.42	0.21627	0.92	0.60142	1.42	0.80912
0.43	0.22469	0.93	0.60734	1.43	0.81179
0.44	0.23315	0.94	0.61317	1.44	0.81442
0.45	0.24164	0.95	0.61892	1.45	0.81701
0.46	0.25017	0.96	0.62459	1.46	0.81955
0.47	0.25872	0.97	0.63018	1.47	0.82206
0.48	0.26729	0.98	0.65568	1.48	0.82452
0.49	0.27587	0.99	0.64110	1.49	0.82694

TABLE 8-3 (*Continued*)

r/z	W_0	r/z	W_0	r/z	W_0
1.50	0.82932	2.00	0.91056	7.50	0.99769
1.51	0.83167	2.02	0.91267	7.80	0.99794
1.52	0.83397	2.04	0.91472	8.00	0.99809
1.53	0.83624	2.06	0.91672	8.20	0.99823
1.54	0.83847	2.08	0.91865	8.40	0.99835
1.55	0.84067	2.10	0.92053	8.60	0.99846
1.56	0.84283	2.15	0.92499	8.80	0.99856
1.57	0.84495	2.20	0.92914	9.00	0.99865
1.58	0.84704	2.25	0.03301	9.20	0.99874
1.59	0.84910	2.30	0.93661	9.40	0.99882
1.60	0.85112	2.35	0.93997	9.60	0.99889
1.61	0.85312	2.40	0.94310	9.80	0.99898
1.62	0.85507	2.45	0.94603	10.00	0.99901
1.63	0.85700	2.50	0.94877	10.20	0.99907
1.64	0.85890	2.55	0.95134	10.40	0.99912
1.65	0.86077	2.60	0.95374	10.60	0.99917
1.66	0.86260	2.65	0.95599	10.80	0.99922
1.67	0.86441	2.70	0.95810	11.00	0.99926
1.68	0.86619	2.75	0.96008	11.20	0.99930
1.69	0.86794	2.80	0.96195	11.50	0.99935
1.70	0.86966	2.85	0.96371	11.80	0.99940
1.71	0.87136	2.90	0.96536	12.00	0.99943
1.72	0.87302	2.95	0.96691	12.20	0.99945
1.73	0.87467	3.00	0.96838	12.50	0.99949
1.74	0.87628	3.10	0.97106	12.80	0.99953
1.75	0.87787	3.20	0.97346	13.00	0.99955
1.76	0.87944	3.30	0.97561	13.20	0.99957
1.77	0.88098	3.40	0.97753	13.50	0.99960
1.78	0.88250	3.50	0.97927	13.80	0.99962
1.79	0.88399	3.60	0.98083	14.00	0.99964
1.80	0.88546	3.70	0.98224	14.20	0.99965
1.81	0.88691	3.80	0.98352	14.50	0.99967
1.82	0.88833	3.90	0.98468	14.80	0.99969
1.83	0.88974	4.00	0.98573	15.00	0.99971
1.84	0.89112	4.20	0.98757	15.40	0.99973
1.85	0.89248	4.40	0.98911	15.80	0.99975
1.86	0.89382	4.60	0.99041	16.00	0.99976
1.87	0.89514	4.80	0.99152	16.40	0.99977
1.88	0.89643	5.00	0.99246	16.60	0.99978
1.89	0.89771	5.20	0.99327	17.00	0.99980
1.90	0.89897	5.40	0.99396	17.40	0.99981
1.91	0.90021	5.60	0.99457	17.80	0.99982
1.92	0.90143	5.80	0.99510	18.00	0.99983
1.93	0.90263	6.00	0.99556	20.00	0.99988
1.94	0.90382	6.20	0.99596	25.00	0.99994
1.95	0.90498	6.40	0.99632	30.00	0.99996
1.96	0.90613	6.60	0.99664	40.00	0.99998
1.97	0.90726	6.80	0.99692	50.00	0.99999
1.98	0.90838	7.00	0.99717	100.00	1.00000
1.99	0.90948	7.30	0.99750	∞	1.00000

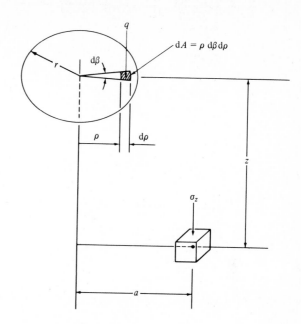

FIGURE 8-13
Vertical stress from loaded
circular area.

The value of N_z can be determined from Fig. 8-14, developed by Foster and Ahlvin. It is based on the assumption that the mass is a semi-infinite elastic medium whose Poisson's ratio is 0.5. It is applicable to points under as well as outside the centerline of a circular footing. Figure 8-13 gives a general configuration of load and stress conditions. For those interested, the analytical solution involves integrating the equation

$$\sigma = \left(\frac{3qr^3}{2\pi}\right) \int_0^{2\pi} \int_0^r \frac{\rho \, d\beta \, d\rho}{(\rho^2 + z^2 + a^2 - 2a\rho \cos \beta)^{5/2}}$$

EXAMPLE 8-5

Given   A circular area, $r = 1.6$ m, induces a soil pressure at the surface of 100 kN/m².

Find   The pressure at a depth of (a) 2 m directly under the center of the circular area, and (b) 2 m below and 2 m away from the center of the circle.

Procedure   (a) $r/z = 1.6/2 = 0.8$. From Table 8-3, $W_0 = 0.52386$. Thus

$$\sigma_z = 100 \times 0.52386$$

Answer

$$\sigma_z = 52.386 \text{ kN/m}^2$$

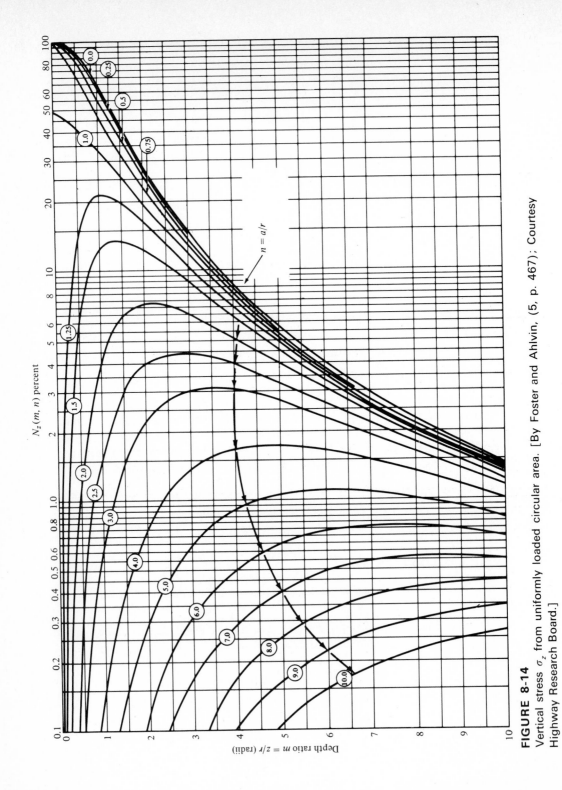

FIGURE 8-14
Vertical stress σ_z from uniformly loaded circular area. [By Foster and Ahlvin, (5, p. 467); Courtesy Highway Research Board.]

Plot axes:
$N_z(m, n)$ percent (horizontal axis, logarithmic, 0.1 to 100)
Depth ratio $m = z/r$ (radii) (vertical axis, 0 to 10)
$n = a/r$

Curve labels: 0.0, 0.25, 0.5, 0.75, 1.0, 1.25, 1.5, 2.0, 2.5, 3.0, 4.0, 5.0, 6.0, 7.0, 8.0, 9.0, 10.0

From Fig. 8-13, for a depth ratio $z/r = 1.25$ radii and an offset distance $= 0$ from the centerline, $a/r = 0$,

$$\% \text{ of total} \approx 52\% \text{ of } 100$$

Answer

$$\sigma_z = 52 \text{ kN/m}^2$$

(b) At a depth $= 2$ m ($m = r/2 = 1.25$) and offset $= 2$ m ($n = a/r = 1.25$)

$$\sigma_z = 22\% \text{ of } 100$$

Answer

$$\sigma_z = 22 \text{ kN/m}^2$$

8-7 PRESSURE CAUSED BY UNIFORMLY LOADED RECTANGULAR AREA

From Boussinesq's equation (8-2a) the vertical stress under a corner of a rectangular area uniformly loaded with a uniform load of intensity q (Fig. 8-15) can be expressed as

$$\sigma_z = \frac{3qz^3}{2\pi} \int_0^a \int_0^b \frac{dy\, dx}{(x^2 + y^2 + z^2)^{5/2}} \tag{a}$$

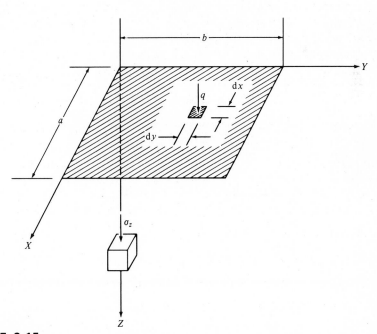

FIGURE 8-15
Vertical stress under *corner* of rectangular area uniformly loaded.

TABLE 8-4
Influence Values $f_z(m, n)$ for Case of Rectangular Area Uniformly Loaded (Boussinesq Solution)

m	0.1	0.2	0.3	0.4	0.5	0.6	0.7	0.8	0.9	1.0	1.2	1.4
0.1	0.00470	0.00917	0.01324	0.01678	0.01978	0.02223	0.02420	0.02576	0.02698	0.02794	0.02926	0.03007
0.2	0.00917	0.01790	0.02585	0.03280	0.03866	0.04348	0.04735	0.05042	0.05283	0.05471	0.05733	0.05894
0.3	0.01324	0.02585	0.03735	0.04742	0.05593	0.06294	0.06859	0.07308	0.07661	0.07938	0.08323	0.08561
0.4	0.01678	0.03280	0.04742	0.06024	0.07111	0.08009	0.08735	0.09314	0.09770	0.10129	0.10631	0.10941
0.5	0.01978	0.03866	0.05593	0.07111	0.08403	0.09472	0.10340	0.11034	0.11584	0.12018	0.12626	0.13003
0.6	0.02223	0.04348	0.06294	0.08009	0.09472	0.10688	0.11679	0.12474	0.13105	0.13605	0.14309	0.14749
0.7	0.02420	0.04735	0.06859	0.08735	0.10340	0.11679	0.12772	0.13653	0.14356	0.14914	0.15703	0.16199
0.8	0.02576	0.05042	0.07308	0.09314	0.11034	0.12474	0.13653	0.14607	0.15370	0.15978	0.16843	0.17389
0.9	0.02698	0.05283	0.07661	0.09770	0.11584	0.13105	0.14356	0.15370	0.16185	0.16835	0.17766	0.18357
1.0	0.02794	0.05471	0.07938	0.10129	0.12018	0.13605	0.14914	0.15978	0.16835	0.17522	0.18508	0.19139
1.2	0.02926	0.05733	0.08323	0.10631	0.12626	0.14309	0.15703	0.16843	0.17766	0.18508	0.19584	0.20278
1.4	0.03007	0.05894	0.08561	0.10941	0.13003	0.14749	0.16199	0.17389	0.18357	0.19139	0.20278	0.21020
1.6	0.03058	0.05994	0.08709	0.11135	0.13241	0.15027	0.16515	0.17739	0.18737	0.19546	0.20731	0.21509
1.8	0.03090	0.06058	0.08804	0.11260	0.13395	0.15207	0.16720	0.17967	0.18986	0.19814	0.21032	0.21836
2.0	0.03111	0.06100	0.08867	0.11342	0.13496	0.15326	0.16856	0.18119	0.19152	0.19994	0.21235	0.22058
2.5	0.03138	0.06155	0.08948	0.11450	0.13628	0.15483	0.17036	0.18321	0.19375	0.20236	0.21512	0.22364
3.0	0.03150	0.06178	0.08982	0.11495	0.13684	0.15550	0.17113	0.18407	0.19470	0.20341	0.21633	0.22499
4.0	0.03158	0.06194	0.09006	0.11527	0.13724	0.15598	0.17168	0.18469	0.19540	0.20417	0.21722	0.22600
5.0	0.03160	0.06199	0.09014	0.11537	0.13736	0.15612	0.17185	0.18488	0.19561	0.20440	0.21749	0.22632
6.0	0.03161	0.06201	0.09016	0.11541	0.13741	0.15617	0.17191	0.18496	0.19569	0.20449	0.21760	0.22644
8.0	0.03162	0.06202	0.09018	0.11543	0.13744	0.15621	0.17195	0.18500	0.19574	0.20455	0.21767	0.22652
10.0	0.03162	0.06202	0.09019	0.11544	0.13745	0.15622	0.17196	0.18502	0.19576	0.20457	0.21769	0.22654
∞	0.03162	0.06202	0.09019	0.11544	0.13745	0.15623	0.17197	0.18502	0.19577	0.20458	0.21770	0.22656

The integral is difficult and far too long to provide a practical benefit here. The integration was performed by Newmark (12) with the following results:

$$\sigma_z = \frac{q}{4\pi} \left[\frac{2mn\sqrt{m^2 + n^2 + 1}}{m^2 + n^2 + 1 + m^2 n^2} \frac{m^2 + n^2 + 2}{m^2 + n^2 + 1} + \sin^{-1} \frac{2mn\sqrt{m^2 + n^2 + 1}}{m^2 + n^2 + 1 + m^2 n^2} \right] \quad (8\text{-}9)$$

where $m = a/z$ and $n = b/z$. Equation (8-9) can also be expressed as

$$\sigma_z = q \cdot f_z(m, n) \quad (8\text{-}9a)$$

where $f_z(m, n)$ is the shape function of the dimensionless ratios m and n. The influence values for various combinations of m and n can be found directly from Table 8-4.

| n | | | | | | | | | | |
1.6	1.8	2.0	2.5	3.0	4.0	5.0	6.0	8.0	10.0	∞
0.03058	0.03090	0.03111	0.03138	0.03150	0.03158	0.03160	0.03161	0.03162	0.03162	0.03162
0.05994	0.06058	0.06100	0.06155	0.06178	0.06194	0.06199	0.06201	0.06202	0.06202	0.06202
0.08709	0.08804	0.08867	0.08948	0.08982	0.09006	0.09014	0.09016	0.09018	0.09019	0.09019
0.11135	0.11260	0.11342	0.11450	0.11495	0.11527	0.11537	0.11541	0.11543	0.11544	0.11544
0.13241	0.13395	0.13496	0.13628	0.13684	0.13724	0.13736	0.13441	0.13744	0.13745	0.13745
0.15027	0.15207	0.15326	0.15483	0.15550	0.15598	0.15612	0.15617	0.15621	0.15622	0.15623
0.16515	0.16720	0.16856	0.17036	0.17113	0.17168	0.17185	0.17191	0.17195	0.17196	0.17197
0.17739	0.17967	0.18119	0.18321	0.18407	0.18469	0.18488	0.18496	0.18500	0.18502	0.18502
0.18737	0.18986	0.19152	0.19375	0.19470	0.19540	0.19561	0.19569	0.19574	0.19576	0.19577
0.19546	0.19814	0.19994	0.20236	0.20341	0.20417	0.20440	0.20449	0.20455	0.20457	0.20459
0.20731	0.21032	0.21235	0.21512	0.21633	0.21722	0.21749	0.21760	0.21767	0.21769	0.21770
0.21509	0.21836	0.22058	0.22364	0.22499	0.22600	0.22632	0.22644	0.22652	0.22654	0.22656
0.22025	0.22372	0.22610	0.22940	0.23088	0.23200	0.23235	0.23249	0.23258	0.23261	0.23263
0.22372	0.22736	0.22986	0.23336	0.23496	0.23617	0.23656	0.23671	0.23681	0.23684	0.23686
0.22610	0.22986	0.23247	0.23613	0.23782	0.23912	0.23954	0.23970	0.23981	0.23985	0.23987
0.22940	0.23336	0.23613	0.24010	0.24196	0.24344	0.24392	0.24412	0.24425	0.24429	0.24432
0.23088	0.23496	0.23782	0.24196	0.24394	0.24554	0.24608	0.24630	0.24646	0.24650	0.24654
0.23200	0.23617	0.23912	0.24344	0.24554	0.24729	0.24791	0.24817	0.24836	0.24841	0.24846
0.23235	0.23656	0.23954	0.24392	0.24608	0.24791	0.24857	0.24886	0.24907	0.24914	0.24919
0.23249	0.23671	0.23970	0.24412	0.24630	0.24817	0.24886	0.24916	0.24939	0.24946	0.24952
0.23258	0.23681	0.23981	0.24425	0.24646	0.24836	0.24907	0.24939	0.24964	0.24972	0.24980
0.23261	0.23684	0.23985	0.24429	0.24650	0.24841	0.24914	0.24946	0.24972	0.24981	0.24989
0.23263	0.23686	0.23987	0.24432	0.24654	0.24846	0.24919	0.24952	0.24980	0.24989	0.25000

When the point at which the stress is desired does not fall below a *corner* of the area, the area is "adjusted" into rectangles as shown in Fig. 8-16, such that corners become located over the point in question. Subsequently the effects are superimposed, as illustrated by Example 8-6. Figure 8-16 shows different locations of the point in question relative to the loaded area (solid boundaries) and the effective load combinations. From this figure the load is calculated as follows:

(a) Load = load as determined directly from Table 8-4.
(b) Load = load from $AFIE + FBGI + GCHI + HDEI$.
(c) Load = load from $AEFD + EBCF$.
(d) Load = load from $GIAE - GHBE - GIDF + GHCF$.

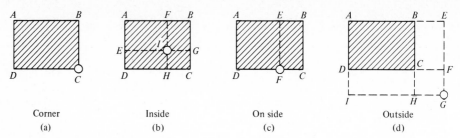

Corner   Inside   On side   Outside
(a)   (b)   (c)   (d)

FIGURE 8-16
Sketch showing the rectangle boundaries for the four basic cases of load superposition.

EXAMPLE 8-6

Given   A rectangular footing 2 m by 3 m carries a total load of 120 metric tons (t) uniformly distributed.

Find   The stress 2.5 m below the footing at a point (a) under one corner, and (b) under the center.

Procedure

$$q = \frac{120}{2 \times 3} = 20 \text{ t/m}^2.$$

(a) Point at corner.

$$m = \frac{a}{z} = \frac{2}{2.5} = 0.8 \qquad n = \frac{3}{2.5} = 1.2$$

From Table 8-4, $f_z(m, n) = 0.16843$. Thus,

$$\sigma_z = (20 \text{ t})(0.16843)$$

Answer

$$\sigma_z = 3.3686 \text{ t/m}^2$$

(b) Point at center.

$$m = \frac{1}{2.5} = 0.4 \qquad n = \frac{1.5}{2.5} = 0.6$$

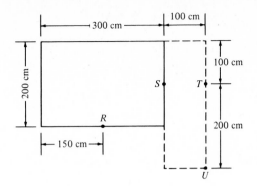

FIGURE 8-17

From Table 8-4, $f_z(m, n) = 0.08009$. Thus

$$\sigma_z = 4(20 \text{ t})(0.08009)$$

Answer

$$\sigma_z = 6.4072 \text{ t/m}^2$$

EXAMPLE 8-7

Given   A rectangular loading as shown in Fig. 8-17. $q = 20 \text{ t/m}^2$.

Find σ_z at $z = 2.5$ m for points (a) R, (b) S, (c) T, (d) U.

Procedure   (a)   Point R.

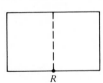

$$m = \frac{1.5}{2.5} = 0.6 \qquad n = \frac{2}{2.5} = 0.8$$

$$f_z(m, n) = 0.12474$$

$$\sigma_z = 2(20 \text{ t})(0.12474)$$

Answer

$$\sigma_z = 4.9896 \text{ t/m}^2$$

(b)   Point S.

$$m = \frac{3}{2.5} = 1.2 \qquad n = \frac{1}{2.5} = 0.4$$

$$f_z(m, n) = 0.10631$$

$$\sigma_z = 2(20 \text{ t})(0.10631)$$

Answer

$$\sigma_z = 4.2524 \text{ t/m}^2$$

(c) Point T. For rectangle $TGAE$,

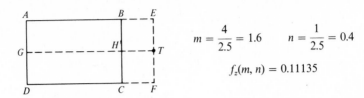

$$m = \frac{4}{2.5} = 1.6 \qquad n = \frac{1}{2.5} = 0.4$$

$$f_z(m, n) = 0.11135$$

For square $THBE$,

$$m = \frac{1}{2.5} = 0.4 \qquad n = \frac{1}{2.5} = 0.4$$

$$f_z(m, n) = 0.06024$$

Superimposing loads (see Fig. 8-16).

$$\sigma_z = 2(20 \text{ t})(0.11135 - 0.06024)$$

Answer

$$\sigma_z = 2.0444 \text{ t/m}^2$$

(d) Point U. For rectangle $UHAE$,

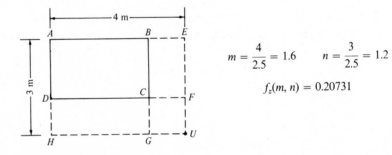

$$m = \frac{4}{2.5} = 1.6 \qquad n = \frac{3}{2.5} = 1.2$$

$$f_z(m, n) = 0.20731$$

For rectangle $UGBE$,

$$m = \frac{1}{2.5} = 0.4 \qquad n = \frac{3}{2.5} = 1.2$$

$$f_z(m, n) = 0.10631$$

For rectangle $UHDF$,

$$m = \frac{4}{2.5} = 1.6 \qquad n = \frac{1}{2.5} = 0.4$$

$$f_z(m, n) = 0.11135$$

For square $UGCF$,

$$m = n = \frac{1}{2.5} = 0.4$$

$$f_z(m, n) = 0.06024$$

Then,

$$\sigma_z = (20\ \text{t})(0.20731 - 0.10631 - 0.11135 + 0.06024)$$

Answer

$$\sigma_2 = 0.9978\ \text{t/m}^2$$

8-8 TOTAL LOAD ON RECTANGULAR AREA IN UNDERSOIL

In the preceding section we have evaluated the vertical stresses induced by a uniformly distributed load over a rectangular area to a point at z depth below the surface. Now we shall evaluate the total load induced on a rectangular area below the surface by a concentrated load applied at the surface. For example, the typical problem may be represented by a wheel load applied at the surface, creating stresses on a buried pipe or culvert, or perhaps on the roof of a relatively shallow tunnel. Although we shall be assuming a horizontal plane, the procedure gives reasonably acceptable results for arch-shaped surfaces by assuming the horizontal plane to be the projection of the circular shape.

Figure 8-18 depicts a rectangular plane, a distance z beneath the surface, subjected to a concentrated surface load Q over one of its corners. The total load on the shaded area is a summation of all the increments of forces induced by the surface load Q.

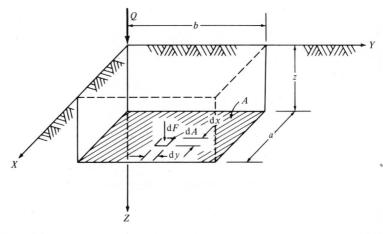

FIGURE 8-18
Load on rectangular area undersoil from concentrated surface load Q over one corner.

The increment of force over a small area may be expressed as the product of the stress and the increment of area, as indicated in Eq. (a):

$$dF = \sigma_z \, dA \tag{a}$$

From Eq. (8-2a), $\sigma_z = 3Qz^3/2\pi R^5$. Thus Eq. (a) becomes

$$F = \int_A \frac{3Q}{2\pi} \frac{z^3}{R^5} \, dA = \frac{3Qz^3}{2\pi} \int_0^a \int_0^b \frac{dx \, dy}{(x^2 + y^2 + z^2)^{5/2}} \tag{b}$$

The integral portion of Eq. (b) is identical to that of Eq. (a) in Section 8-7; i.e., except for the constants, the equations are the same. Hence Eq. (8-9a) could be written as,

$$F = Q f_z(m, n) \tag{8-10}$$

where $f_z(m, n)$ is the influence coefficient or shape function which can be obtained directly from Table 8-4. Examples 8-8 and 8-9 illustrate this application.

EXAMPLE 8-8

Given   A concentrated surface load $Q = 100$ t is situated directly over the center of an area 2 m by 3 m, 2.5 m below the surface.

Find   The total load on the area.

Procedure   We first divide the area such that the load falls over one of the corners, as shown in Fig. 8-19. Thus from Table 8-4, for $m = 1.5/2.5 = 0.6$ and $n = 1/2.5 = 0.4$,

$$f_z(m, n) = 0.08009$$

Then from Eq. (8-9), the total load is

$$F_{tot} = (4 \text{ areas}) \times Q \times f_z(m, n)$$

$$F = 4 \times 100 \times 0.08009$$

Answer

$$F = 32.036 \text{ t}$$

EXAMPLE 8-9

Given   A surface load as shown in Fig. 8-20.

Find   The total load on the shaded area *ABCD*.

Procedure   For the rectangle *FBCE*

$$m = \frac{3}{2.5} = 1.2 \qquad n = \frac{3}{2.5} = 1.2 \qquad f_z(m, n) = 0.19584$$

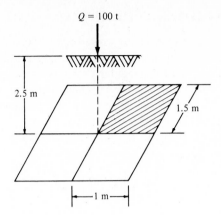

FIGURE 8-19

For the rectangle $FADE$,

$$m = \frac{3}{2.5} = 1.2 \qquad n = \frac{1}{2.5} = 0.4 \qquad f_z(m, n) = 0.10631$$

$$F_{ABCD} = F_{FBCE} - F_{FADE}$$

or

$$F = 100 \times (0.19584 - 0.10631)$$

Answer

$$F = 8.953 \text{ t}$$

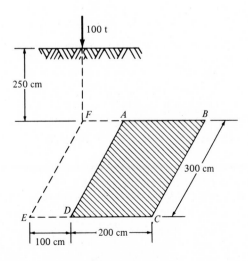

FIGURE 8-20

8-9 NEWMARK'S INFLUENCE CHART

The procedures outlined in the preceding sections for the determination of vertical stresses σ_z induced by uniformly loaded rectangular or circular areas are rather clumsy when applied to irregularly shaped areas. Newmark devised a graphical procedure for computing stresses induced by irregularly shaped loaded areas (12, 13).

Newmark's procedure evolves from the expression for the vertical stress under the center of a loaded circular area, given by Eq. (8-7) or (8-7a). From these expressions the ratio of σ_z/q equals W_0, the influence coefficient given in Table 8-3. That is, Eq. (8-7) can be written as

$$\frac{\sigma_z}{q} = W_0 = \left[1 - \frac{1}{(r^2/z^2 + 1)^{3/2}} \right] \tag{a}$$

The relationship between σ_z/q and r/z may be illustrated by extracting a few values from Table 8-3, as shown in Table 8-5. For convenience ten equal increments of σ_z/q between $\sigma_z/q = 0$ and $\sigma_z/q = 1$ will be selected.

The values of r/z represent concentric circles of relative radii. Plotted for a selected scale for z, these circles are shown in Fig. 8-21, with the last circle not shown since $r/z = \infty$.

Now divide the circles by evenly spaced rays emanating from the center, for convenience say, 20. Thus we obtain a total of (10 circles) (20 rays) = 200 influence units. Thus the influence value IV is

$$IV = \frac{1}{(\text{no. circles})(\text{no. rays})} \tag{b}$$

In our case

$$IV = \frac{1}{10 \times 20} = 0.005$$

The stress at a depth z for a specific point is

$$\sigma_z = q \times IV \times (\text{number of influence units}) \tag{8-11}$$

To use this chart, one draws an outline of the loaded surface to a scale such that the distance AB from Fig. 8-21 equals the depth of the point in question. The point beneath the loaded area for which the vertical stress is sought is then located over the center of the chart. Hence the area will encompass a number of influence units on the chart (in our case each unit has a value of 0.005). Thus by counting the number of

TABLE 8-5
Values of r/z for Selected Values of σ_z/q

σ_z/q	0	0.10	0.20	0.30	0.40	0.50	0.60	0.70	0.80	0.90	1.00
r/z	0	0.27	0.40	0.52	0.64	0.77	0.92	1.11	1.39	1.91	∞

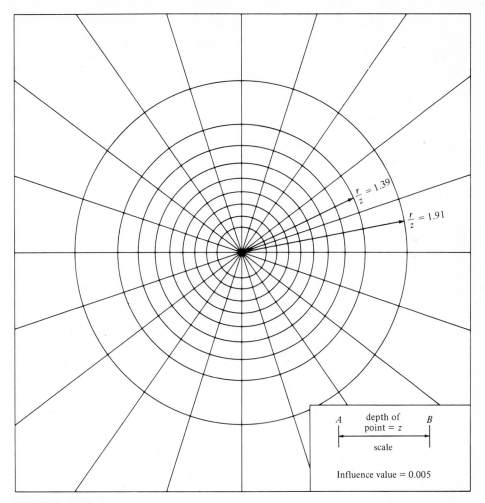

FIGURE 8-21
Newmark influence chart for vertical stress at any depth $z = AB$.

influence units and by using Eq. (8-11), one may proceed to determine the stress at the given point. Example 8-9 may serve to illustrate this procedure.

One may note that while the values for r/z indicated in Table 8-5 remain fixed for the selected values of σ_z/q, the scale for the influence chart was arbitrarily chosen and can therefore be altered as needed. Similarly the number of rays or the number of radii may also vary as desired, thereby varying the influence values for these charts.

EXAMPLE 8-10

Given   The T-shaped foundation is loaded with a uniform load of 100 kN/m² (≈ 2 kips/ft²) (see Fig. 8-22).

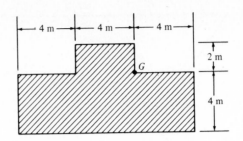

FIGURE 8-22

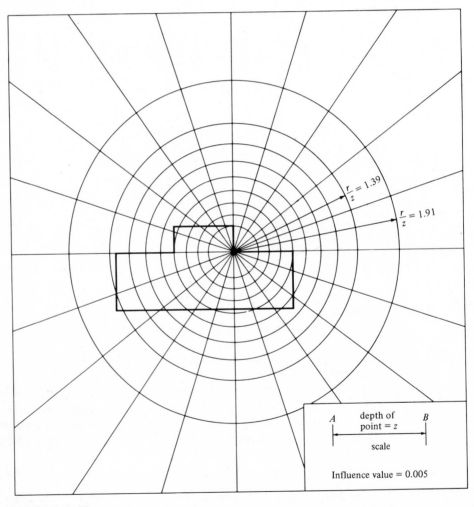

FIGURE 8-23

Find   The pressure at 6 m below point G.

Procedure   The scale is determined such that the distance AB in Fig. 8-23 represents 6 m. Hence the T-shaped area is redrawn to scale with point G placed over the center of the chart. The number of "squares" encompassed by Fig. 8-23 is 66. Hence the total pressure at G is $0.005 \times 66 \times 100$ kN/m^2.

Answer

$$\text{Total pressure} = 33 \text{ kN/m}^2$$

8-10 APPROXIMATE ESTIMATE OF VERTICAL STRESS

Approximate estimates of the average vertical stress under a uniformly loaded area at a given depth z can be made by assuming that the applied surface load spreads downward to a horizontal plane which is enveloped by four planes sloping from the edges of the loaded area at an angle of 30° with the vertical. Another method is to assume a slope of 2:1, as shown in Fig. 8-24. The methods are approximate but rather easy and expedient, and they are quite commonly used for estimating *average* stresses. Generally this approach yields values of σ_z slightly lower than those obtained by previously discussed methods for shallow depths but of comparable magnitude at greater depths.

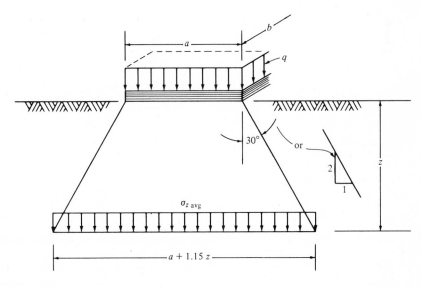

FIGURE 8-24
Approximate $\sigma_{z \, avg}$ on plane at depth z.

For a uniformly loaded surface area $a \times b$, $\sigma_{z\,avg}$ is approximately as follows (see Fig. 8-24). For a 30° slope,

$$\sigma_{z\,avg} = q \frac{a \times b}{(a + 1.15z)(b + 1.15z)} \qquad (8\text{-}12)$$

For a 2:1 slope,

$$\sigma_{z\,avg} = q \frac{a \times b}{(a + z)(b + z)} \qquad (8\text{-}12a)$$

EXAMPLE 8-11

Given   A 120-t load is uniformly distributed on a surface area of 3 m by 2 m.

Find   The approximate $\sigma_{z\,avg}$ at $z = 2.5$ m.

Procedure   From Eq. (8-12),

$$\sigma_{z\,avg} = \frac{120}{(2 + 1.15 \times 2.5)(3 + 1.15 \times 2.5)}$$

$$\sigma_{z\,avg} = \frac{120}{4.88 \times 5.88}$$

Answer

$$\sigma_{z\,avg} = 4.18 \text{ t/m}^2$$

From Eq. (8-12a)

$$\sigma_{z\,avg} = \frac{120}{(2 + 2.5)(3 + 2.5)}$$

Answer

$$\sigma_{z\,avg} = 4.85 \text{ t/m}^2$$

As a matter of comparison, from Example 8-6, at a corner,

$$\sigma_z = 3.37 \text{ t/m}^2 \text{ (via Boussinesq) } (0.690 \text{ kip/ft}^2)$$

and at the center

$$\sigma_z = 6.41 \text{ t/m}^2$$

The average σ_z from corner and center is then 4.83 t/m². Hence it appears that our "estimates" are not unreasonable for this depth.

PROBLEMS

8-1. Calculate the vertical stress at a depth of 3 m due to the weight of the soil (geostatic stress) if $\gamma = 19.3$ kN/m^3.
   (a) The water table is below the 3 m depth.
   (b) The water table is 1 m below the ground surface. Assume total saturation for the "wet" stratum.

8-2. (a) Calculate the vertical stress directly under a concentrated surface load of 1000 kN at depths of 0.5, 1, 1.5, 2, 2.5, 3, 3.5, 4, 4.5, and 5 m.
   (b) Plot the results of part (a), using a convenient scale. (See Example 8-1.)

8-3. Calculate the vertical stress due to a concentrated surface load of 1000 kN at a depth of 2 m at points directly under the load, and at 0.5, 1, 1.5, 2, 2.5, 3, 3.5, and 4 m away. (See Example 8-2.)

8-4. Calculate the vertical stress from a concentrated surface load of 1000 kN at a point whose coordinates are $x = 1$ m, $y = 2$ m, and $z = 3$ m. Surface load coordinates are (0, 0, 0).

8-5. (a) Calculate the vertical stress from a concentrated surface load of 1000 kN at radial distances of $r = 0$, 0.5, 1, 1.5, 2, 2.5, 3, 3.5, and 4 m and depths $z = 1$, 2, 3, and 4 m.
   (b) Plot the values for σ_z determined in part (a) and draw several pressure bulbs similar to those of Fig. 8-7, Section 8-4. Load coordinates are (0, 0, 0).

8-6. A wall footing 10 m long induces a surface load of 300 kN/m. Assume this to be a line load parallel to the y axis and the one end of the footing to be located at coordinates $x = 0$, $y = 0$, and $z = 0$. Determine the vertical stress at points whose coordinates are $x_1 = 2$, $y_1 = 0$, $z_1 = 3$ m and $x_2 = 2$, $y_2 = 5$, and $z_2 = 3$ m. (See Example 8-4.)

8-7. A 700-kN load is supported by a circular footer (at surface) whose diameter is 2 m. Assume the surface pressure to be uniform over the area. (a) Determine the vertical stress at a depth of 3 m directly under the center and at radial distances of 1, 2, 3, and 4 m from the center. (b) Plot the values of σ_z from part (a) and connect these points with a smooth curve.

8-8. A circular footing 3 m in diameter applies a uniform surface pressure of 150 kN/m^2. (a) Via Table 8-3 determine the vertical stress under the center of the footing at $z = 1$, 1.5, 2.5, 3, 3.5, and 4 m. (b) Plot the values for σ_z from part (a) and connect the points with a smooth curve.

8-9. Rework Problem 8-8 using Fig. 8-14.

8-10. A column load of 2000 kN is supported via a 3 m × 4 m footing whose bottom is 1 m below the surface. Assuming that the footer distributes uniform pressure, determine the vertical stress at a depth of 4 m at a point (a) under the center of the footing, and (b) under one of the corners.

8-11. For the footing of Problem 8-10, determine the vertical stress at four points whose (x, y) coordinates are (2, 0), (0, 3), (3, 3), (4, 4), all at 4-m depth. Load coordinates are (0, 0, 0).

8-12. A concentrated surface load of 2000 kN is applied at a point whose cartesian coordinates are $x = 0$, $y = 0$, and $z = 0$. Determine the total load on a rectangular area which lies on the x–y plane, 4 m below the surface, and whose corners have the following (x, y) coordinates: (0, 0), (3, 0), (3, 2), and (0, 2).

8-13. Determine the total load on the area of Problem 8-12 if the concentrated surface load is applied at coordinates $x = -2$, $y = -2$, and $z = 0$.

8-14. A concentrated surface load of 2000 kN is applied at a point whose cartesian coordinates are $x = 0$, $y = 0$, $z = 0$. Determine the total load on a rectangular area which lies parallel to the x–y plane, 5 m below the surface, and whose corners have the following (x, y, z) coordinates: $(1, 1, -5)$, $(4, 1, -5)$, $(4, 3, -5)$, and $(1, 3, -5)$.

8-15.   Rework Problem 8-14 if the concentrated load was applied at coordinate point (1, 2, 0).

8-16.   The footing shown in Fig. P8-16 exerts a uniform load to the soil of 300 kN/m². Determine the pressure at a point 4 m directly under point (a) A and (b) B.

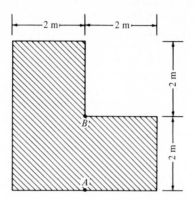

FIGURE P8-16

8-17.   A rectangular footing as shown in Fig. P8-17 (shaded) exerts a uniform pressure of 420 kN/m². Determine the vertical stress at point A for a depth of 3 m.

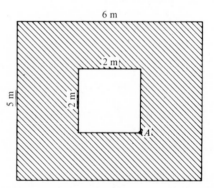

FIGURE P8-17

8-18.   Rework Problem 8-16 using Newmark's influence chart.

8-19.   Rework Problem 8-17 using Newmark's influence chart.

8-20.   A footing shaped as an equilateral triangle of 10 m sides is loaded with a uniform load of 500 kN/m². Determine the pressure at a point located 5 m directly under the middle of one of its sides.

8-21.   For Problem 8-8: (a) Determine the vertical stresses at the given depths using an approximate method. (b) Using the results of Problem 8-8 as the basis of comparison, give the percentage difference in σ_z values for each of these depths.

8-22.   (a)   Construct a Newmark influence chart using 24 rays and values of $\sigma_z/q = 0$, 0.08, 0.16, 0.24, etc.

         (b)   Rework Problem 8-20 using the influence chart developed in part (a) of Problem 8-22.

8-23.   The base pressure under a circular footing, 12 m in diameter, was calculated to be 400 kN/m². Determine the pressure 6 m directly under the center of the footing.

(a) Use influence values given in Table 8-3 for circular footing.

(b) Use the graph by Foster and Ahlvin.

(c) Use Newmark's graph.

8-24. Assume the area shown in Fig. P8-24 to be 7 m below the surface. Determine the total load on this area induced by a concentrated surface load of 2200 kN acting directly over point A.

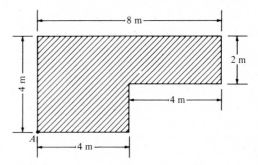

FIGURE P8-24

8-25. Determine the total stress 8 m below point A due to the line pressures shown in Fig. P8-25.

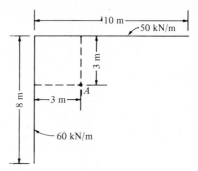

FIGURE P8-25

8-26. Two bearing walls meet at a point A as shown in Fig. P8-26. Assume their loads to act as line loads. Determine the stress at a point 3 m below point A.

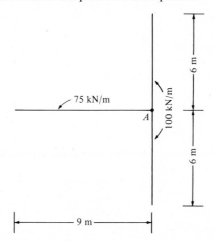

FIGURE P8-26

BIBLIOGRAPHY

1. Acum, W. E. A., and L. Fox, "Computation of Load Stresses in a Three Layer Elastic System," *Geotechnique*, vol. 2, 1951.
2. Ahlvin, R. G., and H. J. Ulery, "Tabulated Values for Determining the Complete Pattern of Stresses, Strains and Deflections Beneath a Uniform Circular Load on a Homogeneous Half Space," *Highw. Res. Board Bull.*, no. 342, 1962.
3. Boussinesq, J., *Application des Potentiels à l'Etude de l'Equilibre et du Mouvement des Solides Elastiques*, Gauthier-Villars, Paris, 1885.
4. Fadum, R. E., "Influence Values for Estimating Stresses in Elastic Foundations," *2nd Int. Conf. Soil Mech. Found. Eng.*, vol. 3, 1948.
5. Foster, C. R., and R. G. Ahlvin, "Stresses and Deflections Induced by a Uniform Circular Load," *Proc. Highw. Res. Board*, vol. 33, 1954.
6. Giroud, J. P., "Stresses under Linearly Loaded Rectangular Area," *ASCE J. Soil Mech. Found. Eng. Div.*, vol. 96, no. SMI, 1970.
7. Gray, H., and I. J. Hooks, "Charts Facilitate Determination of Stresses under Loaded Areas," *Civ. Eng.*, vol. 18, no. 6, 1948.
8. Ho, M. M. K., and R. Lopes, "Contact Pressure of a Rigid Circular Foundation," *ASCE J. Geotech. Eng. Div.*, vol. 103, May 1969.
9. Huang, Y. H., "Stresses and Displacements in Nonlinear Soil Media," *ASCE J. Geotech. Eng. Div.*, vol. 94, Jan. 1968.
10. Koning, H., "Stress Distribution in a Homogeneous Anisotropic, Elastic Semi-infinite Solid," *4th Int. Conf. Soil Mech. Found. Eng.*, vol. 1, 1957.
11. Milovic, D. M., and J. P. Tournier, "Stresses and Displacements Due to Rectangular Load on a Layer of Finite Thickness," *Soils Found.*, vol. 11, 1971.
12. Newmark, N. M., "Simplified Computation of Vertical Pressures in Elastic Foundations," *University of Illinois Eng. Exp. Stn. Circ.* no. 24, vol. 33, no. 4, 1935.
13. Newmark, N. M., "Influence Charts for Computation of Stresses in Elastic Foundations," *University of Illinois Bull.* no. 338, 1942.
14. Newmark, N. M., "Influence Charts for Computation of Vertical Displacements in Elastic Foundations," *University of Illinois Eng. Exp. Stn. Bull.* no. 367, 1947.
15. Skopek, J., "The Influence of Foundation Depth on Stress Distribution," *5th Int. Conf. Soil Mech. Found. Eng.*, vol. 1, 1961.
16. Urena, R. de, J. S. Piquer, F. Muzas, and J. M. Sanz Saracho, "Stress Distribution in Cross-Anisotropic Media," *1st Cong. Int. Soc. Rock Mech.*, Lisbon, Portugal, vol. 1, 1966.
17. Westergaard, H. M., "A Problem of Elasticity Suggested by a Problem in Soil Mechanics: A Soft Material Reinforced by Numerous Strong Horizontal Sheets," *Mechanics of Solids*, S. Timoshenko, 60th Anniversary vol., Macmillan, New York, 1938.

9
Consolidation and Settlement of Structures

9-1 INTRODUCTION

All materials undergo deformation when subjected to loads. The deformation may be uniform for a given load, or it may be progressive and cumulative, depending upon the magnitude of the load, the type of load (static or dynamic), the duration of load, and, of course, the mechanical properties of the material itself. How materials behave under load is one of the most fundamental considerations in structural analysis and design. Indeed, the consolidation of a soil stratum and the subsequent settlement of the superstructure play no less than a major role in foundation engineering.

Many of the materials used in engineering are reasonably homogeneous and isotropic and obey Hooke's law to an acceptable degree. For example, a steel bar subjected to a tensile load will elongate a predictable amount; within the elastic limit of the steel bar, it will double its elongation if the load is doubled; etc. Concrete behaves somewhat similarly, but the relationship between load and deformation is not nearly as well defined as it is for steel, nor are the homogeneity and mechanical properties as consistent as in the case of steel. Generally speaking, soil, as we have seen in the previous discussions, is even less uniform in characteristics and behavior. Hence by inference, we may well expect that soils subjected to foundation pressures will behave with even less predictability than steel or concrete. Yet, as soil engineers we must be able to predict with a reasonable degree of accuracy the deformation of a soil mass subjected to load, and subsequently relate the effect of that deformation to the structure it supports. Despite the knowledge that the material is not elastic, is not homogeneous and isotropic, and is not of easily predictable characteristics, the engineer must, nevertheless, provide a safe and economical design.

In general the engineer is concerned with total *magnitude* and *rate* of consolidation, and ultimately with the settlement of a structure. Furthermore the engineer is not only interested in the amount of settlement of a given structure but, perhaps more importantly, in the relative or *differential settlement* of the structure (22, 35). For example, differential settlement of supports for continuous beams, frames, arches, etc., may prove highly damaging or perhaps even disastrous to the structure. Similarly, a high rate of settlement during construction may be bearable, but large cumulative settlement of the completed structure may be totally intolerable; depending on the structure, the net result may be masonry cracking, leaning of the structure, or perhaps failure in the various components of the structure.

Within the restrictions imposed by various considerations (e.g., economics, structure characteristics, soil properties, design criteria) the engineer's objective is to minimize if not actually eliminate the settlement of a structure. This can sometimes be accomplished by minimizing the intensity of the loads (stress) transmitted to the stratum, by preconsolidating the stratum, by stabilizing the stratum, or by a combination thereof.

9-2 COMPRESSIBILITY OF SOIL

The reduction of volume of a soil mass could be attributed to three possible factors: (1) the escape of water and air from the voids; (2) compression of the solid particles; (3) compression of water and air within the voids. The compression produced by the decrease of the solid particles as a percentage of the total amount is quite negligible; so is that due to the compression of water. Hence virtually all of the compression is attributed to a reduction in the volume of the voids within the soil. If the voids are entirely filled with water, the compression produced is due almost entirely to the escape of water from the voids. For a partially saturated soil, however, the compression of air within the voids may permit appreciable compression of the soil mass even if the escape of water was minimal.

Cohesionless soil such as sand and gravel will, generally, compress during a relatively short time (15, 40). Furthermore the compression of such soils by induced vibrational effects, etc., could be obtained with much greater ease than for the cohesive soils. In fact, quite frequently most of the anticipated settlement of a foundation resting on cohesionless soil has taken place during the construction phase of the structure.

Unlike cohesionless soils, a saturated cohesive soil mass (e.g., fine silts, clays) with low permeability will compress quite slowly since the expulsion of the pore water from such soils is at a significantly slower rate than from cohesionless ones. In other words, because considerably longer time is required for the water to exit from the voids of a cohesive material, the total compression of a saturated clay stratum takes place during a longer period of time than for cohesionless soils (see Fig. 9-2).

The total deformation of a compressible stratum is generally regarded to be the result of a two-phase process: (1) *primary consolidation*, which results from the expulsion of air and pore water from the voids of the soil mass, and (2) *secondary compression*, or *secondary consolidation*, or *creep*, which is speculated to be due to the plastic deformation of the soil as a result of some complex colloid-chemical processes whose roles in this regard are mostly hypothetical at this point. Of the two types of deformations, the *primary* is normally the much larger, easiest to predict, occurs at a faster rate, and is generally the more important of the two. In many problems, however, the *secondary consolidation* is the more important, particularly in highly organic soils (2, 4, 6, 31, 53, 54).

9-3 ONE-DIMENSIONAL CONSOLIDATION; RHEOLOGICAL MODEL

The terms compressibility and consolidation appear synonymous since they both represent a decrease in soil volume. Hence the two terms are frequently used interchangeably in the process of describing the phenomenon of volume decrease, particularly during the evaluation of test data as detailed in subsequent sections. However, *consolidation* is the rate of volume decrease with time. In that sense we view consolidation as a time-related process, involving compression, stress transfer, and of course drainage.

As we have noted in the previous chapter, surface loading results in undersoil stresses in all directions. Related consolidation will, therefore, also result in both the horizontal as well as the vertical directions. However, one notes that the vertical compression or consolidation is the largest, and from a design point of view, it is frequently the most important component. Hence it is the deformation in the vertical direction that we will be focusing upon in this chapter.

In 1923 Karl Terzaghi advanced a rigorous mathematical solution of the process of consolidation of soils. It constitutes one of his greatest contributions to the science of soil mechanics. This theory with all of its underlying assumptions will be developed in subsequent sections of this chapter. At this time, however, let us attempt to illustrate the process of one-dimensional consolidation by means of a model.

Figure 9-1a represents the assembly of a mechanical system used to explain the process of consolidation of a saturated clay. This assembly is analogous to a saturated clay stratum where the volume of the vessel represents the voids of the soil mass with a porosity of n. All of the voids of the soil mass are fully saturated with water. The spring represents the solid particles, while the petcock is analogous to the seepage in the soil through which the water exists. The gauge measures only *hydrostatic excess porewater pressure*. At the instant the petcock is open the total load Q is still carried by the water until some water escapes and subsequently compresses the spring. As more water escapes and the spring begins to compress, more and more of the total load Q is assumed by the spring and less and less by the water, as indicated in Fig. 9-1b. Eventually the hydrostatic excess pressure is reduced to 0, with all the applied load carried by the spring. At this point the deformation would cease, analogous to a complete consolidation.

In correlation with Fig. 9-1b one may represent the total stress σ, resulting from the excess load Q, as a constant which is composed of a *neutral stress*, u, induced by the hydrostatic excess pressure, and of an *effective stress* $\bar{\sigma}$, induced as the soil particles are brought into closer contact with each other. Symbolically we have

$$\sigma = \bar{\sigma} + u \tag{9-1}$$

A more detailed description of this phenomenon was given in Section 8-2.

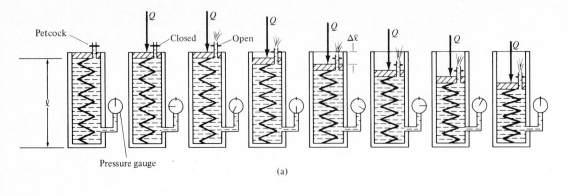

(a)

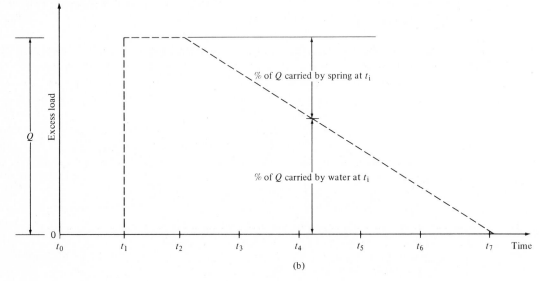

(b)

Corresponding Components Used in the Mechanical Analogy Model

Soil Mass	Equivalent Mechanical Model
Soil	Spring
Saturated voids	Water
Porosity n	Volume of cylinder
Permeability	Petcock
Hydrostatic excess	
Pore-water pressure	Pressure gauge

FIGURE 9-1
Model used to explain by analogy the process of consolidation of a saturated clay.

The rate at which the water exits depends on the size of the petcock. Analogous to this system, for a soil with larger pores the water escapes faster, and therefore the solids (as the spring in the analogy model) will assume some of the externally applied load sooner than in the case of fine-grain soil where the water exits at a slower rate. Of course the cumulative deformation or consolidation will thus be realized later for a cohesive clay with relatively poor drainage characteristics. This behavior is typified by Fig. 9-2.

Figure 9-3 depicts the development of consolidation with time for a cohesive soil. The *initial* segment of consolidation is attributed to the early expulsion of air from the voids. It usually takes place during a relatively short time after the load has been applied. With time the water escapes from the pores, gradually dissipating the excess hydrostatic pore-water pressure. The resulting deformation is known as *primary consolidation*. It is the largest of the three phases and the one singled out in importance for most inorganic clays.

Evidence exists which indicates that some additional consolidation in clays can occur even after the excess pore-water pressure appears to have dissipated. This is generally referred to as *secondary compression* (or *secondary consolidation*). As mentioned previously, its cause is somewhat speculative—perhaps colloid-chemical processes, some small residual excess pore pressure, plastic lag, or some other cause. In most cases involving inorganic soils, the secondary consolidation is rather small and perhaps negligible; in some highly organic soils, such as peat or some colloidal clays, however, the secondary consolidation may be significant indeed.

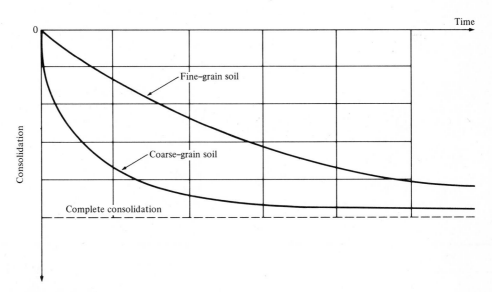

FIGURE 9-2
Time-consolidation relationship under constant load.

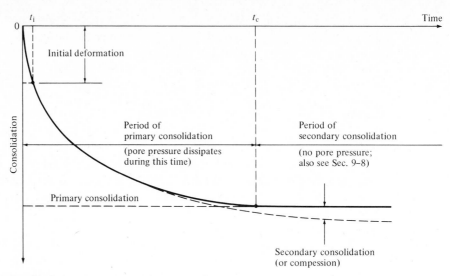

FIGURE 9-3
Components of total settlement for a cohesive soil under constant load.

9-4 LOAD–DEFORMATION CHARACTERISTICS OF SOILS

Unlike some of the engineering materials, such as steel, which display a reasonably well defined stress–strain relationship up to a point (proportional limit), soils do not. The typical soil does not follow Hooke's law. Furthermore, any degree of proportionality between stress and strain that one may construe from test results of one soil will not be applicable to another, even if the index properties of the two appear to be alike. In other words, the *modulus of elasticity* for soil is neither a constant for a given soil nor a predictable variable for different soil masses.

The load–deformation relationship for clays is generally represented in terms of the *pressure–void ratio*, as depicted by Fig. 9-4 which represents the results of a typical laboratory one-dimensional consolidation test. The pressure is plotted to natural scale in Fig. 9-4a and to logarithmic scale in Fig. 9-4b. The logarithmic plotting of the pressure is the more desirable of the two methods.

The first cycle of loading produces a somewhat straight and flat curve (on semilog scale) up to a certain pressure. This is followed by a somewhat smooth but rather pronounced transition into a steep and relatively straight line *CD*. If the pressure is removed, the soil will tend to *rebound* or *swell* back, but not to the original shape; a permanent deformation will have resulted. Then if the load is reapplied, the cycle will form a *hysteresis* loop, with the initial segment of the new curve being somewhat parallel to the first. This curve will gradually blend into the straight line *CDE* shown in Fig. 9-4b. The point of transition from a somewhat flat to a somewhat steep curve (e.g., point near *B*) corresponds to a state of pressure known as *preconsolidation pressure* P_c.

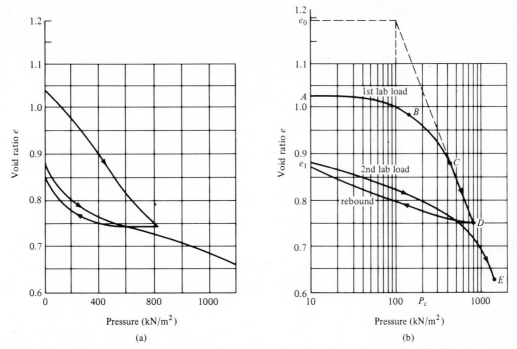

FIGURE 9-4
Pressure–void ratio curves of an undistributed precompressed clay soil. (a) Arithmetic scale. (b) Logarithmic scale.

One would expect the original or in situ void ratio of the soil to be different from the void ratio at the start of the laboratory test. That is, the extraction process obviously relieves the soil sample from in situ pressures, thereby permitting some expansion (i.e., increase mostly in the void ratio) analogous to the effect depicted by the "rebound" curve shown in Fig. 9-4b. Hence the very first laboratory load of Fig. 9-4b, in fact, represents the equivalent of a *recompression* curve of an in situ situation.

In Fig. 9-4b pressure P_c represents the maximum probable pressure the soil experienced in situ. It might be due to either of the two causes:

1. The effects of the overburden conditions existent at the time the sample was obtained
2. The superposition of existing as well as additional imposed stresses

When the effective pressure induced by the existent overburden, commonly denoted by P_0, equals P_c, the clay is said to be *normally consolidated*. When P_c is larger than P_0, the clay is regarded as *overconsolidated*. The ratio of P_c to P_0 is known as the *overconsolidated ratio* (OCR), that is, OCR $= P_c/P_0$.

Overconsolidation of the clay may have been caused by any or a combination of loads:

1. Previous "building" loads which have since been removed
2. Thicker soil overburden which has since been removed or eroded
3. Glacial ice sheets which have since disappeared
4. Fluctuation of the water table
5. Shrinkage stresses

From a practical point of view this implies that if any additional pressures, say from a new building, were to be imposed on the compressible stratum, the settlement would be rather minimal if the resultant effective pressures on the clay layer did not exceed the preconsolidation pressure, but perhaps it would be quite significant if they did (10). Hence it becomes important that one detect the point of preconsolidation in order to properly assess the effects of the induced load on the stratum in question.

Several methods have been suggested for determining the preconsolidation load. The most widely used procedure was proposed by Casagrande (9); it will be explained here. Assume that point B in Fig. 9-5 represents the point of maximum curvature (smallest radius) on an isolated segment of the curve shown in Fig. 9-4b. At point B draw a tangent to the curve and a horizontal line. Bisect the angle α formed by the tangent and the horizontal line. The preconsolidation load P_c is given by the point where a line which bisects this angle meets the tangent to the virgin curve (e.g., CDE in Fig. 9-4b).

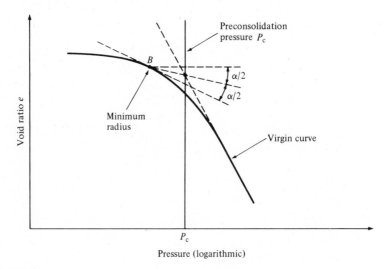

FIGURE 9-5
Casagrande procedure for determining preconsolidation pressure.

The steep straight line of the portion of the pressure–void ratio curve (*CDE* in Fig. 9-4b) assumes the form

$$y = mx + c \tag{a}$$

or

$$e = e_0 - C_c \log_{10} \frac{P}{P_0} \tag{9-2}$$

where C_c is termed the *compression index*. It is the slope of the straight-line portion of the curve on a semilogarithmic plot, with a minus sign indicating a negative slope. In Eq. (9-2) e_0 represents the void ratio at a corresponding effective pressure, say, P_0. The compression index can be roughly approximated via an expression proposed by Terzaghi and Peck (52) related to the liquid limit of the clay:

$$C_c \approx 0.009(\text{L.L.} - 10) \tag{9-3}$$

EXAMPLE 9-1

Given Figure 9-4b is assumed to represent the relationship of the pressure–void ratio as determined from a one-dimensional consolidation test; $P_0 = 100 \ \text{kN/m}^2$.

Find The void ratio at any given pressure for these experimental data parameters.

Procedure At $P_0 = 100 \ \text{kN/m}^2$, $e_c = 1.19$ (by projecting backwards) and at $P = 700 \ \text{kN/m}^2$, $e = 0.78$. Thus

$$C_c = \left(\frac{1.19 - 0.78}{\log(7/1)} \right) = \frac{0.410}{0.8451} = 0.4852$$

Answer

$$e = 1.19 - 0.4852 \log_{10}\left(\frac{P}{100} \right)$$

9-5 ONE-DIMENSIONAL CONSOLIDATION THEORY

A general consolidation theory of soils would take into consideration stress and strain conditions in three dimensions. However, the problem associated with such an undertaking is extremely complex (6, 14, 46). For example, soils are not isotropic, homogeneous, elastic, or "ideal" in any way. Hence to account for the wide variations in properties of the typical soil is perhaps mathematically impossible. Indeed, because of this difficulty many attempts at three-dimensional consolidation analysis focus on numerical or finite-element procedures (12, 17). On the other hand, the *one-dimensional consolidation* theory (1) simplifies the effort greatly, and (2) satisfies most common needs related to settlement.

In the development of the one-dimensional consolidation theory the following assumptions are made:

1. The pressure–void ratio relationship is well defined, that is, $a_v = \partial e / \partial u$, where a_v = *coefficient of compressibility.*
2. The seepage flow is restricted to one direction.
3. Darcy's law applies.
4. The soil mass is completely saturated.
5. The deformation is one dimensional (i.e., in the direction of flow).
6. The water and solid soil particles are incompressible.
7. The soil mass is homogeneous.
8. The coefficient of permeability is constant.

Figure 9-6 shows an element taken from a compressible soil mass. The two piezometers record a difference in head dh as shown; the direction of flow is upward. The hydraulic gradient i is

$$i = \frac{\partial h}{\partial z} = \frac{1}{\gamma_w} \cdot \frac{\partial u}{\partial z} \tag{a}$$

The *change* in gradient over the distance dz is

$$\frac{\partial i}{\partial z} = \frac{1}{\gamma_w} \cdot \frac{\partial^2 u}{\partial z^2} \tag{b}$$

Applying Darcy's law (assumption 3), the quantity of water entering the element q_{in}, for a constant coefficient of permeability k, in the z direction is

$$q_{in} = v\, dA = ki\, dA = \frac{k}{\gamma_w} \frac{\partial u}{\partial z}\, dx\, dy \tag{c}$$

The quantity of water exiting the element q_{out} is

$$q_{out} = \frac{1}{\gamma_w}\left(k + \frac{\partial k}{\partial z}\, dz\right)\left(\frac{\partial u}{\partial z} + \frac{\partial^2 u}{\partial z^2}\, dz\right) dx\, dy \tag{d}$$

But $\partial k / \partial z = 0$ since k is constant. Thus Eq. (d) becomes

$$q_{out} = \frac{k}{\gamma_w}\left(\frac{\partial u}{\partial z} + \frac{\partial^2 u}{\partial z^2}\, dz\right) dx\, dy \tag{d}$$

The change in quantity $\Delta q = q_{out} - q_{in}$ is

$$\Delta q = \frac{k}{\gamma_w} \frac{\partial^2 u}{\partial z^2}\, dx\, dy\, dz \tag{e}$$

The product dx dy dz represents the volume of the element. Also, by definition q = volume/time; hence Δq represents the *time rate of change of volume.*

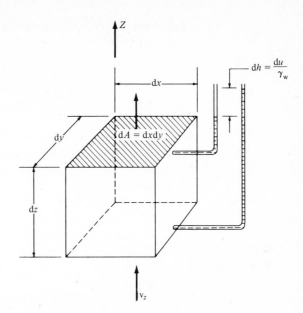

FIGURE 9-6
Vertical inflow and outflow at
a point in a soil mass.

Now to revert to some basic definitions: The volume of the voids $V_v = [e/(1 + e)]$ dx dy dz, and the volume of the solids $V_s = [1/(1 + e)]$ dx dy dz. (The expressions $V_v = eV/(1 + e)$ and $V_s = V/(1 + e)$ were derived in Example 4-1.)

Since there is no change in the volume of water or of the solid particles (assumption 6), any change in volume must be in V_v. Generally such a change, with respect to time, may be given by

$$\frac{\partial V_v}{\partial t} = \frac{\partial}{\partial t}\left(\frac{e}{1 + e}\right) dx\ dy\ dz \qquad (f)$$

But since $V_s = [1/(1 + e)]$ dx dy dz = constant, Eq. (f) becomes

$$\frac{\partial V_v}{\partial t} = \left(\frac{1}{1 + e}\right)\frac{\partial e}{\partial t} \qquad (g)$$

From the assumption of total saturation (assumption 4) a change in flow Δq [Eq. (e)] would be resulting in a change in volume of the element [Eq. (g)]. Hence equating Eq. (e) to Eq. (g), we have

$$\Delta q = \frac{\partial V_v}{\partial t}$$

or

$$\frac{k}{\gamma_w}\frac{\partial^2 u}{\partial z^2} = \left(\frac{1}{1 + e}\right)\frac{\partial e}{\partial t} \qquad (9\text{-}4)$$

From assumption 1, $\partial e = a_v \cdot \partial u$. Hence Eq. (9-4) becomes

$$\frac{k}{\gamma_w}\left(\frac{1+e}{a_v}\right)\frac{\partial^2 u}{\partial z^2} = \frac{\partial u}{\partial t} \tag{9-5}$$

or

$$C_v \frac{\partial^2 u}{\partial z^2} = \frac{\partial u}{\partial t} \tag{9-6}$$

Equation (9-6) is Terzaghi's *consolidation equation*, where C_v, called the *coefficient of consolidation*, in square centimeters per second, is

$$C_v = \frac{k}{\gamma_w}\left(\frac{1+e}{a_v}\right) \tag{9-7}$$

and $\partial^2 u / \partial t^2$ = change in hydraulic pressure gradient, in N/cm^4

$\partial u / \partial t$ = rate of change of hydrostatic excess pore-water pressure, in N/cm$^2 \cdot$ s

The solution to Eq. (9-6) for a constant initial pore pressure u_0 satisfying the following conditions:

$$\begin{array}{lll} \text{at} & z = 0, & u = 0 \\ \text{at} & z = 2H, & u = 0 \\ \text{at} & t = 0, & u = u_0 \end{array}$$

is

$$u = \sum_{m=1}^{\infty} \frac{2u_0}{M}\left(\sin\frac{Mz}{H}\right)e^{-M^2 T} \tag{9-8}$$

where $M = (\pi/2)(2m+1)$, m = any integer, 1, 2, 3, ...

H = one-half thickness of consolidating stratum and

$$T = \frac{C_v t}{H^2} = \text{time factor} \tag{9-9}$$

The *percentage consolidation* U at a given depth and time t is

$$U = \left(\frac{u_i - u}{u_i}\right) \times 100 = \left(1 - \frac{u}{u_i}\right) \times 100 \tag{9-10}$$

where u = hydrostatic excess pressure at time t and depth z

u_i = initial hydrostatic excess pressure at time $t = 0$ and depth z

The basic U–T relationship can be developed by combining Eqs. (9-8) and (9-10). Thus we have

$$U = 1 - \sum_{m=0}^{\infty} \frac{2}{M}\left(\sin\frac{Mz}{H}\right)e^{-M^2 T} \tag{9-11}$$

---

* For a derivation of this expression, see Taylor (51).

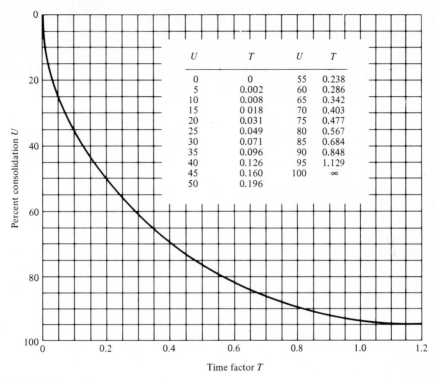

U	T	U	T
0	0	55	0.238
5	0.002	60	0.286
10	0.008	65	0.342
15	0 018	70	0.403
20	0.031	75	0.477
25	0.049	80	0.567
30	0.071	85	0.684
35	0.096	90	0.848
40	0.126	95	1.129
45	0.160	100	∞
50	0.196		

FIGURE 9-7
Average percent consolidation versus time factor.

One may obtain a numerical relationship between U and T by substituting successive values of m from 0 to ∞, as indicated in Fig. 9-7.

The relationship shown in Fig. 9-7 is for a case of constant initial consolidating pressure. Since T, in fact, depends on the type of pressure distribution (e.g., rectangular, trapezoidal, sinusoidal), the above numerical relationship is not quite correct. However, the influence of such variations is rather small, making the use of Fig. 9-7 reasonably accurate.

It has been found that the following empirical relationships can be used, with acceptable accuracy, in place of Eq. (9-11):

$$T = \frac{\pi}{2} U^2 \qquad \text{when } U < 60 \qquad (9\text{-}12a)$$

and

$$T = 0.933 \log_{10}(1 - U) - 0.851 \qquad \text{when } U > 60 \qquad (9\text{-}12b)$$

9-6 DETERMINATION OF COEFFICIENT OF CONSOLIDATION C_v

Only two methods for determining the coefficient of consolidation will be discussed here, one by Casagrande and another by Taylor. Although empirical in nature, they are rather simple and expedient, and give results that are comparable to some developed from theoretical expressions (41).

Figures 9-8 and 9-10 represent the same data from a laboratory test, plotted to two different scales. That is, both represent strain or percent consolidation of the same clay sample. These graphs will be used to determine the coefficient of consolidation by the two different methods as described below.

The Logarithm-of-Time Fitting Method

Figure 9-8 depicts the relationship between strain (percent consolidation) and log time for one stage of load.

The first part of the curve approximates a parabola. Hence the 0 percent consolidation coordinate may be obtained in the following manner: select a point of time t at the early stage of consolidation. Next determine the difference in the ordinate between that point and the one corresponding to $4t$. This is designated as δ in Fig. 9-8. Then lay off this δ value above the curve at point t to obtain the corrected 0 point.

Casagrande suggested that the point of 100 percent primary consolidation may be determined by the intersection of the tangent to the straight-line portion of the primary

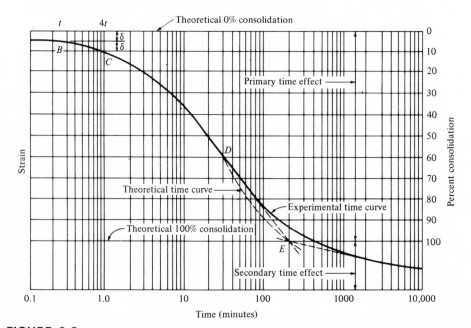

FIGURE 9-8
Time versus consolidation.

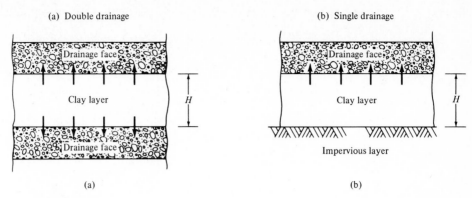

FIGURE 9-9
(a) Double drainage. (b) Single drainage.

consolidation curve and the tangent of the straight-line portion of the secondary consolidation portion of the curve. With the 0 and 100 percent consolidation points established, one may divide the ordinate into ten equal parts to translate the dial readings into percent consolidation.

With the 0 and 100 percent primary compression points located, the coefficient of consolidation may be computed from Eq. (9-8) and at the 50 percent point by noting that the time factor at the 50 percent consolidation corresponds to the value of 0.196. Hence corresponding with this time factor, Eq. (9-9) may be rearranged to express the coefficient of consolidation given by Eqs. (9-13a) and (9-13b) for the cases of drainage shown in Fig. 9-9.

For double drainage (Fig. 9-9a),

$$C_v = \frac{0.196H^2}{4t_{50}} \tag{9-13a}$$

and for single drainage (Fig. 9-9b),

$$C_v = \frac{0.196H^2}{t_{50}} \tag{9-13b}$$

The Square-Root-of-Time Fitting Method

Proposed by D. Taylor, the 0 percent and the 100 percent consolidation can be determined by first plotting the dial readings on the ordinate axis to a natural scale and then the corresponding values on the abscissa as the square root of time, as indicated in Fig. 9-10.

Taylor observed that the first portion of the curve, perhaps past 50 to 60 percent of consolidation, is essentially a straight line, while at 90 percent consolidation the abscissa is 1.15 times the abscissa of the straight line intersecting the 90 percent consolidation. Hence Taylor suggested the following method for determining the

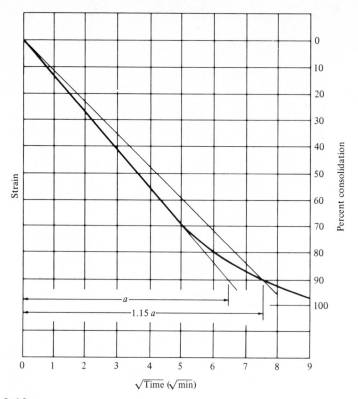

FIGURE 9-10
Time versus consolidation.

points of 0 percent and 100 percent consolidation, and for subsequently determining the coefficient of consolidation.

For the data plotted as shown in Fig. 9-10 draw a tangent to the straight-line portion of the curve. A second line, coinciding with the point of 0 time, is drawn such that the abscissa of this line is 1.15 times as great as that of the first line. The 90 percent consolidation is assumed to be the point of intersection of the time–consolidation curve with the second straight line drawn. Hence a scale for the percent consolidation could be drawn by dividing the distance between 0 and 90 percent into nine equal parts, as shown in Fig. 9-10.

The coefficient of consolidation for the curve may therefore be determined from Eq. (9-9). Noting that T_{90} from Fig. 9-7 has a value of 0.848, the coefficient of consolidation expressions are given by Eqs. (9-14a) and (9-14b).

For double drainage (Fig. 9-9a),

$$C_v = \frac{0.848H^2}{4t_{90}}$$

(9-14a)

For single drainage (Fig. 9-9b),

$$C_v = \frac{0.848H^2}{t_{90}} \qquad (9\text{-}14b)$$

Again note that the time–consolidation relationships indicated in Figs. 9-8 and 9-10 represent the behavior of a specific soil sample subjected to only one load range. Hence the coefficient of consolidation will be altered by different time–consolidation curves.

EXAMPLE 9-2

Given   Figures 9-8 and 9-10 represent time–consolidation relationships for the same clay sample, 3 cm thick, subjected to a given pressure range under double drainage.

Find   (a)   The coefficient of consolidation C_v for the sample by the *two* methods described,
       (b)   The time necessary for 60 percent consolidation of the same sample if it were 2 m thick. Assume an average C_v.

Procedure   (a)   From Eq. (9-14a),

$$C_v = \frac{0.848H^2}{4t_{90}} = \frac{0.848(3)^2}{4t_{90}}$$

From Fig. 9-10, at 90 percent consolidation, $\sqrt{t} = 7.6$ and $t = 57.76$ min. From Fig. 9-8, $t_{50} = 18$.
       Thus based on the time-fitting method,

$$C_{v90} = \frac{0.848(3)^2}{4 \times 57.76 \times 60}$$

Answer

$$\boxed{C_{v90} = 5.505 \times 10^{-4} \text{ cm}^2/\text{s}}$$

Based on the logarithm-of-time method,

$$C_{v50} = \frac{0.196(3)^2}{4 \times 18 \times 60}$$

Answer

$$\boxed{C_{v50} = 4.084 \times 10^{-4} \text{ cm}^2/\text{s}}$$

(b) Average

$$C_v = \frac{5.505 + 4.084}{2} \times 10^{-4} = 4.795 \times 10^{-4} \text{ cm}^2/\text{s}$$

From Fig. 9-7 (tabulated values), $T_{60} = 0.286$. Thus,

$$t_{60} = \frac{T_{60} H^2}{4C_v} = \frac{0.286(200)^2}{4 \times 4.795 \times 10^{-4}}$$

$$t_{60} = 5.97 \times 10^6 \text{ s} = 9.94 \times 10^4 \text{ min}$$

Answer

$$t_{60} = 69 \text{ days}$$

9-7 ONE-DIMENSIONAL CONSOLIDATION TEST

The compressive properties of a soil are usually determined in the laboratory by a consolidation test (ASTM D-2435-70). The basic apparatus, known as *consolidometer*, is shown schematically in Figs. 9-11 and 9-12.

Briefly, a consolidometer consists of a ring into which a clay sample, undisturbed or remolded, is carefully trimmed and fitted. On top and at the bottom of the sample are porous stones which permit vertical drainage of the water expelled from the soil sample. A load is applied to the specimen via a swivel head in order to provide uniform pressure to the soil sample. A dial gauge measures the deformation of the sample.

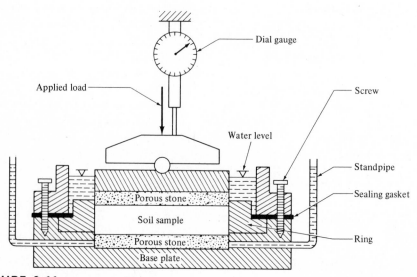

FIGURE 9-11
Cross section of typical fixed-ring consolidometer.

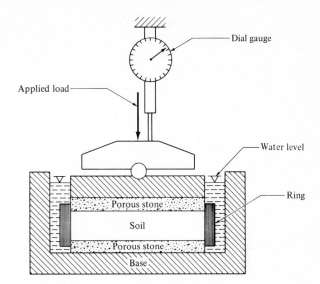

FIGURE 9-12
Cross section of floating-ring
consolidometer.

The soil sample is totally submerged in water during the test, simulating assumed total saturation of the in situ sample. It is the fully saturated clay stratum, in situ, that displays the lowest strength and greatest compressibility (3).

The typical soil sample may vary from about 6 to 10 cm in diameter and from 2 to 5 cm in thickness. The thicker samples are used infrequently, however, since they require a long period of time for complete consolidation. Also, the thickness of the soil sample is usually small relative to its diameter (say, one-half or one-third of the diameter) in order to minimize the effects of friction between the soil and the ring. In Fig. 9-11 the ring remains fixed relative to the base, while in Fig. 9-12 the ring floats relative to the base of the apparatus.

Although various load increments may be used (32–34, 44, 45, 55), it is customary to use a sequence of loads of $\frac{1}{4}, \frac{1}{2}$, 1, 2, 4, and 8 kg/cm^2; that is, each load increment is double the preceding load. Although it is desirable that every load increment be sustained for 24 h, it is often satisfactory to load the soil every 4 to 5 h for small loads and adopt the 24-h recommendations only for the increments for which more reliable results are required. Dial readings are recorded at the beginning of the test and usually at 0.25, 1, 4, 9, 16, 25, 36, 49, 64, 100, and 1440 min. Sometimes more frequent readings are recorded (see Table 9-1 in Example 9-3). After the largest load for the test has been applied, the load is usually decreased in two to three increments, permitting a plotting of pressure with the void ratio during the rebounding phase of the test (see pressure–void ratio plot from Example 9-3). One may consult almost any manual on soil testing for a more detailed description of the test.

Before and after the consolidation test begins, the water content and the weight of the sample are determined. Subsequently the specific gravity of the soils is also determined, permitting the determination of the void ratio for each increment of load.

The calculations and plotting of the results for a typical soil sample are illustrated by Example 9-3.

It is quite apparent that the sample tested is far from being totally undisturbed, i.e., the sample is extracted from the stratum (usually via a Shelby-tube sampler), is further extracted from the sampler (usually pushed out), is handled during the trimming process, is further deformed during the fitting into the consolidation ring, etc. It is with this in mind that the author has found the testing of samples cut from Shelby tubes, tested in the cut section of the tube, a most desirable approach. Here the Shelby tube is cut to a desired length with a clay sample inside it. Instead of extracting the sample, trimming it, and fitting it into a ring, the soil is tested within the Shelby tube, with the Shelby tube acting as a floating ring. This method (1) permits an accurate reading of the thickness; (2) eliminates the necessity for extracting, trimming, and general handling; (3) permits consolidation testing of coarser textures (mixes of sand, clays, and silts) which frequently fall apart during the extraction and trimming processes; (4) reduces the sample disturbance; and (5) appears more accurate as well as more expedient than those described above. In essence, the apparatus is identical to that shown in Fig. 9-12 for the floating-ring consolidometer. The only difference is that the ring is a Shelby-tube section which is dispensed with after the completion of the test.

EXAMPLE 9-3

Given   Data from the laboratory consolidation test of a clay sample, Table 9-1.

Find   (a)   Void ratio–pressure relationship.
          (b)   Coefficient of consolidation for the 2- to 4-kg/cm^2 pressure range.

Procedure   Sample calculations for Table 9-1 are carried out as follows:

$$V_i = \text{volume at any time } i = \pi/4(10.16)^2 H_i = 81.073 H_i$$

where H_i = sample thickness at time i

$$V_s = \frac{W_s}{\gamma_s} = \frac{365.6}{2.70} = 135.41 \text{ cm}^3$$

Initial thickness = $3 + (3.0 - 2.597) \times 10^{-4}$ in $\times$ 2.54 cm/in = 3.102 cm

Then

$$V_i = 81.073 \times 3.102 = 251.49 \text{ cm}^3$$

$$V_{vi} = 251.49 - 135.41 = 116.08 \text{ cm}^3$$

$$e = \frac{V_{vi}}{V_3} = \frac{116.08}{135.41} = 0.857$$

The results for (a) and (b) are given in Table 9-2.

TABLE 9-1
Data from Laboratory Consolidation Test of Clay Sample

$G = 2.7$; dry weight = 365.6 g; diameter = 10.16 cm; $H = 3$ cm (measured at dial reading = 2.597)

Time t (min)		Load Stage (kg/cm²) and Dial Readings ($R \times 10^4$ in)													
		0.5		0.5–1.0		1.0–2.0		2.0–4.0		4.0–8.0		8.0–2.0		2.0–0.1	
		R	R	R	R	R	R	R	R	R	R	R	R	R	R
0	0	3000	0	2930	0	2681	0	2400	0	1845	0	1175	0	1880	0
0.25	0.5	2994	6	2915	15	2664	17	2373	27	1805	40				
1	1	2990	10	2899	31	2647	34	2345	55	1763	82				
2.25	1.5	2987	13	2884	46	2631	50	2318	82	1720	125				
4	2	2983	17	2868	62	2614	67	2290	110	1680	165				
6.25	2.5	2980	20	2853	77	2597	84	2261	139	1639	206				
9	3	2977	23	2837	93	2580	101	2231	169	1599	246				
12.25	3.5	2973	27	2822	108	2564	117	2201	199	1558	287				
16	4	2970	30	2807	122	2547	134	2165	235	1518	327				
25	5	2967	33	2791	138	2530	151	2125	275	1477	368				
36	6	2963	37	2776	153	2512	169	2085	315	1435	410				
49	7	2960	40	2761	168	2495	186	2045	355	1394	451				
64	8	2956	44	2745	184	2477	204	2010	390	1353	492				
100	10	2951	49	2730	199	2459	222	1970	430	1310	535				
225	15	2947	53	2714	215	2441	240	1930	470	1267	578	1440	157	1892	705
400	20	2943	57	2698	231	2422	259	1890	510	1225	620				
1440	38	2930	70	2681	248	2400	281	1845	555	1175	670				

TABLE 9-2
Results of Example 9-3 Calculations

Load Stage (kg/cm²)	Void Ratio e at Initial Load	Thickness of Sample (cm)	H/z	Fitting Time (min) (see Figs. 9-8, 9-10)		C_v(cm²/s $\times 10^4$)	
				t_{90}	t_{50}	$\dfrac{0.848\,H^2}{4t_{90}}$	$\dfrac{0.196\,H^2}{4t_{50}}$
0.5	0.888	3.154	1.577				
0.5–1	0.878	3.137	1.568				
1–2	0.839	3.071	1.535				
2–4	0.796	3.000	1.5000	57.76	18	5.505	4.084
4–8	0.763	2.945	1.473				
8–	0.610	2.689	1.344				
8–	0.656	2.751	1.376				
2–0.1	0.716	2.868	1.434				

9-8 SECONDARY COMPRESSION

As was pointed out in Section 9-3, additional deformation can occur in clays for a considerable period of time after the primary consolidation has ceased. This occurs at a much slower rate, at a very small seepage velocity, and presumably under zero excess pore pressure. Actually some excess pore pressure (perhaps too small to be measured) must exist to cause the outward seepage.

The amount of secondary compression can be estimated by maintaining a constant load on a clay long enough past the point of primary consolidation to establish a relationship between secondary deformation and time. An example of this is given by the lower portion of the curve in Fig. 9-13. This is also depicted in Fig. 9-8. The slope of this portion of the curve is called the *coefficient of secondary compression* C_α given by Eq. (9-15):

$$C_\alpha = \frac{\varepsilon_1 - \varepsilon_2}{\log(t_2/t_1)} \tag{9-15}$$

where ε_1 and ε_2 are *strains* at times t_1 and t_2, respectively. Some common ranges of C_α are given in Table 9-3. The amount of settlement associated with secondary compression is given in Section 9-10.

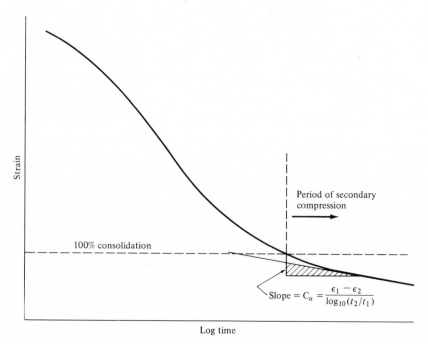

FIGURE 9-13
Determination of the coefficient of secondary consolidation.

TABLE 9-3
Some Common Values of C_{α}

Overconsolidated clays	0.0005–0.0015
Normally consolidated clays	0.005 –0.03
Organic soils, peats	0.04 –0.1

9-9 THREE-DIMENSIONAL CONSOLIDATION

In geotechnical engineering it is common to predict the magnitude and progress of consolidation and subsequent settlement based upon Terzaghi's one-dimensional theory of consolidation, discussed in the preceding section. The assumptions upon which the theory is based have been listed in Section 9-5. The validity of some of these is questionable, and therefore may deserve a qualitative scrutiny.

Briefly, in the one-dimensional theory the strains (vertical deformation) are assumed to be directly proportional to the change in the vertical stress, which is assumed to be, at any one time, equal to the excess pore pressure since dissipated. Furthermore the total vertical stress is constant and uniformly distributed and is equal to the applied load. Also, the lateral strains are assumed to be zero, and the flow of water is assumed going in only one direction. Actually the geometric and stratigraphic conditions never permit such boundary conditions.

A more realistic assessment is that the total stress is not time independent, the effective stress and normal strains are not really linearly tied, and the excess pore pressure also varies with time. Pore pressures dissipate at a rate faster than that predicted by the one-dimensional theory. For zero lateral strain either total lateral physical restraint must be provided, or the load must be distributed uniformly over an infinite area. Also, the dissipation of pore pressure is not only one dimensional, but is likely to be, in many cases, as large or larger in the horizontal direction than in the vertical direction.

The above constitutes apparent simplifications on which the one-dimensional theory rests. Hence three-dimensional consolidation theories have been proposed to account for a more general set of boundary conditions, nonlinear σ–ε relations, layered systems, and large strains (6, 14, 27, 42, 44–46). They constitute much more complex mathematical formulations and are beyond the scope of this text.

9-10 SETTLEMENT PREDICTIONS BASED ON ONE-DIMENSIONAL CONSOLIDATION

Although based on perhaps oversimplified assumptions, as briefly described in the previous section, Terzaghi's one-dimensional consolidation theory is relatively simple mathematically and has proven of value in many practical applications. Hence it remains a commonly used basis for estimating settlement. Here we shall briefly focus on settlement predictions for (1) normally consolidated clays and (2) over-consolidated clays.

Normally Consolidated Clay

Figure 9-14 represents a schematic of a homogeneous totally saturated clay mass. For the sake of illustration, the total volume is divided into totally saturated voids and solids. As mentioned in the previous discussion, the relative amount of deformation of the solids is negligible. Hence it is the change in the void volume that is assumed to be the cause of the settlement. Thus the total settlement S equals the total *change* of stratum thickness ΔH:

$$S = \Delta H = H_i - H_f = h_i - h_2 = (h_1 - h_2)\frac{H_i}{H_i} = H_i\left(\frac{h_1 - h_2}{h_s + h_1}\right)$$

or

$$S = \Delta H = H\frac{(h_1 - h_2)/h_s}{(h_s + h_1)/h_s} = H_i\left(\frac{e_i - e_f}{1 + e_i}\right) \tag{9-16}$$

where the quantities H_i, H_f, h_1, h_2, h_s are shown in Fig. 9-14. In terms of pressure, assuming a straight-line relationship (semilog scale) between e and P, we have from Eq. (9-2),

$$e_0 - e \cong e_i - e_f = C_c \log_{10}\frac{P}{P_0} = C_c \log_{10}\frac{P_0 + \Delta P}{P_0}$$

Substituting in Eq. (9-16),

$$S = \Delta H = \left(\frac{H}{1 + e_i}\right)C_c \log_{10}\left(\frac{P_0 + \Delta P}{P_0}\right) \tag{9-17}$$

Either Eq. (9-16) or Eq. (9-17) is a suitable expression for settlement for a normally consolidated clay, although Eq. (9-17) is perhaps the more convenient of the two since e_f is not needed. P_0 and ΔP can be calculated via methods detailed in Chapter 8.

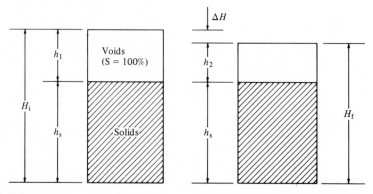

FIGURE 9-14
Soil phase diagram.

The expression for secondary settlement ΔH_s can be estimated from Eq. (9-18):

$$\Delta H_s = HC_\alpha \frac{t_1 + \Delta t}{t_1} \tag{9-18}$$

where C_α is given by Eq. (9-15).

For a layered system (more than one consolidation layer) the problem becomes significantly more complicated. This becomes apparent when one uses the relative permeabilities of the different strata and the different coefficients of compressibility. The excess pore-pressure dissipation of the various strata is difficult to evaluate in a meaningful manner, since one stratum serves as a drainage medium for the other, etc.

Overconsolidated Clay

When the soil is overconsolidated (i.e., $P_c > P_0$), the compression index C_c is different from that used in Eq. (9-17). An approach for determining C_c for such a case was developed by Schmertmann (47). Referring to Fig. 9-15, the procedure is as follows:

1. Plot the laboratory one-dimensional consolidation curve as shown.
2. Compute P_0 and e_0. Point a is the coordinate point for these values.
3. Determine P_c. One method was described in Section 9-4, Fig. 9-5.
4. Determine the slope C_r of the rebound-curve segment de. Use that slope to draw line ab. Thus point b is determined.
5. From an ordinate value of $0.4e_0$ draw a horizontal line to intersect the extension of the laboratory curve. This determines point c.
6. Connect points b and c with a straight line. The slope of this line is C_c for overconsolidated clays, as suggested by Schmertmann.

EXAMPLE 9-4

Given   The data shown in Fig. 9-16.

Find   Estimated average settlement from primary consolidation of clay layer under center of footing.

Procedure   The midplane of the clay thickness will be assumed to represent the average behavior of the clay stratum. Thus P_0 is the sum of the following:

$$
\begin{array}{rcl}
3 \text{ m} \times 19.62 \text{ kN/m}^3 & = & 58.86 \text{ kN/m}^2 \\
7 \text{ m} \times 9.80 \text{ kN/m}^3 & = & 68.60 \text{ kN/m}^2 \\
\underline{2 \text{ m} \times 9.60 \text{ kN/m}^3} & = & \underline{19.20 \text{ kN/m}^2} \\
& & P_0 = 146.66 \text{ kN/m}^2
\end{array}
$$

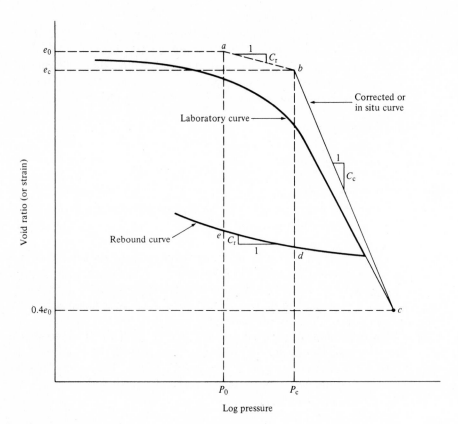

FIGURE 9-15

Determination of C_c for overconsolidated clays. [After Schmertmann (47).]

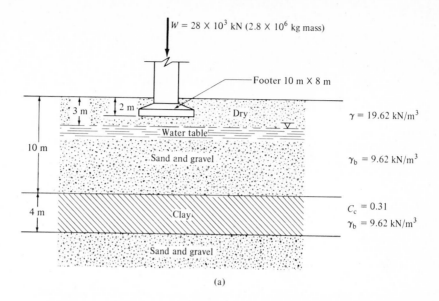

(a)

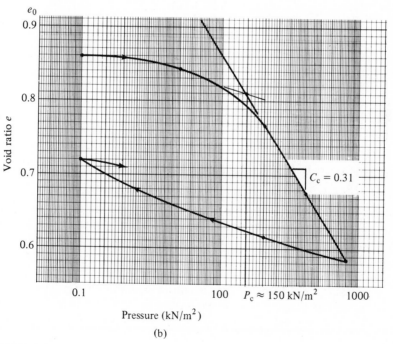

FIGURE 9-16

From Fig. 9-16b, $P_c \approx 150$ kN/m². Thus clay is normally consolidated. Also, $e_0 \approx 0.81$. Now determine ΔP. From Table 8-4, for $z = 14$,

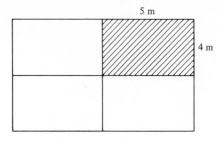

5 m

4 m

$$m = \tfrac{4}{14} = 0.286, \quad n = \tfrac{5}{14} = 0.351,$$

$$\text{and} \quad f_z(m, n) \approx 0.04$$

Hence

$$\Delta P = 4 \times 0.04 \times \frac{28{,}000}{8 \times 10} = 56 \text{ kN/m}^2$$

From

$$S = \Delta H = \frac{H}{1 + e_0} C_c \log_{10} \frac{P_0 + \Delta P}{P_0}$$

we have

$$\Delta H = \frac{4 \text{ m}}{1 + 0.81} (0.31) \log_{10} \frac{146.66 + 56}{146.66} = 0.096 \text{ m}$$

Answer

$$\Delta H = 9.6 \text{ cm}$$

9-11 SETTLEMENT DURING CONSTRUCTION

The change in the effective pressure ΔP is normally time dependent rather than instantaneous. During the construction of most buildings, the application of load, and therefore ΔP, may be realized over a relatively long period of time—perhaps many months or sometimes years. Figure 9-17a shows a typical load–time relationship during a construction period. The dashed line indicates the assumed load rate. Figure 9-17b shows the time–settlement relationship (calculated from theory) for instantaneous loading. This is shown by curve $0CBE$. The "corrected" settlement could be obtained by a method proposed by Terzaghi and further refined by Gilboy by assuming that: (1) the actual settlement at the end of the construction time would be about the same as that resulting from the total load acting for half of the loading time, and (2) the load–time relationship is linear as indicated in Fig. 9-17a. Thus based on these two assumptions, the settlement at the end of the loading period is equal to the settlement on the instantaneous-load curve corresponding to one-half of the total loading period. At any other time t the actual settlement is equal to t/t_p of the settlement AC. Hence a number of points could be determined in this manner, thereby

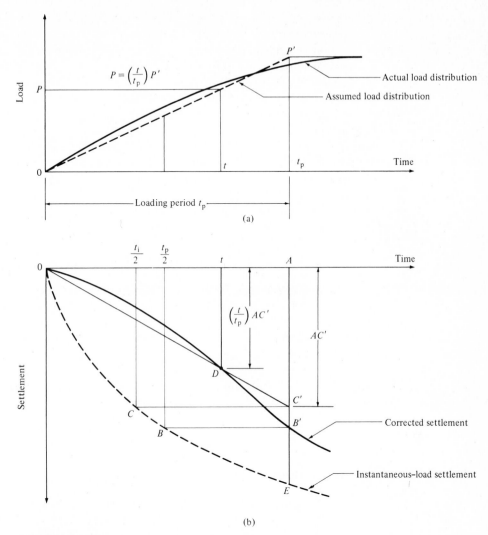

FIGURE 9-17
Procedure for determining settlement during construction period.

yielding the "corrected settlement" curve shown in Fig. 9-17b. The corrected settlement curve beyond the completion of construction is assumed to be the same as the instantaneous-load curve, offset to the right by one-half the construction period.

9-12 TOLERABLE SETTLEMENTS IN BUILDINGS

The problem of settlement has plagued engineers and builders for a very long time. Some of the largest rates and total magnitudes of settlement are found in Mexico City, where rates of settlement of 20 to 30 cm per year and total settlements in excess of 1 m over the years are not uncommon. The "leaning" Tower of Pisa in Italy is perhaps the most famous example of a structure which, according to recent measurement, has settled 2.8 m on one side and 0.8 m on the other. This corresponds to a differential settlement of 2 m and a tilt in excess of 5° (36). Undoubtedly many of the architectural masterpieces of long ago and somewhat more recent ones, such as the Campanile Tower in Italy, collapsed due primarily to high differential settlement. Indeed, the general problem of settlement and the establishing of tolerable settlement limits in buildings are still plaguing us.

Subsequent to his evaluation of various data and a review of building codes in a number of countries, Feld (18) in essence concluded that no answer may be expected as to how much settlement can be tolerated by a given structure without the availability of further pertinent data which would reflect various rigidities and different rates of settlement. Skempton, Peck, and McDonald (49) reviewed 98 case histories and concluded, in part, that slopes of $\delta/L = 1/300$ may cause cracking of panels in frame buildings and slopes of $1/150$ may cause structural damage to columns and beams (where δ represents the differential settlement over a horizontal distance L). Grant, Christian, and Vanmarcke (22) combined the data from 95 additional buildings with those of Skempton, Peck, and McDonald. They concluded that at deflection slopes δ/L exceeding $1/300$ buildings will probably experience some damage. These and numerous other studies* reflect caution in using any hard and fast figures without due regard to many variables (e.g., type of construction, functional and structural tolerances, soil conditions). In other words, a value of $\delta/L \approx 1/300$ should be viewed merely as a broad estimate; each site and building should be evaluated individually. See Section 14-10 for further discussion in this regard.

PROBLEMS

9-1.   Define: (a) normally consolidated clay, (b) overconsolidated clay, (c) overconsolidated ratio.

9-2.   Define and explain the nature of: (a) primary consolidation and (b) secondary compression.

9-3.   The graph of e–log p shown in Fig. P9-3 represents the lab loading results of a one-dimensional consolidation test. Determine the compression index: (a) if $P_0 = P_c$, (b) if $P_0 = 90$ kN/m^2. Use Cassagrande's method for determining P_c.

9-4.   Assume that the e–log p relationship shown in Fig. P9-3 is representative of the clay stratum shown in Fig. P9-4. Determine: (a) P_0 and P_c. (b) OCR.

9-5.   Assume that the e–log p relationship shown in Fig. P9-3 is representative for the clay shown in Fig. P9-4. Determine:
   (a) C_c via Schmertmann's method.
   (b)   An estimate of the settlement under the center of the footing.
   (c)   An estimate of the settlement under a corner of the footing.

9-6.   Assume that the e–log p relationship shown in Fig. P9-6 represents the laboratory results of a one-dimensional consolidation test. The clay stratum from which the test sample was extracted was 3.5 m thick, totally saturated. Assume $P_0 = P_c$. How much additional

* For a rather comprehensive list of related investigations, see (22).

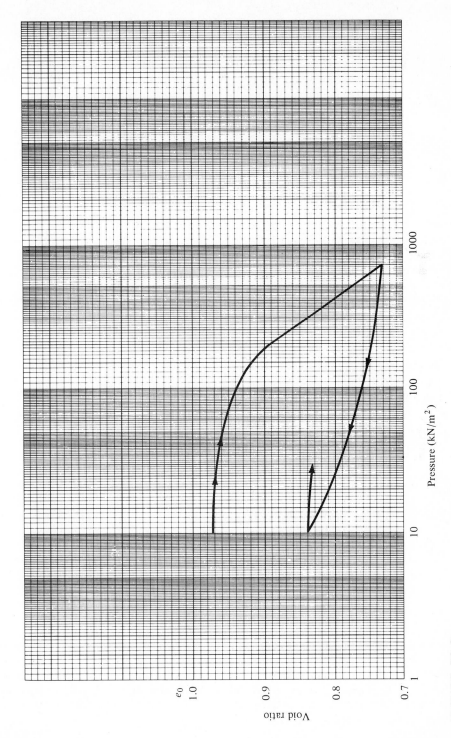

FIGURE P9-3
$e - \log p$ test.

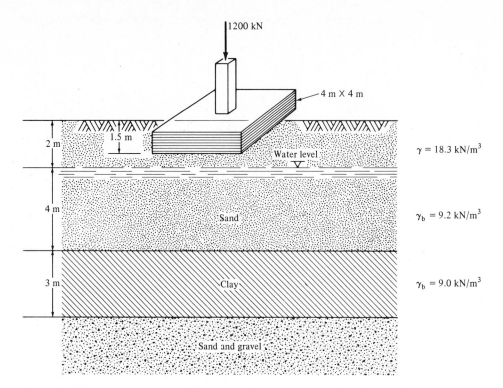

FIGURE P9-4

effective pressure can the clay withstand if the ultimate expected settlement is not to exceed 12 cm?

9-7. (a) Repeat Problem 9-6 assuming $P_0 = 0.8P_c$.

(b) For $P_0 = 0.8P_c$ determine how much additional effective pressure the clay can withstand if the expected settlement is not to exceed 24 cm. Is ΔP from part (b) twice that from (a)? Explain.

9-8. Refer to Fig. P9-8.

(a) Determine the points of 0 and 100 percent consolidation.

(b) Determine the percent consolidation scale from the results of part (a).

9-9. Determine the coefficient of consolidation in Fig. P9-8, assuming that the soil sample connected with this graph was 2.7 cm thick, tested with a top and a bottom porous stone.

9-10. Determine the time for 50 percent consolidation of the sample in Problem 9-9 for various thicknesses. (a) 1 m. (b) 3 m. (c) 8 m. State any assumptions.

9-11. (a) Repeat Problem 9-10 for 30, 40, 60, and 70 percent consolidation.

(b) Draw a graph of percent consolidation versus time for a thickness of 3 m. Does the shape resemble that of Fig. 9-7? Should it? Explain.

9-12. The column–footing combination of a parking garage supports an 8-MN (8×10^6 newtons) load and rests on layered strata as shown in Fig. P9-12. Assuming that the clay compressibility is represented by Fig. P9-3, determine the total settlement of the footing. State your assumptions.

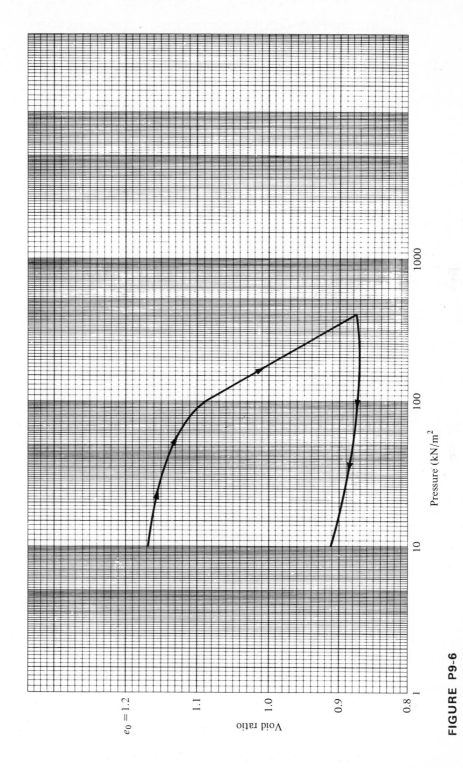

FIGURE P9-6

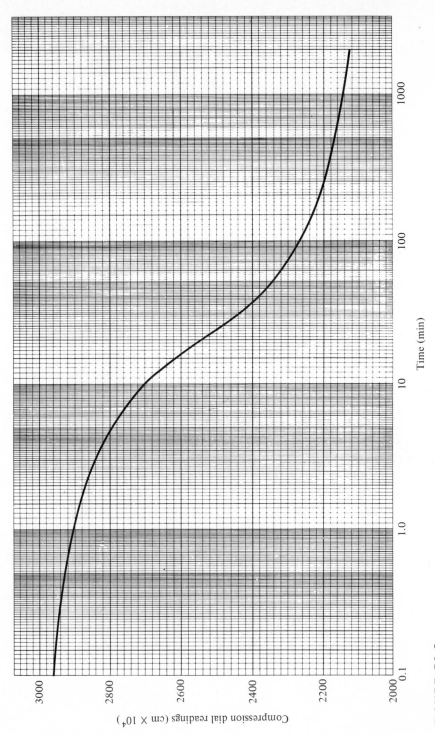

FIGURE P9-8

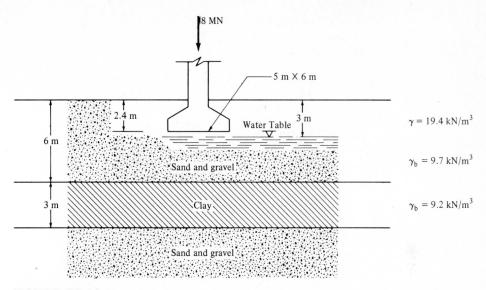

FIGURE P9-12

9-13. Determine the size of the footing for Problem 9-12 if the tolerable settlement is to be limited to 10 cm.

9-14. Determine the total settlement for the foundation condition shown in Fig. P9-14. Assume the submerged average unit weight of the sand and gravel as well as that of clay to be 9 kN/m^3.

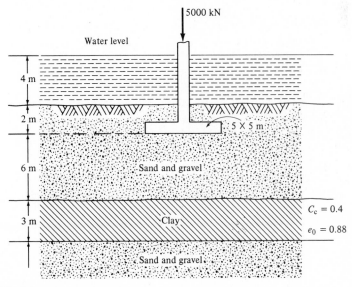

FIGURE P9-14

9-15.   A reasonably homogeneous layer of clay, 4 m thick, is expected to experience an ultimate settlement of 160 mm. After 3 years the average settlement was measured to be 55 mm. How much longer will it take for the average settlement to double (to 110 mm)?

9-16.   Column A shown in Fig. P9-16 is part of an existing building which had been erected long, long ago. A Mr. Doe comes along and decides to build a new building with a basement as shown. About a year later, measurements indicate that column A has settled. Assume normally consolidated conditions.

   (a)   Explain the reason for this settlement.

   (b)   Determine the theoretical maximum the column might have settled.

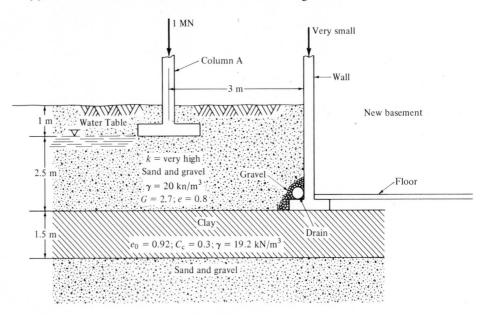

FIGURE P9-16

BIBLIOGRAPHY

1.   Baligh, M. M., and N. F. Fuleihan, "Consolidation Theory with Stress Reduction Due to Settlement," *ASCE J. Geotech. Eng. Div.*, vol. 104, May 1978.

2.   Barden, L., "Primary and Secondary Consolidation of Clay and Peat," *Geotechnique*, vol. 18, no. 1, Mar. 1968.

3.   Barden, L., A. O. Madedor, and G. R. Sides, "Volume Change Characteristics of Unsaturated Clay," *ASCE J. Geotech. Eng. Div.*, vol. 95, Jan. 1969.

4.   Barden, L., "Time Dependent Deformation of Normally Consolidated Clays and Peats," *ASCE J. Geotech. Eng. Div.*, vol. 95, Jan. 1969.

5.   Berry, P. L., and B. Vickers, "Consolidation of Fibrous Peat," *ASCE J. Geotech. Eng. Div.*, vol. 101, Aug. 1975.

6.   Biot, M. A., "General Theory of Three-Dimensional Consolidation," *J. Appl. Phys.*, 1941.

7. Bolt, G. H., "Physico-Chemical Analysis of the Compressibility of Pure Clays," *Geotechnique*, vol. 6, 1956.

8. Bond, D., "Influence of Foundation on Size of Settlement," *Geotechnique*, vol. 11, no. 2, June 1961.

9. Casagrande, A., "The Determination of the Preconsolidation Load and Its Practical Significance," *1st Int. Conf. Soil Mech. Found. Eng.*, Cambridge, Mass., 1936.

10. Casagrande, L., "Effect of Preconsolidation on Settlements," *ASCE J. Geotech. Eng. Div.*, vol. 90, Sept. 1964.

11. Cavounidis, S., and K. Hoeg, "Consolidation during Construction of Earth Dams," *ASCE J. Geotech. Eng. Div.*, vol. 103, Oct. 1977.

12. Christian, J. T., and J. W. Boehmer, "Plane Strain Consolidation by Finite Elements," *ASCE J. Geotech. Eng. Div.*, vol. 96, July 1970.

13. Crawford, C. B., "Resistance of Soil Structure to Consolidation," *Can. Geotech. J.*, vol. 2, 1965.

14. Cryer, C. W., "A Comparison of the Three-Dimensional Consolidation Theories of Biot and Terzaghi," *Quart. J. Mech. Appl. Math.*, vol. 16, 1963.

15. D'Appolonia, D. J., E. E. D'Appolonia, and R. F. Brissette, "Settlement of Spread Footings on Sand," *ASCE J. Geotech. Eng. Div.*, vol. 94, May 1968.

16. D'Appolonia, D., H. G. Poulos, and C. C. Ladd, "Initial Settlement of Structures on Clays," *ASCE J. Geotech. Eng. Div.*, vol. 97, Oct. 1971.

17. Domaschuk, L., and P. Valliappan, "Nonlinear Settlement Analysis by Finite Element," *ASCE J. Geotech. Eng. Div.*, vol. 10, July 1975.

18. Feld, J., "Tolerance of Structures to Settlement," *ASCE J. Soil Mech. Found. Eng.*, vol. 91, no. SM3, 1965.

19. Freeze, R. A., "Probabilistic One-Dimensional Consolidation," *ASCE J. Geotech. Eng. Div.*, vol. 103, July 1977.

20. Gibson, R. E., "The Progress of Consolidation in a Clay Layer Increasing in Thickness with Time," *Geotechnique*, vol. 8, 1958.

21. Gibson, R. E., G. L. England, and M. J. L. Husey, "The Theory of One-Dimensional Consolidation of Saturated Clays. I. Finite Non-Linear Consolidation of Thin Homogeneous Layers," *Geotechnique*, vol. 17, 1967.

22. Grant, R., J. T. Christian, and E. H. Vanmarcke, "Differential Settlement of Buildings," *ASCE J. Geotech. Eng. Div.*, vol. 100, Sept. 1974.

23. Hamilton, J. J., and C. B. Crawford, "Improved Determination of a Sensitive Clay," *Spec. Tech. Publ. 254*, Am. Soc. Test. Mater., Philadelphia, Pa., 1959.

24. Hassib, M. H., "Consolidation Characteristics of Granular Soils," Columbia University, New York, 1951.

25. Holtz, W. G., and J. W. Hilf, "Settlement of Soil Foundations Due to Saturation," *5th Int. Conf. Soil Mech. Found. Eng.*, Paris, France, vol. 1, 1961.

26. Hwang, C. T., N. R. Morgenstern, and D. W. Murray, "On Solutions of Plane Strain Consolidation Problems by Finite Element Methods," *Can. Geotech. J.*, vol. 8, no. 1, Feb. 1971.

27. Koppula, S. D., and N. R. Morgenstern, "Consolidation of Clay Layer in Two Dimensions," *ASCE J. Geotech. Eng. Div.*, vol. 98, Jan. 1972.

28. Lambe, T. W., "Methods of Estimation Settlement," *ASCE J. Geotech. Eng. Div.*, vol. 90, Sept. 1964.

29. Leonards, G. A., and P. Girault, "A Study of the One-Dimensional Consolidation Test," *5th Int. Conf. Soil Mech. Found. Eng.*, Paris, France, vol. 1, 1961.

30. Leonards, G. A., and A. G. Altschaeffl, "Compressibility of Clay," *ASCE J. Geotech. Eng. Div.*, vol. 90, Sept. 1964.

31. Lo, K. Y., "Secondary Compression of Clays," *ASCE J. Soil Mech. Found. Eng. Div.*, vol. 87, 1961.

32. Lowe, J., P. F. Zaccheo, and H. S. Feldman, "Consolidation Testing with Back Pressures," *ASCE J. Geotech. Eng. Div.*, vol. 90, Sept. 1964.

33. Lowe, J., *et al.*, "Controlled Gradient Consolidation Test," *ASCE J. Soil Mech. Found. Eng. Div.*, vol. 95, no. SM1, 1969.

34. Lowe, J. "New Concepts in Consolidation and Settlement Analysis," *ASCE J. Geotech. Eng. Div.*, vol. 100, no. GT6, 1974.

35. McKinley, D., "Field Observation of Structures Damaged by Settlement," *ASCE J. Geotech. Eng. Div.*, vol. 90, Sept. 1964.

36. Mitchell, J. K., V. Vivatrat, and T. W. Lambe, "Foundation of Tower of Pisa," *ASCE J. Geotech. Eng. Div.*, vol. 103, 1977.

37. Moore, P. J., and G. K. Spencer, "Settlement of Building of Deep Compressible Soil," *ASCE J. Geotech. Eng. Div.*, vol. 95, May 1969.

38. Olsen, R. E., and C. C. Ladd, "One-Dimensional Consolidation Problems," *ASCE J. Geotech. Eng. Div.*, vol. 105, Jan. 1979.

39. Osterman, J., and G. Lindskog, "Settlement Studies of Clay," *Swed. Geotech. Inst. Publ.*, no. 7, 1964.

40. Pyke, R., H. B. Seed, and C. K. Chan, "Settlement of Sands under Multi-Directional Shaking," *ASCE J. Geotech. Eng. Div.*, vol. 101, Apr. 1975.

41. Rowe, P. W., "Measurement of the Coefficient of Consolidations of Lacustrine Clay," *Geotechnique*, vol. 9, no. 3, 1959.

42. Rowe, P. W., "The Calculation of the Consolidation Rates of Laminated, Varved or Layered Clays, with Particular Reference to Sand Drains," *Geotechnique*, vol. 14, 1964.

43. Salas, J. A. Jimenez, and J. M. Serratosa, "Compressibility of Clays," *3rd Int. Conf. Soil Mech. Found. Eng.*, vol. 1, 1953.

44. Schiffman, R. L., "Consolidation of Soil under Time-Dependent Loading and Variable Permeability," *Proc. Highw. Res. Board*, vol. 37, 1958.

45. Schiffman, R. L., and R. E. Gibson, "Consolidation of Non-Homogeneous Clay Layers," *Proc. ASCE*, vol. 90, no. SM5, 1964.

46. Schiffman, R. L., A. T. F. Chen, and J. C. Jordan, "An Analysis of Consolidation Theories," *ASCE J. Geotech. Eng. Div.*, vol. 95, Jan. 1969.

47. Schmertmann, J. H., "The Undisturbed Consolidation Behavior of Clay," *Trans. ASCE*, vol. 120, 1955.

48. Seed, H. B., "Settlement Analyses, A Review of Theory and Testing Procedures," *ASCE J. Geotech. Eng. Div.*, vol. 91, Mar. 1965.

49. Skempton, A. W., R. B. Peck, and D. H. McDonald, "Settlement Analyses of Six Structures in Chicago and London," *Proc. Inst. Civ. Eng.*, July 1955.

50. Smith, R. E., and H. E. Wahls, "Consolidation under Constant Rates of Strain," *ASCE J. Geotech. Eng. Div.*, vol. 95, Mar. 1969.

51. Taylor, D. W., *Fundamentals of Soil Mechanics*, Wiley, New York, 1948.

52. Terzaghi, K., and R. B. Peck, *Soil Mechanics in Engineering Practice*, 2nd ed., Wiley, New York, 1967.

53. Wahls, H. E., "Analysis of Primary and Secondary Consolidation," *ASCE J. Soil Mech. Found. Eng. Div.*, vol. 88, no. SM6, Dec. 1962.

54. Weber, W. G., "Performance of Embankments Constructed over Peat," *ASCE J. Soil Mech. Found. Eng. Div.*, vol. 95, 1969.

55. Wissa, A. E. A., J. T. Christian, E. H. Davis, and S. Heiberg, "Consolidation at Constant Rate of Strain," *ASCE J. Geotech. Eng. Div.*, vol. 97, Oct. 1971.

56. Yen, B. C., and B. Scanlon, "Sanitary Landfill Settlement Rates," *ASCE J. Geotech. Eng. Div.*, vol. 101, May 1975.

10
Shear Strength of Soil

10-1 INTRODUCTION

In many soil mechanics problems the shear strength of the soil emerges as one of its most important characteristics. Indeed, this problem group may include:

1. Stability of slopes (e.g., hillsides, cuts, embankments, earth dams)
2. Ultimate bearing capacity of a soil
3. Lateral pressure against retaining walls, sheeting, or bracing
4. Friction developed by piles

Only a small number of construction problems are not in some manner related to the shear strength of the soil.

The shear strength of the soil may be attributed to three basic components:

1. Frictional resistance to sliding between solid particles
2. Cohesion and adhesion between soil particles
3. Interlocking and bridging of solid particles to resist deformation

It is neither easy nor practical to clearly delineate the effects of these components upon the shear strength of the soil. This becomes more apparent when one, relates these components to the many variables which directly or indirectly influence them, not to mention the lack of homogeneity and uniformity of the characteristics that typify most soil masses. For example, these components may be influenced by changes in the moisture content, pore pressures, structural disturbance, fluctuation in the groundwater table, underground water movement, stress history, time, perhaps chemical action or environmental conditions.

There are a number of tests, both laboratory and field, which are used to obtain a measure of the shear strength of a given soil. These are coupled with various theoretical considerations. A combination of the theoretical with the experimental efforts provides the background and tools for the basic understanding of and dealing with many of the phenomena related to the shear strength of the soils.

10-2 BRIEF REVIEW OF SOME ELEMENTS FROM BASIC MECHANICS

With some appropriate modifications, many of the concepts used in some of the basic mechanics courses can be employed to explain various phenomena in soils. This has already been partially demonstrated in connection with groundwater, pressure distribution, etc., discussed in some previous chapters. In this chapter we shall focus on the soil strength and some related parameters. By way of introduction, however, it may be advisable to review some of the relevant basic ideas presumably presented to all students as prerequisites to this course.

Friction between Solid Bodies

A block of weight W is placed on a horizontal plane as shown in Fig. 10-1. The forces acting on the block in Fig. 10-1a are its gravity and the reaction of the plane. Although a friction force is "available," it does not come into play since there is no horizontal force applied to the block. Suppose now that a horizontal force P_1 is applied to the block, as shown in Fig. 10-1b. If P_1 is small, the block will not move; the resultant R_a makes an angle α, commonly referred to as the *angle of obliquity* of force P_1. For equilibrium, a resisting horizontal force F_1 must therefore exist to balance P_1. For this case, the angle α is less than the angle of friction denoted by φ. N is the component of the weight normal to the plane.

If the applied horizontal force is increased to P_2, as shown in Fig. 10-1c, the friction force also increases to a value of F_2. It reaches a maximum value when the angle of obliquity α is equal to the angle of friction φ. One notes that at this point a state of impending motion (slip) exists. For the range of values $0 < \alpha < \varphi$ no slip or sliding occurs. At the point of impending motion, the maximum or available frictional force F could be related to the normal resisting force by a coefficient of friction μ, as given by Eq. (10-1):

$$F = \mu N \tag{10-1}$$

The coefficient of friction μ is independent of the area of contact. It is, however, strongly dependent on the nature of the surface in contact—the type of material, the condition of the surface, etc. Furthermore, in most materials the coefficient of static friction is somewhat larger than the kinetic coefficient.

If the horizontal force is increased to a value of P_3 as shown in Fig. 10-1d, such that the angle α is greater than φ, the block will start sliding. The frictional force cannot exceed the value given by $F = \mu N$, and therefore the block will accelerate to the right.

A similar analogy is indicated in Fig. 10-2. The block resting on a plane inclined at an angle α_1 will remain at rest provided the angle α_1 is less than the friction angle φ. The frictional force F is only as large as necessary to balance the component of the weight parallel to the plane. On a steeper slope, as indicated in Fig. 10-2b, we have a case of impending motion when the angle of the plane α_2 equals the angle of friction φ.

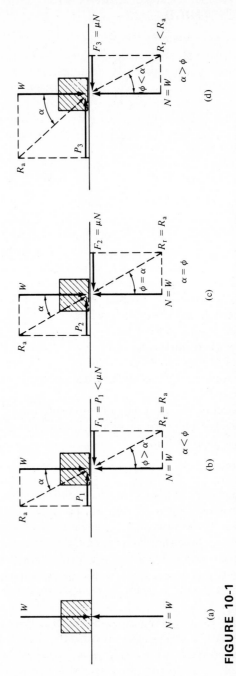

FIGURE 10-1
Development of friction on horizontal plane. (a) No friction. (b) Development of friction. (c) Friction, impending motion. (d) Friction, motion (acceleration).

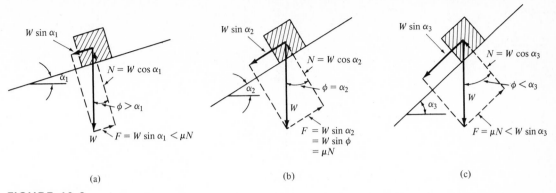

FIGURE 10-2
Development of friction on inclined plane. (a) Friction, no motion. (b) Friction, impending motion. (c) Friction, motion (acceleration).

It is equal to the coefficient of friction times the normal force, as indicated by Eq. (10-1). Finally, if the plane is inclined at an angle α_3 which is greater than angle φ, the block will start sliding. The maximum available frictional force is smaller than the component of the weight parallel to the plane.

From either Fig. 10-1c or Fig. 10-2b, both conditions of impending motions, one notes that

$$\tan \varphi = \frac{F}{N} = \mu \tag{10-2}$$

Although the frictional resistance in sands and other cohesionless granular materials resembles that discussed above, there are some important differences. In such soils the frictional resistance consists of both sliding and rolling friction, coupled with a certain degree of interlocking of the solid particles. In addition, various other factors, such as degree of saturation, particle size and shape, consistency, or intergranular pressure, have a pronounced effect on the shear strength of the soil.

Stress at a Point

One may recall from mechanics that, in general, both *normal* and *shear* stresses act on planes passing through a point in an element of mass which has been subjected to external loading. Figure 10-3 depicts these stresses on a small element containing point A which is located within the mass shown in Fig. 10-3a. The stresses normal to a plane are designated by σ and those parallel to any given plane by τ. Thus nine components of stress define the state of stress at a point. If such stresses on all possible planes through a point are investigated, three planes at right angles to each other will be found to be subjected to normal stresses only, as shown in Fig. 10-3c. These stresses, in our case σ_1, σ_2, and σ_3, are known as *principal stresses*, and the planes on

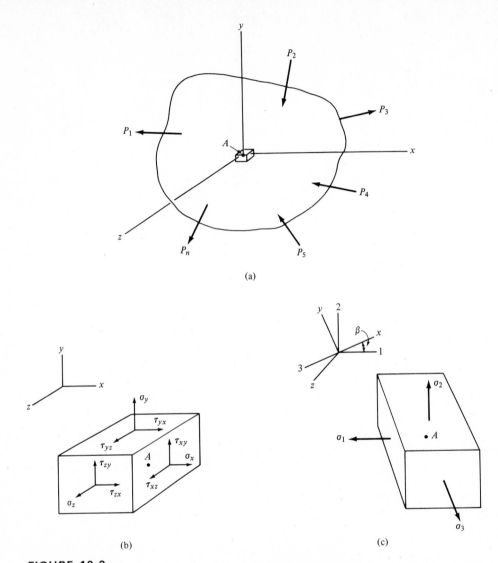

FIGURE 10-3

Stress at a point; case of triaxial stresses. (a) Point A in element of mass subjected to load. (b) General state of stress at point A. (c) Principal stresses at point A.

which they act are known as *principal planes*. Sometimes these planes are further defined as the major, the intermediate, and the minor principal planes, tied to the magnitudes of the respective stresses acting on them. Of course such planes are at a different orientation from those shown in Fig. 10-3b since shear stresses do not exist on principal planes.

It is convenient, and in the case of soil studies usually sufficient, to illustrate some of the above-mentioned relationships by considering the case of biaxial stresses, as shown in Fig. 10-4. The element in Fig. 10-4b is an enlarged version of the element indicated in Fig. 10-4a. As mentioned previously, the general case of stress consists of both normal and shear stresses. Now let us assume the element to be cut by a plane passing through m–n, and place the left segment in equilibrium, as shown in Fig. 10-4c. If the inclined plane is oriented θ degrees relative to the y axis, and if ΔA_n represents the area on the inclined plane, the areas of the vertical and horizontal faces of the elements are $\Delta A_n \cos \theta$ and $\Delta A_n \sin \theta$, respectively. The components of

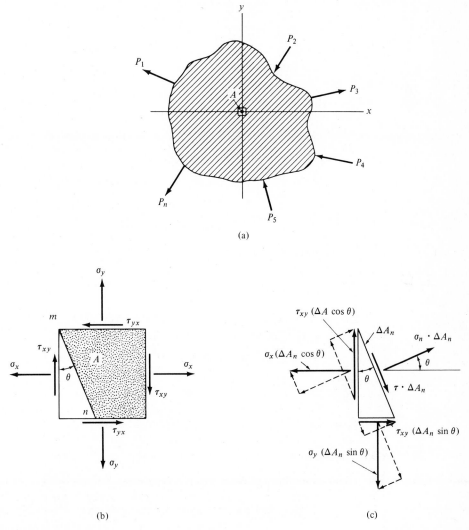

(a)

(b)   (c)

FIGURE 10-4
Stresses at a point; case of biaxial stresses.

these forces, parallel and perpendicular to the m–n plane, are shown by the dashed lines. Summing the force in the direction normal to the plane, we have

$$\sigma_n \Delta A_n = \sigma_x (\Delta A_n \cos \theta) \cos \theta - \tau_{xy}(\Delta A_n \cos \theta) \sin \theta$$
$$+ \sigma_y(\Delta A_n \sin \theta) \sin \theta - \tau_{xy}(\Delta A_n \sin \theta) \cos \theta$$

which reduces to

$$\sigma_n = \sigma_x \cos^2 \theta + \sigma_y \sin^2 \theta - 2\tau_{xy} \sin \theta \cos \theta$$

From the trigonometric identities,

$$\cos^2 \theta = \frac{\cos 2\theta + 1}{2} \; ; \qquad \cos \theta \sin \theta = \frac{\sin 2\theta}{2} \; ; \qquad \sin^2 \theta = \frac{-\cos 2\theta + 1}{2}$$

Thus after rearranging terms, in terms of double angles, the above equation becomes

$$\sigma_n = \left(\frac{\sigma_x + \sigma_y}{2} \right) + \left(\frac{\sigma_x - \sigma_y}{2} \right) \cos 2\theta - \tau_{xy} \sin 2\theta \qquad (10\text{-}3)*$$

Summing forces in the direction parallel to the plane, we have

$$\tau \Delta A_n = \sigma_x(\Delta A_n \cos \theta) \sin \theta + \tau_{xy}(\Delta A_n \cos \theta) \cos \theta$$
$$- \sigma_y(\Delta A_n \sin \theta) \cos \theta - \tau_{xy}(\Delta A_n \sin \theta) \sin \theta$$

which reduces to

$$\tau = (\sigma_x - \sigma_y) \sin \theta \cos \theta + \tau_{xy}(\cos^2 \theta - \sin^2 \theta)$$

and, in terms of double angles,

$$\tau = \left(\frac{\sigma_x - \sigma_y}{2} \right) \sin 2\theta + \tau_{xy} \cos 2\theta \qquad (10\text{-}4)*$$

The orientation of the principal plane is determined by setting the derivative $d\sigma_n/d\theta = 0$. From Eq. (10-3),

$$-(\sigma_x - \sigma_y) \sin 2\theta - 2\tau_{xy} \cos 2\theta = 0$$

from which

$$\tan 2\theta = \frac{-\tau_{xy}}{(\sigma_x - \sigma_y)/2} \qquad (10\text{-}5)*$$

Mohr's Circle for Stress

The normal and shear stresses given by Eqs. (10-3) and (10-4), respectively, may be represented graphically by an extremely useful device known as *Mohr's circle for stress*. It is named after the German engineer Otto Mohr who devised it in 1882.

* For a more comprehensive treatment, see J. N. Cernica, *Strenth of Materials*, 2nd ed., Holt, Rinehart and Winston, New York, 1977.

The Mohr circle represents these equations in a manner that makes them more easily understood and remembered and brings out their physical significance more lucidly.

It is a rather simple matter to show that these are the equations of a circle in a σ_n–τ plane. For convenience, Eqs. (10-3) and (10-4) are rewritten below, Eq. (10-3) having been slightly rearranged:

$$\sigma_n - \left(\frac{\sigma_x + \sigma_y}{2}\right) = \left(\frac{\sigma_x - \sigma_y}{2}\right) \cos 2\theta - \tau_{xy} \sin 2\theta \tag{10-3}$$

$$\tau = \left(\frac{\sigma_x - \sigma_y}{2}\right) \sin 2\theta + \tau_{xy} \cos 2\theta \tag{10-4}$$

Squaring both sides of each equation, we get

$$\left(\sigma_n - \frac{\sigma_x + \sigma_y}{2}\right)^2 = \left(\frac{\sigma_x - \sigma_y}{2}\right)^2 (\cos 2\theta)^2 - (\sigma_x - \sigma_y)(\tau_{xy}) \cos 2\theta \sin 2\theta$$
$$+ \tau_{xy}^2 (\sin 2\theta)^2$$

and

$$\tau^2 = \left(\frac{\sigma_x - \sigma_y}{2}\right)^2 (\sin 2\theta)^2 + (\sigma_x - \sigma_y)(\tau_{xy}) \cos 2\theta \sin 2\theta + \tau_{xy}^2 (\cos 2\theta)^2$$

Hence adding the two equations,

$$\left(\sigma_n - \frac{\sigma_x + \sigma_y}{2}\right)^2 + \tau^2 = \left(\frac{\sigma_x - \sigma_y}{2}\right)^2 + \tau_{xy}^2 \tag{10-6}$$

This is the equation of a circle of the form

$$(\sigma_n - a)^2 + \tau^2 = R^2$$

where the radius is

$$R = \sqrt{\left(\frac{\sigma_x - \sigma_y}{2}\right)^2 + \tau_{xy}^2}$$

and $a = (\sigma_x + \sigma_y)/2$. The center of the circle lies at a point $[(\sigma_x + \sigma_y)/2, 0]$. Note that the center of the circle always lies on the σ_n axis.

The above discussion may be more meaningfully complemented and expanded by an actual construction of Mohr's circle. As a start, some basic steps shall be outlined:

1. The *normal* stresses are plotted as horizontal coordinates. The *tensile* stresses are considered positive (plotted to the right of the origin); the *compressive* stresses are considered negative (plotted to the left of the origin). However, in soil mechanics it is customary to indicate *compressive* stresses as positive.

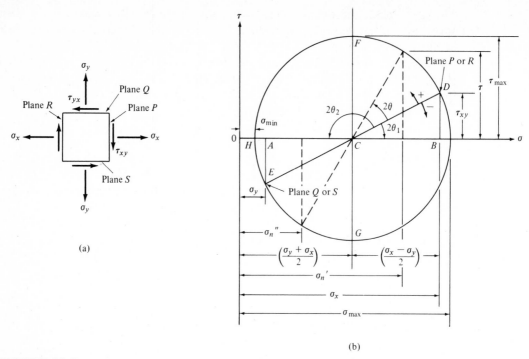

FIGURE 10-5

2. The *shear* stresses are plotted as vertical coordinates. Positive shear stresses

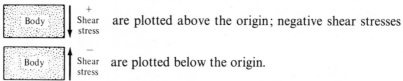

are plotted above the origin; negative shear stresses

are plotted below the origin.

3. Positive angles* on the circle are obtained when measured in the counterclockwise sense; negative angles on the circle are obtained in the clockwise sense. An angle of 2θ on the circle corresponds to an angle of θ on the element.

Figure 10-5b shows Mohr's circle for the assumed stresses given in Fig. 10-5a at the stressed point, assuming that $\sigma_x > \sigma_y > 0$. The normal and shear stresses on vertical plane P give the coordinates of one point on the circle, as indicated in Fig. 10-5b by point D. Plane R, also vertical and 180° from P, is located at $2\theta = 360°$ on the circle from plane P, resulting in the *same* point on the circle. Similarly, the stresses on horizontal planes Q and S plot a different single point E. This shows that the stresses on parallel planes of an element oriented at any angle are correspondingly equal.

* This sign convention should not be confused with that for shear stresses.

The point of intersection of a line DE connecting these two points and the horizontal axis σ_n gives the location of the center of the circle. The diameter of the circle is the length of line DE.

The circle proves to be indeed a valuable visual aid. For example, the maximum shear stress is readily apparent to be equal to the radius of the circle. The orientation of this stress with respect to the original plane can either be measured or easily calculated with the help of trigonometry. The principal stresses σ_{max} and σ_{min} can likewise be spotted and determined easily, and the orientation can easily be measured or calculated. Furthermore, on the principal planes (at σ_{max} and σ_{min}) the shear stresses are zero.

Figure 10-7 shows a number of additional basic relationships which can be obtained from Mohr's circle, such as that for the element of Fig. 10-6. Observed directly from Fig. 10-7, some of these basic relationships are:

1. The maximum shear stress is equal to the radius of the circle. Furthermore the maximum shear stresses are on planes that make an angle of 45° with the principal planes.
2. The resultant stress on any plane, e.g., S, has a magnitude of $\sqrt{\sigma^2 + \tau^2}$. Its magnitude is equal to the distance $0S$ shown in Fig. 10-7. Furthermore its angle of obliquity is equal to $\tan^{-1}(\tau/\sigma)$.
3. The maximum angle of obliquity is constructed by the line $0T$ in Fig. 10-7, such that the line is tangent to the circle at point T. Correspondingly, the stresses, normal and shear, which correspond to point T on the circle, represent the stresses on the plane of maximum obliquity. Note that the shear stress on this

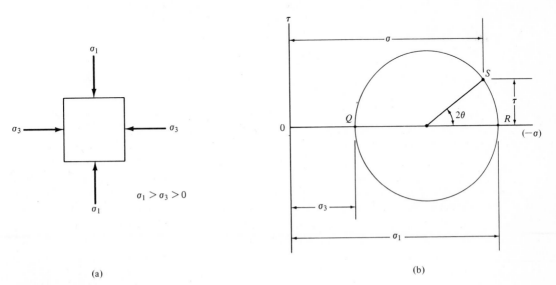

(a)

(b)

FIGURE 10-6
Mohr's circle for case of principal stresses.

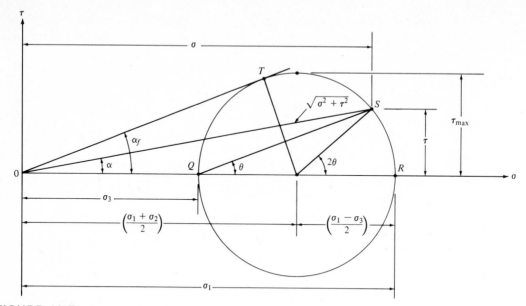

FIGURE 10-7
Some basic relationships, Mohr's circle for element of Fig. 10-6a.

plane is less than τ_{max}. Interestingly one notes that slippage occurs at the point of maximum obliquity and not at the angle where τ_{max} occurs. It is therefore α_f which assumes the more prominent position regarding slip failure.

10-3 MOHR'S THEORY OF FAILURE

The shear strength of soil is generally regarded as the resistance to deformation by continuous shear displacement of soil particles along *surfaces of rupture*. That is, the shear strength of the soil is not regarded solely in terms of its ability to resist peak stresses, but it must be viewed in the context of deformation which may govern its performance. In that light, therefore, shear failure is necessarily viewed as the state of deformation when the functional performance of the soil mass is impaired.

There are a number of different theories as to the nature and extent of the state of stress and deformation at the time of failure. Failure of a soil mass, particularly cohesionless soil which develops its strength primarily from solid frictional resistance between and interlocking of grains, appears to be best explained by *Mohr's rupture theory*. According to Mohr's theory, the shear stress in the plane of slip reaches at the limit a maximum value which depends on the normal stress acting in the same planes and the properties of the material. This represents the combination of normal and shear stresses which results in a maximum angle of obliquity α_f (see Fig. 10-7).

Hence, based on Mohr's theory, the shear strength of a given soil is a function of the normal stress and the soil properties (e.g., strength parameters c and φ). For

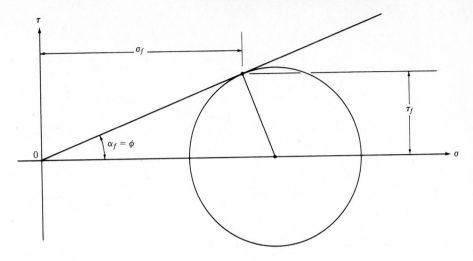

FIGURE 10-8
Relationship between shear strength, normal stress, and angle of obliquity α.

cohesionless soils, for example, the angle of obliquity reaches a maximum or limiting value α_f equal to the friction angle φ, as indicated in Fig. 10-8. Hence, for this special case, where the cohesion intercept is zero ($c = 0$), the shear strength could be expressed as

$$s = \tau_f = \sigma_f \cdot \tan \varphi \qquad (10\text{-}7)$$

where s represents a convenient and widely used symbol to denote shear strength in soils. One may note that for $\alpha_f > 0$, the value for s is always less than τ_{max}.

Figure 8-9a depicts a general s–σ relationship while Fig. 8-9b illustrates a linear s–σ relationship. The linearity may not be quite correct (for example, see Fig. 10-17), but is reasonably accurate and mathematically quite convenient.

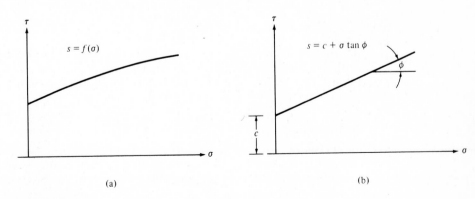

FIGURE 10-9
Shear strength according to Mohr and Coulomb.

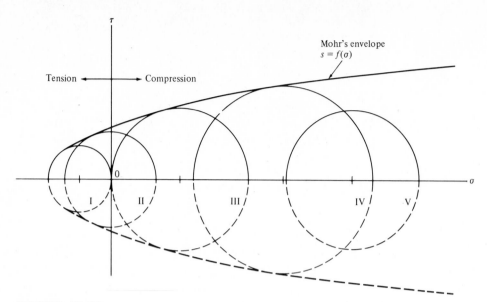

FIGURE 10-10
Mohr's circles for various cases of stress: I, simple tension; II, pure shear; III, simple compression; IV and V, biaxial compression.

Now let us assume that the function $s = f(\sigma)$ represents the shear strength of the soil at impending failure. This may be represented by the heavy line in Fig. 10-10, designated as *Mohr's envelope*. Based on Mohr's theory, any combination of stresses that falls within this envelope represents a stable condition. All five Mohr circles represented in Fig. 10-10 fall in this category. It is to be noted, however, that circles I–IV are tangent to the Mohr envelope, thereby depicting a condition of impending failure. The corresponding points of tangency represent the resultant's shear and normal stresses for the respective angles of obliquity. Also, the corresponding shear strengths for the various cases are the shear stresses at the points of tangency. Circle V, on the other hand, is well within Mohr's envelope. The maximum shear stress for this condition is below the shear strength of the soil.

The friction angle decreases slightly in most soils, and particularly granular soils, with increasing confining stresses. Hence Mohr's envelope is slightly curved, a best fitted curve, as indicated in Fig. 10-10. However, the variation from a straight line in most instances is relatively small. Thus it is mathematically convenient to represent the Mohr envelope by a straight line, sloped at an angle φ. This is shown in Fig. 10-9b as first proposed by Coulomb in 1776 in connection with his investigations of retaining walls.

Figure 10-11c shows a Mohr circle and the corresponding strength envelope for an element stressed as shown in Fig. 10-11a. For this case the expression for the shear strength s may be given as

$$s = c + \sigma \tan \varphi \tag{10-8}$$

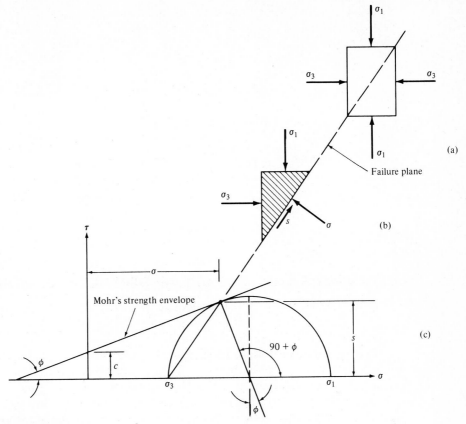

FIGURE 10-11
(a) Element; $\sigma_1 > \sigma_3$. (b) σ_n and τ stresses on plane of failure. (c) Mohr's circle for stress condition shown in (a).

where c = *cohesion* or cohesion intercept
 φ = angle of internal friction

Also from Fig. 10-11c, the orientation of the failure plane with reference to the plane of major principal stresses in a granular soil may be readily established as $2\theta_{cr} = 90 + \varphi$, or

$$\theta_{cr} = 45° + \frac{\varphi}{2} \qquad (10\text{-}9)$$

The normal and shear stresses on the plane of failure are shown in Fig. 10-11b.

   Although it is difficult to measure it very accurately, pore-water pressure has long been recognized as an important factor in shear strength evaluation. Hvorslev's work (21) provided relevant data to support the use of effective stress parameters in describing the shear strength of soils. Sometimes referred to as the Mohr–Coulomb–Hvorslev equation, the relationship is

$$s = \bar{c} + \bar{\sigma} \tan \bar{\varphi} \qquad (10\text{-}10)$$

where $\bar{c}$ = effective cohesion value
$\quad \bar{\sigma}$ = effective normal stress, $\sigma - u$
$\quad u$ = pore-water pressure
$\quad \bar{\varphi}$ = effective angle of internal friction

Equation (10-10) does not imply, however, any unique relationship between effective stress and shear strength. Many other factors enter into the picture. Void ratio at failure, time, stress history, grain structure (i.e., flocculated or dispersed), environmental conditions (i.e., pore water, water table fluctuation, temperature), degree of saturation, conditions that formulate the formation of soil, capillary tension, and effective stresses in the direction normal to the plane of greatest shear distortion are some of the factors that may have a significant effect, particularly for clay soils.

10-4 DETERMINATION OF THE SHEAR STRENGTH OF SOILS

In the preceding section the strength envelope was developed primarily from theoretical considerations. Little was said about the practical and the experimental aspects. That is, while the theoretical presentation serves as a satisfactory tool in defining failure in soils, its validity and usefulness must be closely tied to the parameters observed experimentally and under field conditions. Indeed, to a foundation designer the very essence of the Mohr envelope is experimental data. The failure criteria for soils, subsequent to due considerations of the related parameters, are usually determined experimentally by one or a combination of the following laboratory tests:

1. Direct shear test
2. Triaxial compression test
3. Unconfined compression test

Details related to testing equipment and procedures are beyond the scope of this text. This information can be found in any number of soil-testing manuals. Instead we shall focus primarily on the more general considerations regarding loading, data measurement and interpretation, and sample selection and representation.

As frequently emphasized in various previous sections, the typical soil stratum is neither continuous, homogeneous, nor isotropic. Hence under the best of conditions, a limited number of tests give but a rough approximation of the stratum characteristics. It is thus rather apparent that a soil sample being tested should be as representative as possible of the sample in the field, and as undisturbed as possible. Also, the loads, restraints, and other conditions of testing should correspond to a reasonable degree to those in situ. Yet it is recognized that the mere act of obtaining a sample from a natural deposit radically alters the state of stress and induced strains.

Direct Shear Test (ASTM D-3080-72)

Purportedly first used by Coulomb in 1776, the direct shear test is a simple and widely used test for determining the shear strength of soils. The apparatus is a shear box consisting of two sections, an upper and a lower section. The lower section is fixed to a frame; the upper section moves horizontally relative to the lower section once the shear load is applied. This is depicted in Fig. 10-12a. The soil sample is placed in the box, with approximately half of the sample within either section. A shear force T pulls the upper section relative to the lower, thereby creating a shear plane and corresponding shear stresses, as indicated in Fig. 10-12a. A normal force N is applied to the plane of shear via a plate placed over the soil specimen. Porous stones may be used at the top and bottom of the shear box to facilitate drainage.

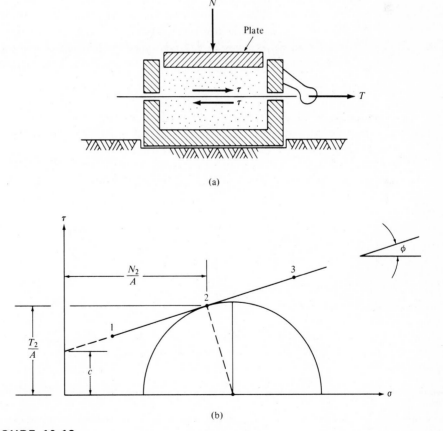

(a)

(b)

FIGURE 10-12
Direct shear test. (a) Schematic of direct shear apparatus. (b) Plotted test results from direct shear test and a superimposed Mohr circle.

The sample is subjected to a shear force and subsequent rupture by increasing the horizontal force *T,* either at a constant speed (*strain-controlled* test) or at a constant load (*stress-controlled* test) until "failure" is induced. This procedure is repeated for several values of normal force *N.* Hence by plotting the normal stresses and corresponding shear stresses from the results of such tests, a failure envelope is obtained, as shown in Fig. 10-12b. The normal and shear stresses are obtained by dividing the normal force and shear force, respectively, by the cross section of the box. Usually the box is a square of sides equal to approximately 5 cm. For example, if the normal and horizontal forces for the second trial are represented by N_2 and T_2, respectively, the coordinate point for test 2 is given as $(N_2/A, T_2/A)$. The corresponding Mohr circle for this case is shown in Fig. 10-12b. The soil parameters φ and c could be measured directly from the graph in Fig. 10-12b.

If the test is run on samples which have not undergone substantial water drainage, it is commonly referred to as *undrained* shear test; i.e., the load is applied soon after the placement of the specimen in the shear box. However, the soil sample may be subjected to a sustained normal load and subsequent consolidation before the test is begun. In that case, if the test is run quickly, thereby developing pore pressures, the test is referred to as *consolidated undrained*; if performed very slowly so that no pore pressure develops, the test is referred to as *consolidated drained*.

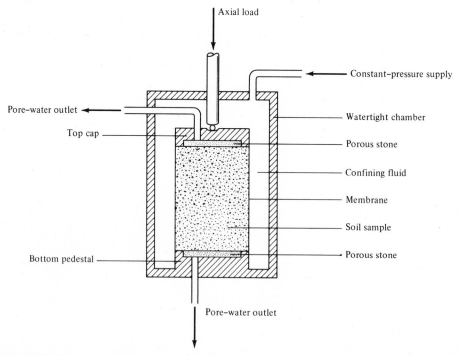

FIGURE 10-13
Schematic diagram of triaxial test apparatus.

Triaxial Test (ASTM D-2850-70)

Figure 10-13 gives the essential features of a triaxial test apparatus (15). A cylindrical sample of soil is enclosed on its curved surface by a thin rubber membrane which extends over a top cap and bottom pedestal. The cell is then filled with water and subjected to a desired pressure. The water pressure, commonly referred to as *confining pressure*, acts horizontally on the cylindrical surface of the sample through the rubber membrane as well as vertically through the top cap. An axial load, commonly referred to as *deviator load*, is then applied and steadily increased until failure of the specimen occurs. The procedure could be repeated for different confining pressures and corresponding deviator loads. We may then construct a Mohr circle for each combination of confining and total axial stresses (σ_3 and σ_1, respectively). The tangent to the resulting circles becomes the Mohr envelope, as shown in Fig. 10-14b.

The lateral pressure produces a normal horizontal stress on all vertical planes of the specimen. Hence any two vertical planes of the specimen experience equal stress σ_3, as shown in Fig. 10-14a. No shear stresses are produced on the horizontal planes or on any two orthogonal vertical planes by either the axial load or the confining pressure. That is, the vertical and horizontal planes of the element shown are subjected only to maximum and minimum (confining) normal stresses σ_1 and σ_3 (principal stresses), respectively.

The imposed conditions in a triaxial test may be manipulated to within a reasonable facsimile of those estimated in situ; that is, the apparatus permits:

1. Control of confining pressure as well as deviator stress
2. Control of pore-water discharge and pressure
3. Reasonably well defined "boundary" conditions

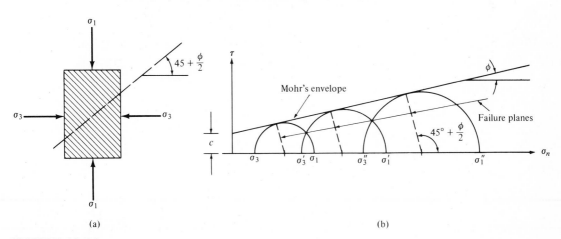

(a)   (b)

FIGURE 10-14

(a) Stressed triaxial specimen. (b) Mohr's circles and Mohr's envelope from three combinations of principal stresses σ_1 and σ_3.

Usually the ratio of length to diameter of the soil specimen ranges between 2 : 1 and 3 : 1.

The maximum principal stress σ_1 is the combination of the deviator stress $\Delta\sigma_1$ and the confining stress σ_3 and is expressed by

$$\sigma_1 = \Delta\sigma_1 + \sigma_3$$

where $\Delta\sigma_1 = P/A$, P being the deviator load, and A the cross-sectional area of the soil specimen.

Unconfined Compression Test (ASTM D-2166-66, 72)

Essentially a special case of the triaxial compression test where the minor principal stress is zero, the *unconfined compression test* is one of the easiest and simplest tests for determining the shear strength of cohesive soil. Indeed, perhaps due most to its simplicity and acceptable results for clay, it has found wide use throughout the world. It is applicable to undisturbed as well as disturbed cohesive soils. Obviously it is not applicable to cohesionless soils since the specimen would fall apart as it is laterally unsupported.

Figure 10-15a shows the stress conditions for an unconfined compression test.

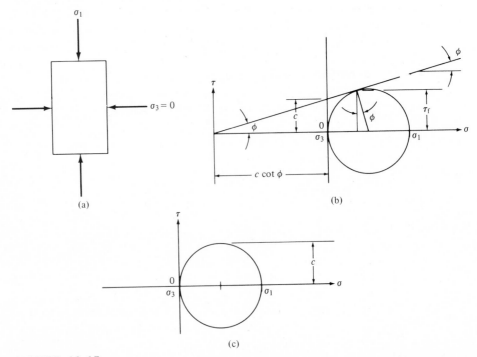

FIGURE 10-15
(a) Unconfined compression test specimen subjected to major principal stress σ_1 and minor principal stress σ_3. (b) Corresponding Mohr's circle for a cohesive soil where $\phi > 0$. (c) Mohr's circle for $\phi = 0$.

Figure 10-15b depicts Mohr's circle for a cohesive soil for which $\varphi > 0$. It is readily apparent that the shear strength of the soil, say τ_f, is a little less than the radius of the circle. For a highly cohesive, true clay, the angle of internal friction is rather small and frequently assumed to be negligibly small, say zero. In that case the corresponding Mohr circle is that indicated by Fig. 10-15c, with the ultimate shear strength of the soil equal to the cohesion c.

EXAMPLE 10-1

Given   The following data from direct shear tests:

Test	Normal Force (kg)	Shear Force (kg)	Area of Sample (cm)
1	4	5.80	5.5 × 5.5
2	8	6.94	5.5 × 5.5
3	12	8.1	5.5 × 5.5
4	16	9.6	5.5 × 5.5

Find   The value for c and φ.

Procedure   The corresponding normal and shear stresses are calculated as follows:

$$\sigma_1 = \frac{P_1}{A} = \frac{(4 \text{ kg})(9.81 \text{ N/kg})(1 \text{ kN/1000 N})}{(5.5 \text{ cm})^2 (0.01 \text{ m/cm})^2}$$

$$\sigma_1 = 13 \text{ kN/m}^2$$

$$\tau_1 = \frac{T_1}{A} = \frac{(5.8 \text{ kg})(9.81 \text{ N/kg})(1 \text{ kN/1000 N})}{(5.5 \text{ cm})^2 (0.01 \text{ m/cm})^2}$$

$$\tau_1 = 17.5 \text{ kN/m}^2$$

The values calculated for tests 1–4 are:

Test	$\sigma (kN/m^2)$	$\tau (kN/m^2)$
1	13	17.5
2	26	22.4
3	39	26.2
4	52	31.0

From the values of σ vs. τ as plotted in Fig. 10-16 the values of c and φ are measured to be approximately 13 kN/m² and 19°, respectively. For the superimposed Mohr circle for test 2, for example, the major and minor principal stresses can be measured as 57 and 10 kN/m², respectively.

Answer

$$c = 13 \text{ kN/m}^2 \quad \text{and} \quad \varphi = 19°$$

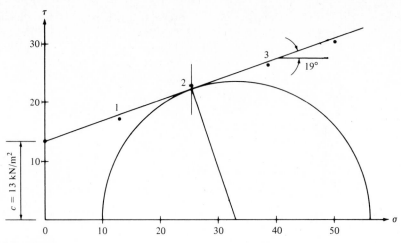

FIGURE 10-16

EXAMPLE 10-2

Given   The following data from triaxial tests:

Test	Confining Stress $\sigma_3(kN/m^2)$	Deviator Stress $\Delta\sigma_1(kN/m^2)$
1	30	57
2	60	79
3	90	92

Find   (a) The values for c and φ.
            (b) The orientation of the failure plane and the shear and normal stresses on the failure plane for test 3.

Procedure The corresponding confining stress σ_3 and the total axial stress σ are calculated as follows:

$$\sigma_3 = 30 \ kN/m^2; \qquad \sigma_3 \text{ values are as given}$$

$$\sigma_1 = \sigma_3 + \text{deviator stress} = 30 + 57 = 87 \ kN/m^2$$

Test	σ_3	σ_1
1	30	87
2	60	139
3	90	182

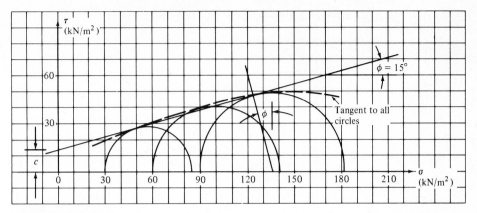

FIGURE 10-17
Mohr's envelope based on three triaxial tests. The solid line represents a straight best-fitted curve: the dashed line is a tangent to all circles.

The Mohr circles for the three tests are plotted in Fig. 10-17. The solid line represents an "averaged" Mohr envelope; the dashed line is perhaps a better fitted envelope. Based on the straight-line slope, the measured values for c and φ are

$$\varphi = 15° \qquad c = 12 \text{ kN/m}^2$$

From the Mohr circle the values for normal and shear stresses on the failure plane for test 3 are measured to be (approximately)

Answer

$$\sigma = 126 \text{ kN/m}^2 \qquad \tau = 44 \text{ kN/m}^2$$

The stresses acting on the faces of the element and the orientation of the failure plane relative to the major principal plane are shown in Fig. 10-18.

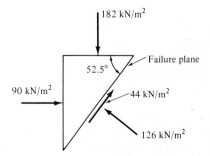

FIGURE 10-18
Stresses acting on element.

10-5 *p-q* DIAGRAMS AND STRESS PATHS

Instead of developing the strength envelope by plotting a series of Mohr circles, then drawing tangents to these circles, as shown in Figs. 10-10 and 10-14 or Example 10-2, it is more convenient to plot the state of stress as a stress point whose coordinates are (p, q), defined as

$$p = \frac{\sigma_1 + \sigma_3}{2} \qquad q = \frac{\sigma_1 - \sigma_3}{2} \tag{10-11}$$

Actually the p and q values represent the coordinates of a point on Mohr's circle, with p representing the center of the circle (always located on the abscissa, or σ axis) and q representing the maximum shear stress, equal to the radius of the circle. The locus of p–q points for a test series is known as a *stress path* (26, 27). Such a graphical depiction is known as the p–q *diagram.*

Figure 10-19 shows the Mohr envelope (φ line) developed from tangents to the circles at points 1, 2, and 3 and line K_f (K_f line), which passes through p–q points A, B, and C—points of maximum shear for the respective circles. Thus the K_f line represents a limiting state of stress at impending failure. The relationship between the φ line and the K_f line may be better illustrated with the aid of Fig. 10-20 for cohesive (but $\varphi > 0$) and cohesionless ($c = 0$) soils.

From either Fig. 10-20a or Fig. 10-20b,

$$\tan \alpha_f = \sin \varphi \tag{a}$$

But from Fig. 10-20a,

$$c \cdot \cot \varphi = d \cdot \cot \alpha_f = \frac{d}{\tan \alpha_f} \tag{b}$$

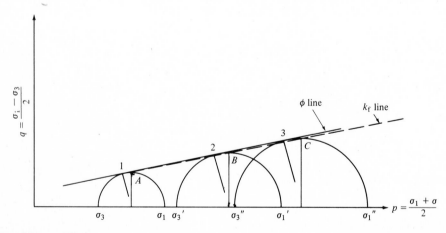

FIGURE 10-19
Stress path for a triaxial test series.

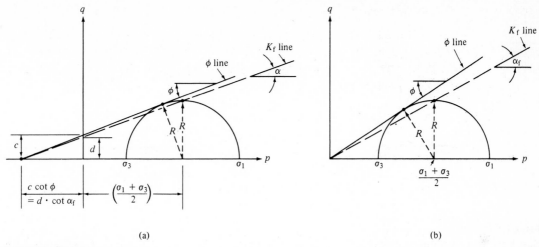

FIGURE 10-20

$p-q$ diagrams and relationship between ϕ line, K_f line, and strength parameters ϕ and c. (a) Cohesive soil. (b) Cohesionless soil.

Combining Eqs. (a) and (b), we get

$$c = \frac{d}{\cos \varphi} \qquad (10\text{-}12)$$

With the K_f line developed and d determined, the parameters c and φ may be either computed or scaled directly from the $p-q$ diagram.

Figure 10-19 depicts the stress path for a series where $\sigma_3 < \sigma_3' < \sigma_3''$ and $\sigma_1 < \sigma_1' < \sigma_1''$. By contrast, Fig. 10-21 shows the stress path for a series where $\sigma_1 > \sigma_3$

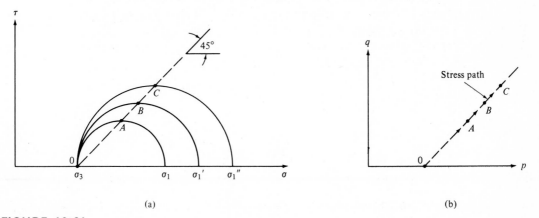

FIGURE 10-21

(a) Series of Mohr's circles where σ_1 increases while σ_3 is held constant. (b) $p-q$ diagram (stress path) corresponding to (a).

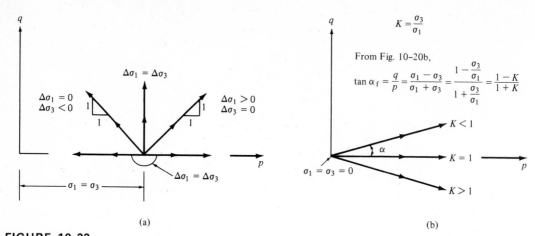

(a)   (b)

FIGURE 10-22
Stress paths for various combinations of stress conditions. (a) Stress paths starting from point where $\sigma_1 = \sigma_3$. (b) Stress paths starting from point where $\sigma_1 = \sigma_3 = 0$.

and σ_3 is held constant. In the same manner, we may hold σ_1 constant while varying σ_3, etc. Thus we may vary σ_1 and σ_3 in various ways to obtain any number of stress paths. Figure 10-22 shows a variety of such paths starting from points $\sigma_1 = \sigma_3$ and $\sigma_1 = \sigma_3 = 0$. Needless to say, the stress increments $\Delta\sigma_1$ or $\Delta\sigma_3$ do not have to vary linearly. Hence the stress paths do not have to be straight lines, and they are not limited to a specific starting point.

Stress paths can be developed for either total or effective stresses. The horizontal distance between total and effective stress paths is the pore pressure u. The pore pressure is positive if the effective stress path is on the left of the total stress path, and vice versa, as shown in Fig. 10-23. Positive pore pressures are usually found in normally consolidated clays, while negative pore pressures may be expected in highly overconsolidated clays.

FIGURE 10-23
General relationship between pore pressure and total and effective stress paths.

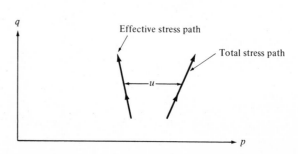

10-6 SHEAR STRENGTH OF COHESIONLESS SOILS

The shear strength of a cohesionless soil may be represented by Eq. (10-13). This is a special case of Eq. (10-8), where cohesion $c = 0$.

$$s = \sigma \tan \varphi \tag{10-13}$$

Generally the value of φ is influenced by:

1. The state of compaction and the void ratio of the soil. The friction angle increases with decreasing void ratio (increasing density), but not linearly.
2. Coarseness, shape, and angularity of the grains. Angular grains interlock more effectively than rounded ones, thereby creating a larger friction angle.
3. Mineralogical content. Hard gravel particles result in higher friction angles than soft grains which may crush more easily, thereby reducing the interlocking or bridging effects. For sand, however, the mineralogical content seems to make little difference except if the sand contains mica. In that case the void ratio is usually larger, thereby resulting in loose interlocking and lower friction angle.
4. Grain size distribution. A dense well-graded sand usually displays a higher friction angle than a dense uniform-sized sand.

Saturated Cohesionless Fine-Grain Soils

As mentioned above, the angle of internal friction of saturated sands and some organic silts is only slightly less than that of the soil in a dry state and of the same relative density. However, the shear strength might be altered significantly by a change in the pore pressures. Hence the shear strength of a saturated cohesionless soil might be given by Eq. (10-14):

$$s = (\sigma - u) \cdot \tan \varphi = \bar{\sigma} \tan \varphi \tag{10-14}$$

where $\bar{\sigma}$ = effective (intergranular) stress
  u = pore-water pressure

Quite apparently, when the pore-water pressure approaches σ, the shear strength approaches zero. When that happens, we may approach impending instability or perhaps motion (e.g., slope failures, boiling). Fluctuation in the water table is a common cause of significant variations in the pore stress and, thereby, in the shear strength of the soil.

Liquefaction of Fine Sand and Inorganic Silts

If a saturated and/or inorganic silt is totally saturated and under hydrostatic neutral stress such that it is not subjected to any effective stress, the mass is in a state of *liquefaction*. Under such circumstances the pore-water pressure u equals the total normal stress σ, thereby reducing the shear strength, as given by Eq. (10-14), to zero. It is quite apparent that in this state the soil is in a *quick condition*, the phenomenon discussed in Section 7-3. Also, if not confined, such a mass will "flow" since it cannot resist stress.

If a submerged fine sand undergoes a sudden decrease in the void ratio, an increase in the pore-water pressure u may result such that the pore pressure may equal or even exceed the value for σ. For example, pile driving, earthquakes, blasts, or other forms of vibration or shock may cause a sudden decrease in the volume, thereby increasing the pore pressure u as a result of a surge in hydrostatic excess pressure (8, 30). Should the value of u reach sufficient magnitude, say $u \geq \sigma$, the shear strength of the soil, as indicated by Eq. (10-14), may be totally lost, resulting in what is known as *spontaneous liquefaction*. Loose fine silty sands are most vulnerable to such effects from shock or dynamic loads or sudden fluctuations in the water table. Compacting a loose sand stratum is frequently a viable option to decrease the possibility of liquefaction.

EXAMPLE 10-3

Given   A cohesionless soil specimen in a triaxial test is subjected to a major and minor principal stress σ_1 and σ_3, respectively.

Find   (a) An expression for φ in terms of σ_1 and σ_3.
(b) An expression for τ and σ on the failure plane in terms of σ_1, σ_3, and φ.

Procedure   The Mohr envelope shown in Fig. 10-24 is for the general stress condition given.

$$\sin \varphi = \frac{(\sigma_1 - \sigma_3)/2}{(\sigma_1 + \sigma_3)/2} = \frac{\sigma_1 - \sigma_3}{\sigma_1 + \sigma_3}$$

Answer
$$\varphi = \sin^{-1}\left(\frac{\sigma_1 - \sigma_3}{\sigma_1 + \sigma_3}\right)$$

From Fig. 10-24,

$$\tau = \frac{\sigma_1 - \sigma_3}{2} \cos \varphi$$

$$\sigma = \frac{\sigma_1 + \sigma_3}{2} - \frac{\sigma_1 - \sigma_3}{2} \sin d$$

Thus

Answer
$$\sigma = \frac{\sigma_1}{2}(1 - \sin \varphi) + \frac{\sigma_3}{2}(1 + \sin \varphi)$$

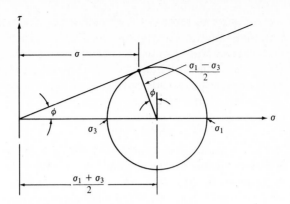

FIGURE 10-24

10-7 SHEAR STRENGTH OF COHESIVE SOIL

A number of noteworthy differences exist between cohesive and noncohesive soils (24, 37):

1. The frictional resistance of cohesive soils is less than that of granular soils.
2. The cohesion of clay is appreciably larger than that of granular soils.
3. Clay is much less permeable than a sandy soil, and the water drainage is, therefore, significantly slower. Hence the pore pressure induced by an increase in load is dissipated very slowly, and the transfer of stress and the corresponding increase in intergranular pressure are likewise much slower.
4. The time-related changes of volume in clays is slower than that for granular material (e.g., consolidation).

Of the number of factors which are recognized to have a direct effect on the shear strength of cohesive soils, most are recognized as individually quite complex. Also, it is recognized that most are significantly interrelated, thereby further increasing the complexity of the problem. Other factors may be still unknown. Generally the degree of consolidation, the drainage, effective stress, and pore pressures are relevant factors to be considered in the strength evaluation of cohesive soils.

The basic procedure and the apparatus for conducting a triaxial compression test have already been described in Section 10-4, but the combination of initial pressures on the specimen, coupled with the moisture conditions, have not (28). These require additional considerations and evaluations. In that regard one notes the partial or extreme initial conditions of pressures and/or the amount of drainage:

1. Complete consolidation
2. Partial consolidation

The degree of saturation may be typified by

1. Drained condition
2. Undrained condition

In order to represent the possible combinations of pressure and drainage, the following nomenclature typifies some of the above-mentioned states of consolidation and degrees of saturation: consolidated drained (CD), consolidated undrained (CU), unconfined undrained (UU). These terms will be used intermittently in subsequent discussions and the reader should be familiar with them.

Normally Consolidated Clays

When the sample is extracted from the ground, the overburden pressure is removed and the pore pressure altered significantly, i.e., negative pore pressures are developed. In order to simulate a somewhat realistic in situ state of stress, the characteristics of saturated, normally consolidated clays extracted from a given stratum are commonly investigated via a consolidated undrained triaxial test. A confining pressure σ_3 and a deviator stress $\Delta\sigma_1$ are applied for undrained conditions. A confining pressure, say σ_0, of the in situ value may be estimated as the overburden pressure for the depth of the stratum from which the sample was extracted. If several such tests are run for varying confining cell pressures, a Mohr envelope may be obtained as indicated in Fig. 10-25a. If the confining pressure is less than the in situ value σ_0, the Mohr envelope depicts a range of preconsolidation of the soil; that is, relative to the confining pressures, the soil specimen appears overconsolidated. The shear strength of the clay specimen tested in this range is higher than that indicated by a straight line through the origin. The relationship between the shear strength and the normal stress in this range is designated by line portion ab, which is slightly curved, but frequently interpreted as a straight line. On the other hand, if the tests are run under confining cell pressures larger than σ_0, the envelope to the ruptured circles is approximately a straight line, represented by segment bc in Fig. 10-25a. Although the effect of pore-water pressure is present for all ranges of normal stress within the Mohr envelope, the pore pressure is larger for the case where the confining pressures are smaller than σ_0. Generally there is an increased drainage of water with increasing confining pressure. However, the effective stress is more significant in the region where the confining pressures are larger than σ_0.

Figure 10-25b shows the general relationship between the shear strength and the effective stress for a case of consolidated drained tests. These tests are appreciably more time consuming than the consolidated undrained tests, and are therefore not as common during routine investigations. The specimen is subjected to a confining pressure, and then the deviator stress is applied at a very slow rate and drainage is permitted at each end of the sample. A series of such tests run under varying confining cell pressures provides a strength envelope as shown in Fig. 10-25b. The strength envelope for such tests is again somewhat curved for the range of confining pressures less than σ_0; for confining pressures greater than σ_0, the Mohr envelope is approximately a straight line. Briefly, a comparison of the two cases shows the following:

1. The corresponding slopes of the straight-line segments of the Mohr envelopes (depicted in Fig. 10-25) are significantly different.

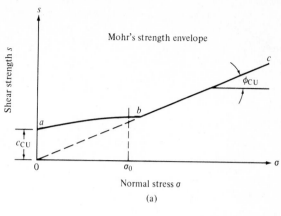

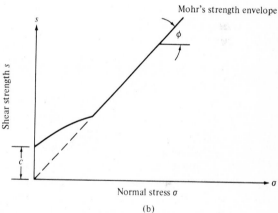

FIGURE 10-25

$s-\sigma$ relationships from consolidated undrained and drained tests. (a) Consolidated undrained (CU) test. (b) Consolidated drained (CD) test.

2. For the drained condition the neutral (pore-water) stress is virtually negligible.
3. The effective friction is significantly larger for the drained case than for the undrained case.

The relationship between angles φ and φ_{CU} may be illustrated by means of Fig. 10-26. For the consolidated drained condition there exists no excess hydrostatic pore-water pressure. Hence the Mohr circle for this case, associated with vertical and confining stresses σ_1 and σ_3, is shown as a solid line. Correspondingly, the strength envelope is represented by the solid line with the corresponding slope of φ. For the consolidated undrained condition there exists a pore pressure u. Thus the total vertical and confining stresses are $\sigma_1 + u$ and $\sigma_3 + u$, respectively. This is shown in Fig. 10-26.

The shear strengths for the two cases are s and s_1, as shown. The actual strength may lie somewhere between that for a consolidated drained and that for an undrained condition since total drainage is unlikely; that is, the actual strength of the sample may range between values of s and s_1.

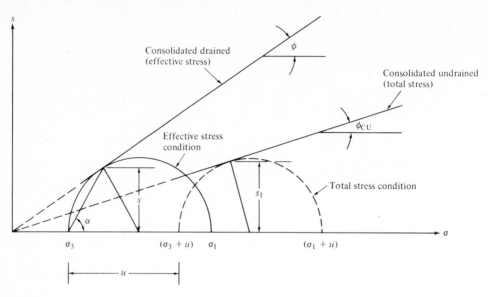

FIGURE 10-26
Consolidated undrained triaxial test results.

Intact Overconsolidated Clays

A soil being evaluated or tested may have been subjected to a great deal of pre-compression induced by loads which since then may have disappeared. For example, pressures induced by excessively thick ice sheets from glacial formations which since have melted, or from soil formations caused by glacial movements which since have eroded may have greatly precompressed an existing stratum. Figure 10-27 is a schematic depicting the formation of *overconsolidating pressures* with time.

Let us assume that at time t_n the present overburden pressure upon a given soil sample is calculated to be P_0. If the soil sample has the history indicated by Fig. 10-27, then the soil at one time has been subjected to some "additional" pressure ΔP. Schematically we will designate the maximum pressure to which the soil sample was subjected as P_c. We shall refer to it as *overconsolidation pressure*, which has been discussed in Chapter 9. For overconsolidated clays the overconsolidation ratio is greater than 1 (that is, OCR > 1).

Figure 10-28 shows the Mohr strength envelope for an overconsolidated clay. One notes that for the range of pressures less than P_c the Mohr envelope deviates from a straight line, i.e., the shear strength is larger than that given by the dotted straight line. Generally preconsolidation causes or results in a smaller void ratio at failure than would otherwise exist, even though the specimen tends to expand as a result of extraction from an in situ condition. Cohesion and general capillary forces tend to resist the volume increase, thereby resulting in a somewhat greater shear strength, as indicated by the curved portion of the envelope. Beyond the preconsolidation

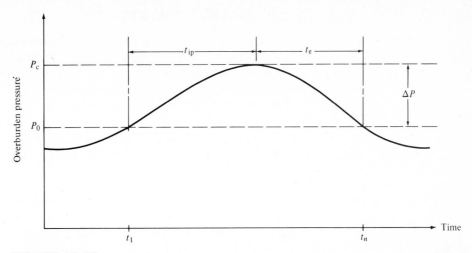

FIGURE 10-27
Development of overconsolidation loads. t_{ip} = period of increasing overburden; t_e = period of erosion or decreasing overburden.

pressure P_c, the effective normal stress and shear strength relationship is given by a reasonably straight line.

The shear strength of clays which have fine discontinuities, hairline cracks, etc., generally referred to as *fissures*, may be appreciably different from that of the typical overconsolidated clay described above. Depending on the magnitude and the orientation of these fissures, test results may be particularly misleading in the overall evaluation. For example, the results from a direct shear test on a sample where fissures are parallel to the shearing force may be appreciably smaller than for one where the orientation of the fissures is 90° to the shear force. On the other hand, the triaxial test yields somewhat more reliable results, improved perhaps by the lateral restraint of the confining pressure.

FIGURE 10-28
Shear strength vs. effective normal stress for overconsolidated clay.

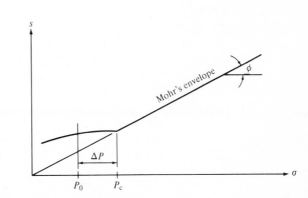

10-8 IN SITU EVALUATION OF SHEAR STRENGTH

As mentioned previously, the extraction of a soil sample and the subsequent changes induced by the extraction process, the handling, and the testing procedures may greatly alter the characteristics of the specimen and therefore the test results. Furthermore it is not always feasible and practical to attempt to duplicate the in situ conditions. Frequently it is both expedient and desirable to test the soil in the in situ condition.

A number of *field tests* are used to estimate the shear strength of the soil. Some of these give results that fit into the theoretically based expressions used to designate the shear strength of the soil. Others are rather empirical in nature and greatly dependent upon the engineer's judgment and experience.

The *vane-shear tester* (ASTM D-2573-72, 78) described in Section 3-7* has found frequent use in determining the shear strength of soft cohesive soils which otherwise may be greatly disturbed during the extraction and testing process. The test is performed at any given depth by first augering to the prescribed depth, cleaning the bottom of the borings, and then carefully pushing the vane instrument into the stratum to be tested. The torque is then applied gradually and the peak value noted. The shear strength of the soil can then be estimated by using Eq. (10-15):

$$\tau = \frac{3T}{28\pi R^3} \tag{10-15}$$

Penetrometers (described in Section 3-7) are sometimes used to test the shear strength of the soil at the surface—lateral or vertical. Their use is primarily applicable to fine-grained soil; coarse and gravelly strata tend to give erroneous results. The procedure for using the penetrometer consists of first cleaning the surface of any loose material, pushing the penetrometer into the stratum to the calibration mark on the head of the penetrometer, and recording the maximum reading on the penetrometer scale. This reading represents the pressure in force per unit area necessary to push the penetrometer to the designated mark. This instrument is also frequently used as a guide for the bearing capacity of the soil. Again, the reliability of the results must be interpreted with a view to the conditions present at the time of testing. For example, a dry clay would test totally differently from a wet clay. This of course is something one would expect. However, the interpretation and subsequent use of the test data must take into account such changes in water content conditions.

The *split-spoon sampler* described in Section 3-7 is still another means used to estimate the shear strength of the soil. The number of blows required to drive the split-spoon sampler (generally referred to as the Standard Penetration Test) is a measure of the soil's resistance to shear. (The Standard Penetration Test is also frequently used to estimate the bearing capacity of the soil.) To a lesser degree the

---

* For a schematic description of the apparatus and derivation of Eq. (10-15) see Section 3-7.

thin-wall tube (frequently referred to as a Shelby-tube sampler) described in Section 3-5 is sometimes used as a check on the shear strength of a soil.

The above are but a few of the many techniques and tools used to determine the shear strength of the soil. However, as mentioned in previous sections, it is to be noted that a soil may have different shear strength characteristics in various planes and for different moisture contents. Hence one should take note of the fact that the above-mentioned techniques may or may not reflect the true shear strength of the soil in all directions.

PROBLEMS

10-1.   How does soil friction (a) differ from and (b) resemble "surface" friction?

10-2.   State Mohr's theory of failure. Briefly, how is it applied to soils?

10-3.   What is Mohr's strength envelope? What does the K_f line represent? What is the significance of each?

10-4.   Name at least three properties which influence the shear strength of cohesionless soils. State how each of these properties affects the strength.

10-5.   Name at least three differences which exist between cohesive and cohesionless soils within the context of shear strength.

10-6.   Briefly, what are the essential features of a triaxial test apparatus?

10-7.   In connection with triaxial testing, what is: (a) deviator stress, (b) confining stress, (c) total axial stress, and (d) the relationship between total axial stress and confining stress?

10-8.   Briefly, describe the physical makeup of a vane-shear tester. How is it used?

10-9.   For each datum in the table:

	Stress	*Compression (kPa)*
Test	σ_1	σ_3
1	100	40
2	200	60
3	300	80
4	400	100

$c = 0$

(a)   Draw a Mohr circle.

(b)   Determine the orientation of the failure plane.

(c)   Determine the normal and shear stresses on the failure plane.

10-10.   For the data given in Problem 10-9.

(a)   Determine Mohr's strength envelope and the corresponding φ values.

(b)   Determine the K_f line and the corresponding α_f values.

10-11. Show the stress paths for each of the following data on a p–q diagram.

Stresses, Compression (kPa)				
	Initial		Final	
Test	σ_1	σ_3	σ_1	σ_3
1	100	100	500	100
2	100	100	100	500
3	100	100	500	500
4	100	100	500	300
5	500	500	500	100
6	500	500	800	600

10-12. A cohesionless soil sample (assume $c = 0$) was placed in a direct shear apparatus and subjected to a normal stress of 180 kN/m². The specimen failed in shear at a shear stress of 122 kN/m². Determine:
(a) The angle of internal friction φ.
(b) The magnitude of the principal stresses, based on Mohr's circle being tangent at (180, 122).
(c) The normal and shear stresses on a properly oriented failure plane.

10-13. From the Mohr circle construction of Problem 10-12, part (b), determine α_f and draw the corresponding stress path.

10-14. A silty sand sample was placed in a direct shear apparatus. When subjected to a zero normal load, it sheared at 26 kN/m². An identical sample was then subjected to a normal stress of 200 kN/m². It failed at 129 kN/m². Determine:
(a) The angle of internal friction φ.
(b) The magnitude of the principal stresses by drawing a Mohr circle tangent at (200, 129).
(c) The normal and shear stresses on a properly oriented failure plane.

10-15. A sand specimen ($c = 0$) was subjected to a drained triaxial test under a chamber pressure of 100 kN/m². A deviator stress of 210 kN/m² resulted in shear failure.
(a) Draw a Mohr's circle for this stress condition.
(b) Determine the angle of internal friction φ.
(c) Draw the K_f line and measure α_f. How does α_f compare with φ?

10-16. A series of triaxial tests were run on similar silty sand specimens obtained from the same source. The test data are:

Stresses (kPa)		
Test	Confining	Deviator
1	30	85
2	60	96
3	90	106

(a) Draw the Mohr envelope for the given data.

(b) Determine an average value for φ.

(c) Based on a best-fitted straight-line Mohr envelope, determine the value of c.

(d) Draw the K_f line, then determine c.

(e) Compare the values of c obtained from parts (c) and (d).

10-17. Draw a p–q diagram for the data of Problem 10-16. Is the slope of this line the same as that of Mohr's strength envelope of Problem 10-16? Explain.

10-18. For test 1 of Problem 10-16, determine:

(a) The orientation of the failure plane.

(b) The shear and normal stresses on the failure plane, and show these on a properly oriented element.

10-19. For test 2 of Problem 10-16, determine:

(a) The orientation of the failure plane.

(b) The shear and normal stresses on the failure plane, and show these on a properly oriented element.

10-20. Define liquefaction. What are the conditions necessary for liquefaction to occur?

BIBLIOGRAPHY

1. Aas, G., "A Study of the Effect of Vane Shape and Rate of Strain on the Measured Values of in-situ Shear Strength of Clays," *6th Int. Conf. Soil Mech. Found. Eng.*, 1965.

2. Ansal, A. M., Z. P. Bazant, and R. J. Krizek, "Viscoplasticity of Normally Consolidated Clays," *ASCE J. Geotech. Eng. Div.*, vol. 105, Apr. 1979.

3. Bishop, A. W., and G. Eldin, "Undrained Triaxial Tests on Saturated Sands and Their Significance in the General Theory of Shear Strength," *Geotechnique*, vol. 2, 1950.

4. Bjerrum, L., N. Simons, and I. Torblaa, "The Effect of Time on the Shear Strength of a Soft Marine Clay," *Proc. Brussels Conf. Earth Pressure Problems*, vol. 1, 1958.

5. Bjerrum, L., S. Kringstad, and O. Kummeneje, "The Shear Strength of Fine Sand," *5th Int. Conf. Soil Mech. Found. Eng.*, Paris, France, vol. 1, 1961.

6. Bjerrum, L., and K. Y. Lo, "Effect of Aging on the Shear-Strength Properties of a Normally Consolidated Clay," *Geotechnique*, vol. 13, no. 2, 1963.

7. Bjerrum, L., and A. Landva, "Direct Simple-Shear Tests on a Norwegian Quick Clay," *Geotechnique*, vol. 16, no. 1, 1966.

8. Blazquez, R., R. J. Krizek, and Z. P. Bazant, "Site Factors Controlling Liquefaction," *ASCE J. Geotech. Eng. Div.*, vol. 106, July 1980.

9. Blight, G. E., "Shear Stress and Pore Pressure in Triaxial Testing," *ASCE J. Geotech. Eng. Div.*, vol. 91, Nov. 1965.

10. Franklin, A. G., L. F. Orozco, and R. Semrau, "Compaction and Strength of Slightly Organic Soils," *ASCE J. Geotech. Eng. Div.*, vol. 99, July 1973.

11. Gibbs, H. J., *et al.*, "Shear Strength of Cohesive Soils," 1st PSC. *ASCE*, 1960.

12. Gibson, R. E., "Experimental Determination of the True Cohesion and True Angle of Internal Friction in Clays," *3rd Int. Conf. Soil Mech. Found. Eng.*, Zurich, Switzerland, vol. 1, 1953.

13. Gibson, R. E., and D. J. Henkel, "The Influence of Duration of Tests at Constant Rate of Strain on Measured 'Drained' Strength," *Geotechnique*, vol. 4, 1954.

14. Hall, E. B., "Shear Strength Determination of Soft Clayey Soil by Field and Laboratory Methods," *Symp. Soil Exploration, Spec. Tech. Publ.* 351, Am. Soc. Test. Mater., Philadelphia, Pa., 1963.

15. Haythornthwaite, R. M., "Mechanics of the Triaxial Test for Soils," *Proc. ASCE*, vol. 86, 1960.

16. Helenelund, K. V., "Vane Tests and Tension Tests on Fibrous Peat," *Proc. Geotech. Conf.*, Oslo, Norway, 1967.

17. Henkel, D. J., "The Effect of Overconsolidation on the Behavior of Clays during Shear," *Geotechnique*, vol. 6, 1956.

18. Henkel, D. J., "The Shear Strength of Saturated Remolded Clays," *Proc. 1960 ASCE Specialty Conf. Shear Strength of Cohesive Soils*, Boulder, Colo.

19. Herrman, H. G., and L. A. Wolfskill, "Residual Shear Strength of Weak Clay," Tech. Rep. 3-699, U.S. Army Eng. Waterw. Exp. Stn., 1966.

20. Hooper, J. A., and F. G. Butler, "Some Numerical Results Concerning the Shear Strength of London Clay," *Geotechnique*, vol. 16, no. 4, 1966.

21. Hvorslev, M. J., "Uber die Festigkeitseigenschaften gestörter Bindiger Böden" (On the strength properties of remolded cohesive soils), Danmarks Naturvidenskabelige, Samfund, Copenhagen, 1937.

22. Kenney, T. C., "Shear Strength of Soft Clay," *Proc. Geotech. Conf.* Oslo, Norway, vol. 2, 1967.

23. Koerner, R. M., "Behavior of Single Mineral Soils in Triaxial Shear," *ASCE J. Geotech. Eng. Div.*, vol. 96, July 1970.

24. Koerner, R. M., "Effect of Particle Characteristics on Soil Strength," *ASCE J. Geotech. Eng. Div.*, vol. 96, 1970.

25. Ladd, C. C., and T. W. Lambe, "The Shear Strength of 'Undisturbed' Clay Determined from Undrained Tests," *Spec. Tech. Publ.* 361, Lab. Shear Testing of Soils, Am. Soc. Test. Mater., Philadelphia, Pa., 1963.

26. Lambe, T. W., "Stress Path Method," *ASCE J. Soil Mech. Eng. Div.*, vol. 93, 1967.

27. Lambe, T. W., and W. A. Marr, "Stress Path Method; 2nd ed.," *ASCE J. Geotech. Eng. Div.*, vol. 105, June 1979.

28. Lundgren, R., J. K. Mitchell, and J. H. Wilson, "Effects of Loading on Triaxial Test Results," *ASCE J. Geotech. Eng. Div.*, vol. 94, Mar. 1968.

29. O'Neil, H. M., "Direct-Shear Test for Effective-Strength Parameters," *Proc. ASCE*, vol. 88, no. SM4, 1962.

30. Peck, R. B., "Liquefaction Potential: Science versus Practice," *ASCE J. Geotech. Eng. Div.*, vol. 105, Mar. 1979.

31. Perlow, J. M., and A. F. Richards, "Influence of Shear Velocity on Vane Shear Strength," *ASCE J. Geotech. Eng. Div.*, vol. 103, Jan. 1977.

32. Prevost, J. H., "Undrained Shear Tests on Clays," *ASCE J. Geotech. Eng. Div.*, vol. 105, Jan. 1979.

33. Rodin, S., "Experiences with Penetrometers, with Particular Reference to the Standard Penetration Test," *5th Int. Conf. Soil Mech. Found. Eng.*, Paris, France, vol. 1, 1961.

34. Rowe, P. W., and L. Barden, "The Importance of Free Ends in Triaxial Testing," *Proc. ASCE*, vol. 90, 1964.

35. Sadasivan, S. K., and V. S. Raju, "Theory for Shear Strength of Granular Materials," *ASCE J. Geotech. Eng. Div.*, vol. 103, Aug. 1977.

36. Skempton, A. W., "Vane Tests in the Alluvial Plain of the River Forth near Grangemouth," *Geotechnique*, vol. 1, no. 2, 1948.

37. Skempton, A. W., and A. E. Bishop, "The Measurement of the Shear Strength of Soils," *Geotechnique*, vol. 2, no. 2, 1950.

38. Taylor, D. W., "A Direct Shear Test with Drainage Control," *Symp. Direct-Shear Testing of Soils, Spec. Tech. Publ.* 131, Am. Soc. Test. Mater., Philadelphia, Pa., 1952.
39. Taylor, D. W., "Review of Research on Shearing Strength of Clay at M.I.T. 1948–1953," Rep. to U.S. Army Corps of Eng. Waterw. Exp. Stn., 1955.
40. Whitman, R. V., and K. E. Healy, "Shear Strength of Sands During Rapid Loading," *Proc. ASCE*, vol. 88, 1962.
41. Wu, T. H., N. Y. Chang, and E. M. Ali, "Consolidation and Strength Properties of a Clay," *ASCE J. Geotech. Eng. Div.*, vol. 104, July 1978.

11

Stability of Slopes

11-1 INTRODUCTION

The term *slope* as used here refers to any earth mass, natural or man-made, whose surface forms an angle with the horizontal. Hills and mountains, river banks and coastal formations are common examples of sloped earth masses formed by nature perhaps as a result of glacial movements, weathering, erosion, deposit buildup and sedimentation, or other factors. Examples of man-made slopes may include *fills* such as embankments, earth dams, levees; examples associated with *cuts* may include highway and railway cuts, canal banks, foundation excavations, and trenches.

Virtually every slope experiences gravitational forces and subsequent changes. Indeed, the combining effects of gravity and water are the primary and direct influences on changes for most of the slopes. Some slopes may possibly also be influenced by such natural phenomena as chemical actions, earthquakes, glacial forces, or wind. In virtually all instances the effect is general flattening of the slope, either suddenly or slowly and cumulatively. Predicting the change with any degree of accuracy may be a difficult task; preventing such change may be an even greater task for the soil engineer.

The stability of a slope is generally viewed in relative terms. As implied above, a slope would not remain unchanged forever. However, from an engineering point of view, we think of a slope as stable if it meets a prescribed need for a fixed period of time with a suitable or acceptable safety factor. For example, we may accept as "safe" a highway cut which may undergo a slight change due to erosion, but which would not experience a mass sliding failure during a given time span. On the other hand, we may not view as acceptable a slope that would appear to be generally stable except for a gradual and progressive erosion from seepage or wave action, etc. It is apparent, therefore, that in the analysis of slopes, *stability* is not a totally descriptive term. Perhaps a more appropriate term would be *functional stability*, which necessarily relates to a specific need and specific governing criteria.

Man-made embankments exist in China, India, and other parts of the world which are centuries old and apparently still stable. Less documentation is available as to failures experienced by the ancient builders. However, the research on slides which occurred in the last century, such as that in Gothenburg, Sweden, in 1916, that which took place during the construction of the Panama Canal, also in the same era, or the large earth flow in quick clay near St. Thuribe, Quebec, of 1898, provided new impetus toward a systematic analysis of slope stability. In this respect much credit is due to a number of Swedish engineers who developed and recommended methods for analyzing the stability of slopes, which are still used today. Some of these methods as well as others developed will be discussed in subsequent sections.

The analytical tools available for evaluating slope stability are generally a combination of theoretical considerations coupled with practical reasoning and observations. As so typical of many soils problems, the factors involved are difficult to evaluate, are many and interrelated, and are subject to change with changing environment. For example, when the shear stresses on a slope equal or exceed the shear strength of the soil, impending failure will occur. However, as discussed in the previous chapter, the evaluation of the magnitude of both the shear stresses and the shear strength is dependent on numerous and complicated factors. Hence predicting the probable extent of gradual slope movement and the subsequent stability of a slope is essentially an evaluation of the induced shear stresses and resisting shear strength. The following sections present several methods related to such an evaluation.

11-2 INFINITE SLOPES

The name *infinite slopes* is given to earth masses of constant inclinations of unlimited extent and uniform conditions at any given depth below the surface. Inherent in this definition, the soil stratum is not necessarily homogeneous and uniform with depth, but has similar characteristics in a given stratum parallel to the surface. The usual plane of failure for such slopes is typified by a relatively significant change in characteristics of the sliding mass to the fixed mass along the plane of failure (e.g., a *clay* mass sliding over a *shale* formation). It is mathematically convenient to delineate the characteristics of the sliding mass into rather specific categories, such as cohesionless or cohesive soils, seepage or no seepage. Indeed, this is perhaps a simplistic view since it is quite unlikely that any slope meets such well-defined conditions to any degree. More realistically, the ordinary slope is quite probably never totally cohesive or cohesionless, never totally dry or totally submerged and seeping. Nevertheless such assumptions and subsequent calculations provide a good basis for the engineer to evaluate the stability. The engineer's judgment will be the determining factor as to where a specific condition may fit between, perhaps, two idealistic extremes.

Cohesive Material—No Seepage

Figure 11-1a shows the forces acting on an element from a slope of infinite extent. No seepage is assumed, and the material above the slip plane is assumed homogeneous and cohesive. The slip plane is parallel to the surface of the slope. A unit thickness of the element is assumed in the direction normal to the page. Forces F_1,

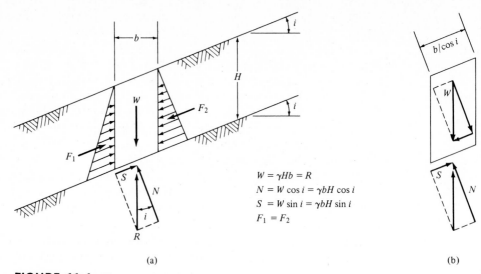

$$W = \gamma H b = R$$
$$N = W \cos i = \gamma b H \cos i$$
$$S = W \sin i = \gamma b H \sin i$$
$$F_1 = F_2$$

(a)   (b)

FIGURE 11-1

Forces on element of infinite slope.

F_2, are assumed equal and opposite, and are therefore ignored in the analysis. Thus the relevant forces acting on the slip plane are shown in Fig. 11-1b. By summing forces in the perpendicular and parallel directions to the slope, respectively, we obtain the resulting normal and shear stresses as indicated in Eqs. (a) and (b). Hence,

$$\sigma = \frac{N}{b/\cos i} = \frac{\gamma b H \cos i}{b/\cos i}$$

$$\sigma = \gamma H \cos^2 i \tag{a}$$

and

$$\tau = \frac{S}{b/\cos i} = \frac{\gamma b H \sin i}{b/\cos i}$$

$$\tau = \gamma H \sin i \cos i \tag{b}$$

The effective unit resistance developed by the soil could be expressed by Eq. (c), where the parameters c_d and φ_d are the developed cohesion and friction angle, respectively. They are equal to or less than the effective cohesion and effective friction angles $\bar{c}$ and $\bar{\varphi}$, respectively. Thus,

$$s = c_d + \sigma \tan \varphi_d \tag{c}$$

Equating Eqs. (b) and (c), one obtains the *critical* depth for the clay stratum as expressed by

$$H_c = \frac{c_d}{\gamma} \left(\frac{\sec^2 i}{\tan i - \tan \varphi_d} \right) \tag{11-1}$$

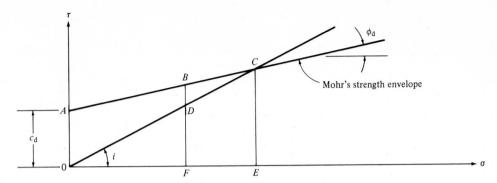

FIGURE 11-2
Mohr's envelope *ABC* representing limiting resistance conditions developed.

Let us assume that line *ABC* in Fig. 11-2 represents limiting resistance conditions (Mohr strength envelope) which may be developed within a soil mass on a sliding plane. Thus for a normal stress $0F$, the shear strength which can be developed in the soil is FB; this is larger than the shear stress FD, potentially developed on the inclined plane. Hence no sliding occurs under such stress conditions. Impending sliding would occur, however, when the normal stress is $0E$ and the corresponding shear stress is CE; obviously the actuating stress at that point equals the shear strength of the soil. Correspondingly depth H, for which the shear stress on the slip plane equals the shear strength of the soil, is commonly referred to as the *critical depth*. In other words, at this depth we have impending shear failure. At any depth greater than the critical depth sliding would theoretically occur. The safety factor F can be obtained from Eqs. (b) and (c), i.e., $F = s/\tau$, or

$$F = \frac{c}{\gamma H \sin i \cos i} + \frac{\tan \varphi}{\tan i} \tag{11-2}$$

Cohesive Soils—Seepage and Water Table at Surface

Figure 11-3 depicts a case where the soil is assumed cohesive, and where seepage coupled with a water table at the surface of the ground is assumed. The pore pressure at a depth H equals $\gamma_w H \cos^2 i$. The effective pressure is $(\gamma - \gamma_w)H \cos^2 i$. Thus the corresponding normal and shear stresses are given by Eqs. (d) and (e), respectively:

$$\bar{\sigma} = \gamma_b H \cos^2 i \tag{d}$$

$$\tau = \gamma H \sin i \cos i \tag{e}$$

The developed shear strength can be written as

$$s = c_d + \gamma_b H \cos^2 i \tan \varphi_d \tag{f}$$

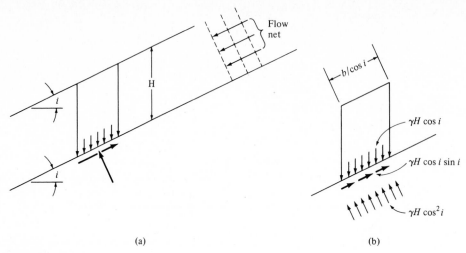

(a)   (b)

FIGURE 11-3

Slope consisting of cohesive soil with seepage, water table at surface. (b) Normal and shear stresses on inclined plane.

Thus from Eqs. (e) and (f) we have

$$c_d + \gamma_b H \cos^2 i \tan \varphi_d = \gamma H \sin i \cos i \left(\frac{\cos i}{\cos i} \right) \qquad \text{(g)}$$

or

$$H_c = \frac{c_d \sec^2 i}{\gamma \tan i - \gamma_b \tan \varphi_d} \qquad \text{(11-3)}$$

The safety factor is

$$F = \frac{c_d}{\gamma H \sin i \cos i} + \left(\frac{\gamma_b}{\gamma} \right) \frac{\tan \varphi_d}{\tan i} \qquad \text{(11-4)}$$

Cohesionless Soils—Seepage and Water Table at Surface

Equation (g) may be utilized with the value of $c_d = 0$. Hence, we obtain the critical slope i for this condition, as expressed by Eq. (11-5):

$$\gamma_b H \cos^2 i \tan \varphi_d - \gamma H \tan i \cos^2 i = 0$$

$$H \cos^2 i (\gamma_b \tan \varphi_d - \gamma \tan i) = 0$$

or

$$\gamma_b \tan \varphi_d - \gamma \tan i = 0$$

from which

$$i = \tan^{-1}\left(\frac{\gamma_b}{\gamma}\right) \tan \varphi_d \qquad (11\text{-}5)$$

(Usually γ_b is about $\frac{1}{2}\gamma$, thus $i \approx \varphi_d/2$—see Example 11-2.) The factor of safety is

$$F = \left(\frac{\gamma_b}{\gamma}\right)\left(\frac{\tan \varphi_d}{\tan i}\right) \qquad (11\text{-}6)$$

Cohesionless Soil—Dry Conditions

For this case Eqs. (a) and (b) apply; Eq. (c) would be altered to reflect a value of $c_d = 0$. Thus equating the expression for shear stress to that for shear strength, the summation of forces parallel to the slope gives Eq. (11-7):

$$\gamma H \sin i \cos i - \gamma H \cos^2 i \tan \varphi_d = 0$$

or

$$\tan i \cos^2 i - \tan \varphi_d \cos^2 i = 0$$

and

$$i = \varphi_d \qquad (11\text{-}7)$$

(As expected, the angle of inclination i is equal to φ_d, the maximum angle that could be developed by the granular dry soil.)

The factor of safety is

$$F = \frac{\tan \varphi_d}{\tan i} \qquad (11\text{-}8)$$

EXAMPLE 11-1

Given   A relatively cohesive soil at a constant infinite slope (Fig. 11-4). Assume negligible seepage and negligible pore pressure. $\gamma = 18.0 \text{ kN/m}^3$; $c = 36 \text{ kN/m}^2$; $\varphi = 14°$; $H = 3 \text{ m}$; $i = 22°$.

Find   (a) The maximum shear stress τ developed.
(b) The maximum shear strength, assuming $c_d = 36 \text{ kN/m}^2$ and $\varphi_d = 14°$.
(c) The critical height H_c for $c_d = c$; $\varphi_d = \varphi$.
(d) The factor of safety with respect to cohesion.
(e) The factor of safety against sliding.

Procedure   (a) From Eq. (b),

$$\tau = \gamma H \sin i \cos i$$

$$\tau = 18.3 \times (\sin 22°)(\cos 22°)$$

Answer

$$\tau = 18.76 \text{ kN/m}^2 \ (0.39 \text{ kips/ft}^2)$$

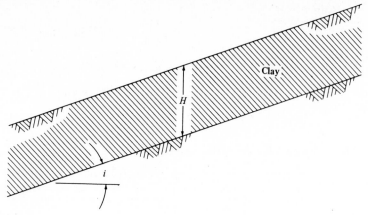

FIGURE 11-4

(b)   From Eq. (c),

$$s = c_d + \sigma \tan \varphi_d = c_d + \gamma H \cos^2 i \tan \varphi_d$$

$$s = 36 + 18.3 \times (\cos^2 22°)(\tan 14°)$$

Answer

$$s = 47.57 \text{ kN/m}^2 \ (0.99 \text{ kips/ft}^2)$$

(c)   From Eq. (11-1),

$$H_c = \frac{c_d}{\gamma} \left(\frac{\sec^2 i}{\tan i - \tan \varphi_d} \right)$$

$$H_c = \frac{36}{18} \left(\frac{\sec^2 22°}{\tan 22° - \tan 14°} \right)$$

Answer

$$H_c = 14.68 \text{ m} \ (47.5 \text{ ft})$$

(d) $c_d = \gamma H \cos^2 i(\tan i - \tan \varphi_d)$

Assuming $\varphi_d = 14°$, the maximum *developed* cohesion is

$$c_d = 18.3 \times (\cos^2 22°)(\tan 22° - \tan 14°) = 7.18 \text{ kN/m}^2$$

$$F = \frac{c}{c_d} = \frac{36}{7.18}$$

Answer

$$F = 5.0$$

(e) From Eq. (11-2),

$$F = \frac{36}{(18)(3)(\sin 22°)(\cos 22°)} + \frac{\tan 14°}{\tan 22°}$$

Answer

$$F = 2.63$$

EXAMPLE 11-2

Given An infinite slope; seepage and water table at surface; and $\gamma = 18 \text{ kN/m}^3$; $\varphi_d = 30°$.

Find The critical-slope angle.

Procedure From Eq. (11-5),

$$i = \tan^{-1}\left(\frac{\gamma_b}{\gamma}\right)\tan \varphi_d$$

$\gamma_b = 18 - 9.82 = 8.18 \text{ kN/m}^3$. Assuming $\varphi_d = \varphi$,

$$i = \tan^{-1}\left(\frac{8.18}{18}\right)\tan 30°$$

Answer

$$i = 14.7°$$

11-3 THE CIRCULAR ARC ANALYSIS

The plane (straight-line) infinite slope discussed in Section 11-2 is not the one usually associated with most slope failures, and in particular for slopes of finite extent. Instead the more typical failure surface is curved, somewhat approximating a circular arc. Generally these surfaces are a mixture of circular arcs and spirals, with the arcs somewhat flat at the ends and sharper in the center. Needless to say, due to the large variation in the soil properties and slope characteristics, the description for such failure surfaces is a general one at best. The actual shape may perhaps be best viewed as an individual feature of every slope failure.

Largely due to the work done by K. B. Petterson in connection with the slope failure at Stigberg Quay in the harbor of Gothenburg, Sweden, in 1916 (45, 46) and by other Swedish engineers, the analysis of slopes based on a circular-arc failure surface became widely used. Their investigations (via borings, etc.) indicated that, in general, the ruptured surface is fairly circular in shape, particularly for a rather homogeneous and isotropic soil condition. Some significant deviation may occur if discontinuities exist in the strata, such as perhaps plane boundaries separating two distinctly different materials (e.g., clay and shale, weak planes).

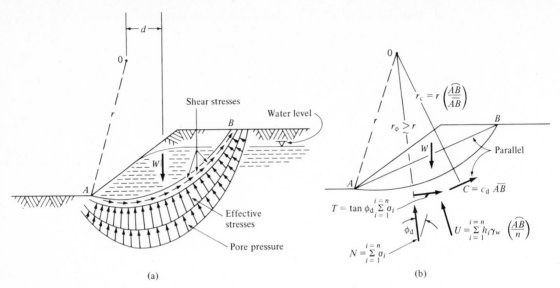

FIGURE 11-5

Forces acting on mass circular failure surface. (a) Stress distribution on failure arc. (b) Resultant forces acting on failure wedge.

Figure 11-5 shows a homogeneous soil mass with an assumed circular failure surface. Generally the forces acting on the mass are induced by (1) gravity (i.e., the weight of the sliding mass); (2) effective stress; (3) shear stress (cohesion and friction); and (4) pore pressure. Figure 11-5a depicts all these conditions. The corresponding resultants for these respective forces are shown in Fig. 11-5b.

The magnitude and orientation of the forces shown in Fig. 11-5 are, to some extent, a matter of conjecture and reasonable assumptions. Generally the pore pressure at any point along the arc may be estimated via the flow net construction described in Chapter 7. Thus we may divide the arc length AB into n equal lengths and compute the average pressure head for each length, as schematically depicted by the pore-pressure diagram in Fig. 11-5a. The average force per unit thickness for any segment i becomes $h_i(\widehat{AB}/n)\gamma_w$. The vector summation of these forces gives the pore-pressure resultant U; it acts through the center of the circle.

The developed cohesion c_d along the arc is usually assumed uniform over the entire length. Thus the total cohesive force is $c_d\widehat{AB}$. We may represent the resultant cohesion force C parallel to the chord $\overline{AB}$ by summing moments about 0 of the cohesion forces, i.e., $c_d\widehat{AB}\,r = c_d\overline{AB}\,r_c$. From this, $r_c = r\,(\widehat{AB}/\overline{AB})$. The resultant cohesion force is $C = c_d\overline{AB}$.

The shear forces along the failure plane are comprised of cohesion and friction components $C + T$ (that is, $s = c_d + \bar{\sigma}\tan\varphi_d$). As mentioned above, the distribution of c_d is usually assumed uniform over the entire arc length. However, the more widely accepted effective stress distribution is zero at the ends of the arc and sinusoidal between. Hence neither the shear nor the normal components of the effective stress is uniform over the length of the arc. Correspondingly, the shear stress varies from c_d at the ends to $(c_d + \bar{\sigma} \tan \varphi_d)$ between.

For homogeneous soils the weight W of the wedge is usually assumed to be the product of the unit soil weight and the wedge volume. Its line of action is, thus, through the centroid of the wedge's cross section. The centroid may be estimated by cutting a piece of cardboard to the shape of the wedge and balancing the cutup on a pin or sharp point.

For an assumed center of failure arc, the actuating or overturning moment is $M_0 = Wd$ (see Fig. 11-5). Thus for a shear strength $S = c_d \widehat{AB} + \tan \varphi_d \sum \bar{\sigma}$, the resisting moment is $M_r = rS$. The pore pressure on the arc does not contribute to the moment since the resultant U passes through the circle center 0. The corresponding safety factor is $F = M_r/M_0$, or

$$F = \frac{r[c_d \widehat{AB} + \tan \varphi_d \sum_{i=1}^{i=n} \bar{\sigma}_i]}{Wd} \tag{11-9}$$

where c_d = average cohesion developed per unit length of arc
s = shear strength per unit length of arc
$S = \sum_{i=1}^{i=n} s(\widehat{AB}/n)$
$\widehat{AB}$ = arc length
r = radius of circle
φ_d = average friction angle developed
$\bar{\sigma}_i$ = effective stress at a point i
W = total weight of wedge
d = moment arm measured from center of rotation to vertical passing through centroid of mass
F = factor of safety

(Refer to Fig. 11-5.)

In Eq. (11-9) the term $\bar{\sigma}_i$ varies over the arc and is not symmetrical. Its value is influenced mainly by the weight of the soil above a point i. This approach leads to the method of slices, discussed in the following section.

11-4 METHOD OF SLICES

The method described in the previous section yields rather inaccurate results if applied to stratified or nonuniform soil properties (for example, c, γ, φ, are not uniform). By subdividing the circular segments into convenient slices, one may be able to account at least partially for the variations in the soil characteristics. Such is the method to be presented here. Pioneered by Swedish engineers, and particularly via the efforts of Fellenius (19) and Petterson (46), the *method of slices* appears to be a more general and perhaps a somewhat more accurate method of stability analysis. It is more applicable in the case of significant variations in soil properties and water conditions.

Ordinary Method of Slices

Figure 11-6 shows a cross section encompassed by a trial failure arc subdivided into a number of vertical sections or *slices*. It is frequently convenient, although not necessary, to divide the cross section into equally wide slices. Also, the thickness of a slice is assumed unity. The stresses acting on a typical slice are the weight of the slice, the normal and shear stresses, the pore pressures, and the shear and normal effective stresses on the interface of the slice induced by deformation. These are shown in Fig. 11-6b. The latter two are rather indeterminate and, for the sake of simplifying

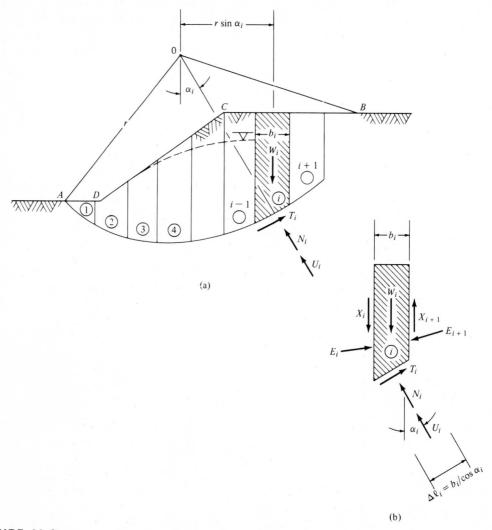

(a)

(b)

FIGURE 11-6
Method of slices. (a) Sliding wedge divided into slices. (b) Forces acting on slice i.

the calculation efforts, are frequently neglected in the analysis. If the geometry of the mass is not greatly changed by deformation, the error introduced in the analysis when the stresses on the vertical faces are neglected is frequently reasonably small, although the assumption does not satisfy statics.

For any given slice the weight could be determined as the product of the cross section of the slice times the unit weight times the unit thickness: $W_i = (A_i)\gamma_i$ (1). It is assumed to act through the midpoint of the area, as shown in Fig. 11-6b. From this figure,

$$N_i = W_i \cos \alpha_i$$

$$\overline{N}_i = W_i \cos \alpha_i - U_i = W_i \cos \alpha_i - u\Delta l_i$$

$$T_i = W_i \sin \alpha_i$$

$$S_i = \overline{N}_i \tan \varphi_d + c_d \Delta l_i = (W_i \cos \alpha_i - U_i) \tan \varphi_d + c_d \Delta l_i$$

The safety factor F is

$$F = \frac{\sum \text{resisting moments}}{\sum \text{overturning moments}} = \frac{r \sum_{i=1}^{i=n} S_i}{r \sum_{i=1}^{i=0} W_i \sin \alpha_i}$$

or, from Fig. 11-6a, in terms of effective stress parameters $\bar{c}$ and $\bar{\varphi}$,

$$F = \frac{\sum_{i=1}^{i=n} [\bar{c} \Delta l + (W_i \cos \alpha_i - U_i) \tan \bar{\varphi}]}{\sum_{i=1}^{i=n} W_i \sin \alpha_i} \qquad (11\text{-}10)$$

Bishop's Simplified Method of Slices

As mentioned previously, the side forces were neglected in the derivation of Eq. (11-10). This violates the equilibrium requirement with respect to translation. Thus the safety factor based on this equation is somewhat in error. Yet to include these forces complicates the approach quite appreciably. Bishop (3) suggested an alternate approach to determine $\overline{N}_i$. He assumed the resultant in the vertical direction of the forces acting on the sides of the slice to be zero. Thus referring to Fig. 11-6b,

$$\overline{N}_i = W_i \cos \alpha_i - U_i = W_i \cos \alpha_i - u_i \left(\frac{b_i}{\cos \alpha_i}\right) \qquad (a)$$

$$T_i = W_i \sin \alpha_i = \frac{\overline{N}_i \tan \bar{\varphi} + \bar{c} \Delta l_i}{F} \qquad (b)$$

From $\sum F_y = 0$ (the summation of the forces in the y direction), we have

$$(U_i + N_i) \cos \alpha_i - W_i + T_i \sin \alpha_i = 0 \qquad (c)$$

Substituting the expression for T_i from Eq. (b) into Eq. (c) and dividing by $\cos \alpha_i$ and rearranging, we have

$$\overline{N}_i + N_i \frac{\tan \alpha_i \tan \bar{\varphi}}{F} = \frac{W_i}{\cos \alpha_i} - U_i - \frac{\bar{c} \Delta l_i}{F} \tan \alpha_i \qquad (d)$$

But $\Delta l_i = b_i/\cos \alpha_i$. Also, $U_i = u_i \Delta l_i = u_i(b_i/\cos \alpha_i)$. Hence substituting in Eq. (d) and solving for $\bar{N}_i$, we have

$$\bar{N}_i = \frac{W_i - u_i b_i - (\bar{c}b_i/F) \tan \alpha_i}{\cos \alpha_i(1 + \tan \alpha_i \tan \bar{\varphi}/F)} \tag{e}$$

From Eq. (a) $\bar{N}_i = (W_i \cos \alpha_i - U_i)$. This is one of the terms in Eq. (11-10). Hence substituting the expression for $\bar{N}_i$ from Eq. (e) into Eq. (11-10), we have

$$F = \frac{\sum_{i=1}^{i=n}\left\{ \dfrac{\bar{c}b_i}{\cos \alpha_i} + \left[\dfrac{W_i - u_i b_i - \bar{c}b_i \tan \alpha_i/F}{\cos \alpha_i(1 + \tan \alpha_i \tan \bar{\varphi}/F)} \right] \tan \bar{\varphi}\right\}}{\sum_{i=1}^{i=n} W_i \sin \alpha_i} \tag{f}$$

Simplifying,

$$F = \frac{\sum_{i=1}^{i=n} [\bar{c}\, b_i + (W_i - u_i b_i) \tan \bar{\varphi}][\sec \alpha_i/(1 + \tan \alpha_i \tan \bar{\varphi}/F)]}{\sum_{i=1}^{i=n} W_i \sin \alpha_i} \tag{11-11}$$

Although Eq. (11-11) is regarded as more compliant to equilibrium requirements than Eq. (11-10), its solution requires an iterative analysis since F appears on both sides of the equation. It lends itself well to a computer analysis, although hand calculations are not out of the question since convergence is rapid. The trial-and-error approach consists of assuming an F on the right-hand side of the equation, then solving for F on the left. If the difference between the assumed F and the computed F is significant, a new F is assumed and the procedure repeated until a satisfactory F is determined. Again, computer programs are generally preferable in this regard (26, 73). If $\bar{\varphi} = 0$, then Eq. (11-11) may be solved directly for F.

Semigraphical Approximation

The method consists of plotting the slope to scale. A trial circle is assumed, and subsequently the wedge is subdivided into convenient slices. The respective weights W_i are calculated and assumed to act through the midpoints of the respective slices. Angles α_i can be measured, and thus normal and tangential components can be calculated. Clockwise and counterclockwise rotational effects of the normal and tangential components are inherent in the rotational equilibrium requirement. Thus,

$$F = \frac{\sum \text{resisting moments}}{\sum \text{overturning moments}}$$

or

$$F = \frac{\sum_{i=1}^{i=n} (\bar{c}\, \Delta l + W_i \cos \alpha_i \tan \bar{\varphi} + T_R)}{\sum_{i=1}^{i=n} T_0} \tag{11-12}$$

where T_R and T_0 represent the tangential components of force *resisting* and *overtiming*, respectively. Example 11-4 may provide some further explanation in this regard.

There are many possible rupture surfaces, and the location of the most critical one (i.e., the surface that results in the least safety factor) is usually determined by trial and error. Generally the failure surfaces for relatively flat slopes (e.g., 1 : 4, 1 : 5, or more) may be rather deep into the base of the slope and somewhat past the toe. For steep (say 1 : 1) and moderately steep (say 1 : 2 or 1 : 3) slopes the circle generally passes through the toe, has a smaller radius and a shorter arc length.

It is usually sufficient to draw three or four arbitrary failure surfaces and subsequently evaluate the safety factor for each. After a few trials a trend would become apparent, thereby pointing toward the probable location of the failure surface near the lowest safety factor. For example, in Fig. 11-7 a number of circles are assumed, and the corresponding safety factors are plotted to a convenient scale at the respective centers of rotation, designated by 0_1, 0_2, and 0_3, for the respective circles 1, 2, and 3 shown. A smooth curve is drawn for the plotted safety factors. The lowest portion of the curve represents the smallest safety factor and provides the center of the circle for the critical failure arc. The corresponding slip surface is shown by the dashed line.

As mentioned above, not all circles pass through the toe; some dip deeper into the base, with failures past the toe. However, for the moderate and steep slope conditions, and relatively cohesive soils, failures through the toe are rather common. As a rough approximation, values of $\beta_A = 25\text{--}30°$ and $\beta_B = 35\text{--}36°$, measured as indicated in Fig. 11-7, provide a starting point for the first center of rotation. Two or three additional points, say equally spaced, for additional trial circles and a subsequent determination of the critical failure surface may be selected as shown in Fig. 11-7.

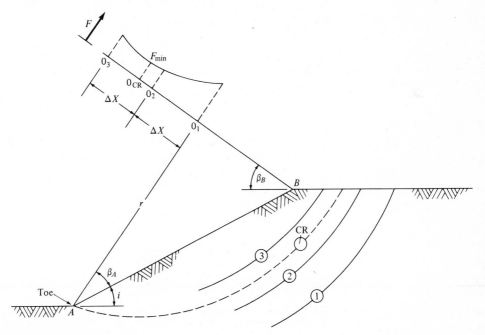

FIGURE 11-7
Trial circles for determining critical failure surface.

In general some guidelines may prove helpful in locating the trial circles for some well-defined and uniform soil properties. (1) For cohesionless soils the critical surface has a radius approaching infinity and coincides with the slope surface itself. (2) For purely cohesive soil and $i < 53°$ the rupture surface is deep seated, with a very large radius (theoretically an infinite radius and infinite depth); the center of the circle is located above the midpoint of the slope. For $i > 53°$ the circle generally passes through the toe, with the center of the circle above the midpoint of the slope, unless the slope is exceedingly steep.

EXAMPLE 11-3

Given   The slope and soil properties, as shown in Fig. 11-8.

Find F via the approximate method of slices for the trial circle shown.

Procedure   The slope was plotted to a convenient scale. For the assumed failure circle and the slices selected, the width of each slice is measured as 2.5 m. For each slice the angle α_i is measured, so is the height of each slice. Unit thickness (1 m) is assumed normal to the page.
The calculations for slice 1 are as follows:

$$W_1 = (area)\gamma = bh\gamma = \left(\frac{2.5 \times 3}{2}\right)19.61 = 73.6 \text{ kN}$$

$$\bar{N}_1 = W_1 \cos \alpha_1 = 73.6 \cos 29° = 64$$

$$\alpha_1 = 29° \text{ (measured value)}$$

$$T_{R_1} = 73.6 \sin 29° = 37$$

$$\bar{N}_1 \tan \varphi = 64 \tan 18° = 20.8$$

The results for the entire slope are given in the following table:

Slice	1	2	3	4	5	6	7	8	9	10	11	Σ
W_i	73.6	196	319	406	479	536	571	544	467	353	142	—
$\bar{N}_i$	64	183	311	405	479	528	540	480	370	236	69	—
T_0 ↱	—	—	—	—	≈ 0	93	186	256	284	220	130	1169
T_R ↰	37	70	42	35	—	—	—	—	—	—	—	184
$\bar{N}_i \tan \varphi$	21	59.6	101	132	156	172	175	156	120	77	22.4	1190

$\sum \Delta L = L$	$L = 2\pi r\left(\dfrac{109.5}{360}\right) = 34.4$	34.4
$\sum c \, \Delta l = cL$	$cL = 28 \times 34 = 963.2$	963.2
F	$F = \dfrac{1190 + 963.2 + 184}{1169} = 2.0$	

Note that slices 1–4 effect a counterclockwise rotation; hence they resist sliding. Slice 5 has a negligible tangential component (say 0).

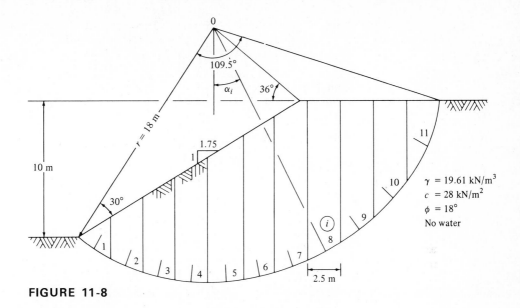

FIGURE 11-8

11-5 STABILITY CHARTS—COUSINS'S APPROACH FOR SIMPLE SLOPES

The method of slices described in the preceding section develops a reasonable basis for estimating slope stability, but the theoretically developed expressions (1) require an iterative procedure for their solution; and (2) do not provide a sufficiently expedient methodology for locating the critical failure surface. Charts and computer programs, however, have helped considerably in this regard (4, 12, 32, 65, 68, 69). Especially expedient are a number of charts for simple slopes developed by Cousins (12); their use will be discussed here.

Figures 11-9 through 11-11 show charts developed by Cousins via a rather extensive computer analysis for specific values of pore-pressure ratios r_u, where $r_u = u/\gamma_w h$. Figures 11-9 and 11-10 relate a slope angle i to a stability number N_F for a number of dimensionless parameters $\lambda_{c\varphi}$. They are defined as $N_F = F\gamma H/\bar{c}$ and $\lambda_{c\varphi} = \gamma H \tan \bar{\varphi}/\bar{c}$. An estimate of the depth factor D (see Fig. 11-12) can also be made directly from the charts for the respective values of i, N_F, and $\lambda_{c\varphi}$.

Figure 11-11 depicts the relationship between slope angle i and coordinates X, Y of the center of the critical slip circle for a number of $\lambda_{c\varphi}$ values and three values of r_u. For convenience the coordinates X and Y are expressed as $(X \tan i)/H$ and $(Y \tan i)/H$. The following examples may enhance the explanation regarding the use of these charts.

Cousins's Stability Charts for Tension Cracks

The preceding discussion ignored the effects of tension cracks. Yet, tension cracks may reduce the safety factor of a slope by perhaps as much as 20 percent, and are

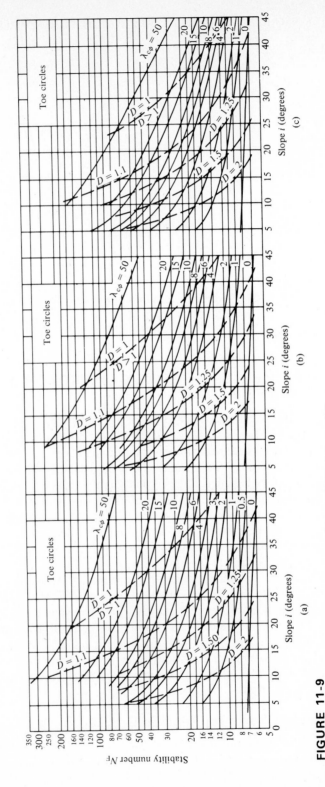

FIGURE 11-9
Stability numbers for toe circles (a) $r_u = 0$. (b) $r_u = 0.25$. (c) $r_u = 0.5$.

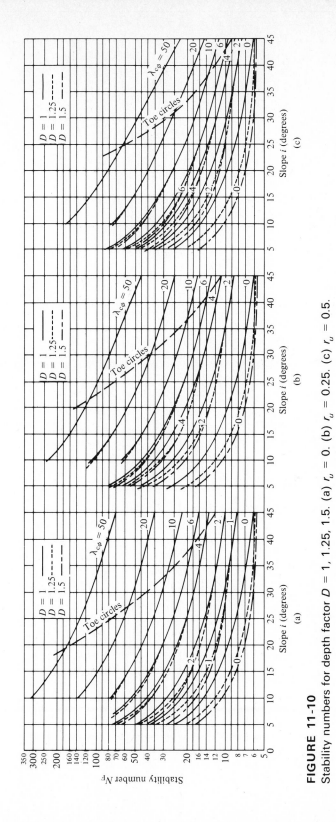

FIGURE 11-10

Stability numbers for depth factor $D = 1$, 1.25, 1.5. (a) $r_u = 0$. (b) $r_u = 0.25$. (c) $r_u = 0.5$.

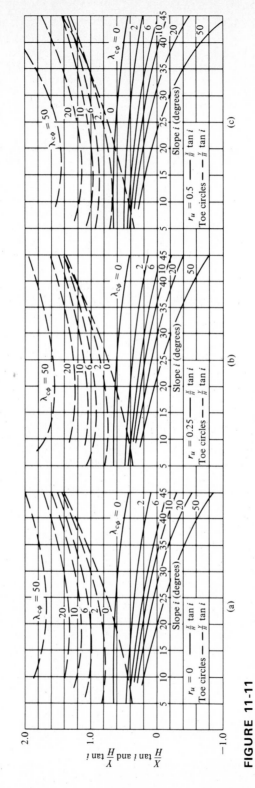

FIGURE 11-11

Coordinates of slip circles for toe circles. (a) $r_u = 0$. (b) $r_u = 0.25$. (c) $r_u = 0.5$.

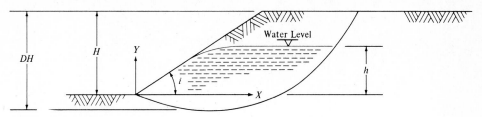

FIGURE 11-12
Notation related to Cousins's charts.

usually regarded as early and important warning signals of impending failure in cohesive soils. Subsequent to the stability charts shown in Figs. 11-9, 11-10, and 11-11, Cousins (13) developed a series of stability charts for toe circles which take into account the effect of tension cracks, both dry and filled with water. These are shown in Fig. 11-13, for pore-pressure ratios $r_u = 0$, 0.25, and 0.5, for a dry tension crack; Fig. 11-14 presents a similar relationship but the tension crack is assumed filled with water, for a case $r_u = 0$.

In developing these charts, Cousins assumed: (1) a totally homogeneous soil and homogeneous pore-pressure ratios, r_u, (2) the depth of the tension cracks was limited to half the height of the slope, and (3) the water filling the tension crack has a unit weight half the bulk unit weight of soil.

Among other observations, Cousins notes that the location of the critical tension crack is quite close to the crest of the slope for the no-water-in-the-tension-crack case and even closer if the tension crack is filled with water. For slope angles $i > 25°$ and $\lambda_{c\varphi} > 3$, the critical tension crack is located at the crest if the crack is filled with water.

EXAMPLE 11-4

Given   A slope as shown in Fig. 11-15. $\bar{\varphi} = 18°$; $\bar{c} = 32$ kN/m²; $\gamma = 19.1$ kN/m²; $\gamma_u = 0.25$.

Find   (a) F. (b) Coordinates X, Y for the center of the critical slip circle.

Procedure   (a) $\bar{c}/\gamma H = 32/19.1 \times 30 = 0.056$

$$\lambda_{c\varphi} = \gamma H \tan \bar{\varphi}/\bar{c} = 19.1 \times 30 \times \tan 18°/32 = 5.82$$

From Fig. 11-9, for $r_u = 0.25$ and $i = 26°$, $N_F = 19$. Thus,

$$F = 19 \times 0.056 = 1.06$$

$$D \approx 1.1$$

Answer
$$F = 1.06$$

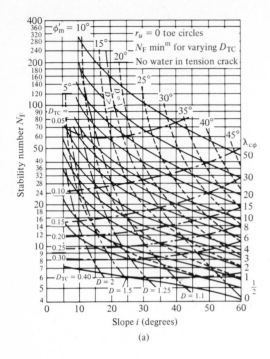

(a)

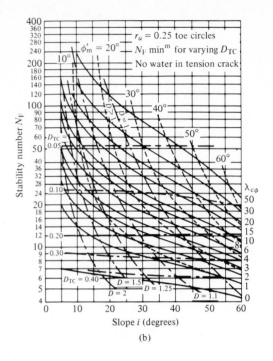

(b)

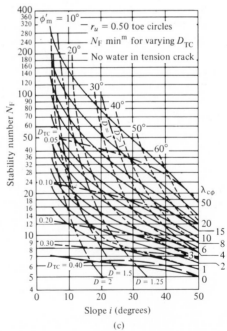

(c)

FIGURE 11-13

Stability charts for no water in tension cracks.

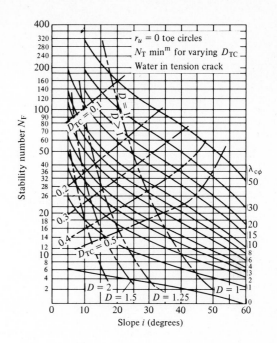

FIGURE 11-14
Stability number N_F for water
in tension crack, $r_u = 0$.

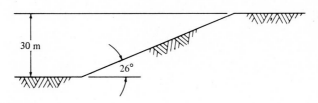

FIGURE 11-15

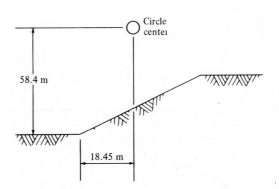

FIGURE 11-16

(b) From Fig. 11-11b,

$$\frac{X}{H}\tan i = 0.3 \qquad \frac{Y}{H}\tan i = 0.95$$

$$\frac{X}{30}\tan 26° = 0.3 \qquad Y = \frac{30 \times 0.95}{0.488}$$

Answer

$$X = 18.45 \text{ m} \qquad Y = 58.40 \text{ m}$$

The center of the critical slip circle is shown in Fig. 11-16.

EXAMPLE 11-5

Given   The data for Example 11-4, except that the slope is unknown.

Find   Slope for a specified $F = 1.4$.

Procedure   From the previously calculated values, $\lambda_{c\varphi} = 5.82$ and $\bar{c}/\gamma H = 0.056$. For $F = 1.4$, $N_F = 1.4/0.056 = 25$. From Fig. 11-9, for $r_u = 0.25$,

Answer

$$i \approx 17°$$

EXAMPLE 11-6

Given   Slope from Example 11-4, but assume $r_u = 0$.

Find F.

Procedure   Assume $r_u = 0$.

$$\bar{c}/\gamma H = 28/19.1 \times 10 = 0.143$$

$$\lambda_{c\varphi} = \gamma H \tan \varphi/\bar{c} = 19.1 \times \tan 18°/28 = 2.28$$

From Fig. 11-9, for $r_u = 0$ and $i = \tan^{-1}(1/1.75) = 29.74°$, $N_F = 13$. Thus,

$$F = 13 \times 0.143 = 1.859$$

$$D \approx 1.1$$

Answer

$$F = 1.859$$

EXAMPLE 11-7

Given   Slope from Example 11-4.

Find F for
(a) No water in tension cracks
(b) Water in tension cracks

Compare smallest F from (a) or (b) with results from Example 11-6.

Procedure   Assume $r_u = 0$. Also, $i = \tan^{-1}(1/1.75) = 29.74°$; $\bar{c}/\gamma H = 28/19.1 \times 10 = 0.143$; $\lambda_{\bar{c}\varphi} = 19.1 \times 10 \times \tan(18°/28) = 2.28$.
(a) From Fig. 11-13 (no-water case, $r_u = 0$), $N_F = 12.5$. Thus,

$$F = 12.5 \times 0.143 = 1.79$$

Answer
$$F = 1.79$$

(b) From Fig. 11-14 (water in tension crack), $N_F = 12.0$. Thus,

$$F = 12 \times 0.143 = 1.72$$

Answer
$$F = 1.72$$

Comparison:

$$\text{Percent difference} = \frac{(1.86 - 1.72)}{1.86} \times 100 = 7.5 \text{ percent less } F$$

Answer
7.5 percent less F

11-6 SLIDING ON INCLINED PLANE

The following method of analysis, proposed by Culmann in 1866, is based on the assumption that failure occurs on a slip plane through the toe of the slope. The general equilibrium conditions are applicable, however, to any wedge-shaped mass sliding on a plane surface that cuts through or above the toe of the slope. Figure 11-17 depicts the typical case analyzed by this method.

Figure 11-17 shows the forces acting on the sliding mass. The expression for the total weight of the wedge, in terms of the dimensions shown in Fig. 11-18a, is

$$W = \tfrac{1}{2}hL\gamma \tag{a}$$

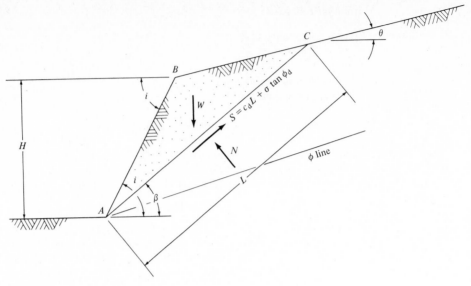

FIGURE 11-17
Culman's slip plane.

From geometry, the expression for h could be determined easily; from Fig. 11-18:

$$AB = \frac{h}{\sin(i - \beta)} = \frac{H}{\sin i}$$

Thus,

$$h = \frac{H \sin(i - \beta)}{\sin i} \qquad \text{(b)}$$

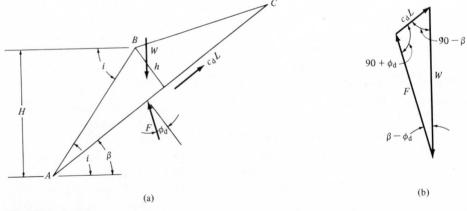

FIGURE 11-18
Forces acting on sliding mass.

Substituting Eq. (b) into Eq. (a), we have

$$W = \tfrac{1}{2}L\gamma H \frac{\sin(i - \beta)}{\sin i} \tag{c}$$

The developed shear strength along plane AC is

$$S = c_d L + W \cos \beta \tan \varphi_d \tag{d}$$

The weight component parallel to the plane AC is $W \sin \beta$. Thus the expression for safety factors becomes

$$F = \frac{c_d + \tfrac{1}{2}\gamma H[\sin(i - \beta)/\sin i] \cos \beta \tan \varphi_d}{\tfrac{1}{2}\gamma H[\sin(i - \beta)/\sin i] \sin \beta} \tag{11-13}$$

Referring to the force polygon shown in Fig. 11-18b and using the *Law of Sines*, we have

$$\frac{c_d L}{\sin(\beta - \varphi_d)} = \frac{W}{\cos \varphi_d} \tag{e}$$

Substituting the expression for W from Eq. (c), we have

$$\frac{c_d L}{\sin(\beta - \varphi_d)} = \frac{\tfrac{1}{2}L\gamma H \sin(i - \beta)}{\sin i \cos \varphi_d}$$

or

$$\frac{c_d}{\gamma H} = \frac{1}{2} \frac{\sin(\beta - \varphi_d) \sin(i - \beta)}{\sin i \cos \varphi_d} \tag{f}$$

where the quantity $c_d/\gamma H$ is known as *stability number*. The condition for impending slip occurs when $c_d/\gamma H$ is a maximum. Thus from Eq. (f) the derivative with respect to β, equated to zero, becomes

$$\frac{d}{d\beta} = \cos(\beta - \varphi_d) \sin(i - \beta) - \sin(\beta - \varphi_d) \cos(i - \beta) = 0$$

from which

$$\frac{\sin(\beta - \varphi_d)}{\cos(\beta - \varphi_d)} = \frac{\sin(i - \beta)}{\cos(i - \beta)}$$

or

$$\tan(\beta - \varphi_d) = \tan(i - \beta)$$

Thus,

$$\beta - \varphi_d = i - \beta$$

Or, solving for β, we have

$$\beta = \tfrac{1}{2}(i + \varphi_d) \tag{11-14}$$

which is the expression for the *critical-slope angle*.

The method gives good results for very steep or vertical slopes; it does not provide satisfactory results for relatively flat slopes.

EXAMPLE 11-8

Given   The section shown in Fig. 11-19.

Find   An expression for H in terms of c_d, i, and φ_d for $F = 1$.

Procedure   From Eq. (11-13),

$$F = \frac{c + \tfrac{1}{2}\gamma H\,[\sin(i - 3)/\sin i]\cos\beta\tan\varphi}{\tfrac{1}{2}\gamma H\,[\sin(i - \beta)/\sin i]\sin\beta}$$

But from Eq. (11-14) $\beta = \tfrac{1}{2}(i + \varphi_d)$. Thus for $F = 1$ and $c = c_d$ and $\varphi = \varphi_d$ we have

$$1 = \frac{c_d + \tfrac{1}{2}\gamma H\,[\sin\tfrac{1}{2}(i - \varphi_d)/\sin i]\cos\tfrac{1}{2}(i + \varphi_d)\tan\varphi_d}{\tfrac{1}{2}\gamma H\,[\sin\tfrac{1}{2}(i - \varphi_d)/\sin i]\sin\tfrac{1}{2}(i + \varphi_d)}$$

or

$$c_d = \tfrac{1}{2}\gamma H\,\frac{\sin\tfrac{1}{2}(i - \varphi_d)}{\sin i}\,[\sin\tfrac{1}{2}(i + \varphi_d) - \cos\tfrac{1}{2}(i + \varphi_d)\tan\varphi_d]$$

and

Answer

$$H = \frac{2c_d\sin i}{\sin\tfrac{1}{2}(i - \varphi_d)[\sin\tfrac{1}{2}(i + \varphi_d) - \cos\tfrac{1}{2}(i + \varphi_d)\tan\varphi_d]}$$

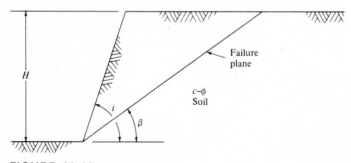

FIGURE 11-19

11-7 SPECIAL TOPICS ON STABILITY

It is rather apparent that the many variables and conditions which may be associated with slope stability are too numerous to be covered within the scope of this text. However, some general guidelines and reflections on various phenomena relevant to some special cases may be worthwhile indeed.

Liquefaction

Slope slides in dry sands do not occur for slope angles equal to or less than the angle of repose. They may occur in loose, saturated sands, or for angles of inclination exceeding the natural angle of repose of the sand. On the other hand, even saturated sands in a dense or reasonably dense state may prove somewhat stable if the angle of inclination is less than the angle of repose, provided the slope is not subjected to excessive disturbances, including *liquefaction* (9, 20, 39, 47). For example, shock waves produced by earthquakes, vibrations, and/or blasting, or a rapid fluctuation of the water table may trigger a condition of liquefaction and subsequent mass slide.

Increasing the density of a sand or stabilizing (see Chapter 16) the mass may be remedies to potential sliding. The increased density may be obtained by vibrating the sand via mechanical tampers, pile driving, etc. Paradoxically the very methods for improving the density may indeed be the cause of such a slide. Thus using perhaps cement or chemical grouting may be an economically viable alternative worth considering in this respect.

Seismic Effects

Earthquakes may induce displacement and subsequent failure of slopes or earth embankments which may be regarded as safe under ordinary conditions (34). Such effects may result in *liquefaction* of the soil, or in a sudden change in pore-pressure and effective stresses in the soil mass, or perhaps in the development of a vertical crack which may cause a reduction in the shear strength and eventual slope failure. Furthermore, the effects of earthquakes may be detrimental to slopes of cohesive as well as cohesionless material (20).

Past practices have employed the use of numerical constants, generally referred to as *seismic coefficients* k_g to account for the effects of gravity. These coefficients were given in percent of gravity. For example, coefficients of 10 percent of the gravitational force (0.1 g) were common. Thus the dynamic forces were treated as static forces. Sometimes referred to as *pseudostatic* analysis, this method appears to be highly empirical and without adequate basis or justification for the selection of the coefficient of 0.1 g. No substantive data exist to suggest that its use gives accurate results or that past slope failures could have been predetermined by its use.

Investigations at the University of California, Berkeley, by Seed (51–60) and others (43, 62, 67) point to deficiencies in the overall analysis when coefficients such as those mentioned above are used without due regard to deformations as well as material makeup and slope geometry. For example, fine loose sands may be subject to liquefaction, while coarser dense sands may not be. The latter appear to be subject to displacements generally concentrated in thin shear zones near the surface of the slope. Also, the effective cohesion and angle of friction in the more cohesive soils may be significantly different under repetitions of seismic loads than those found from general routine laboratory tests. Indeed it is perhaps oversimplistic and dangerous to generalize a very complex physical relationship by the use of some highly empirical and loosely based coefficients in slope-stability analysis. It is perhaps best that one confronted with such problems, cognizant of the limitations of such analyses, research the problem from various points of view before resting with a final answer.

Drawdown

Figure 11-20 depicts a slope subjected to a lowering of the water table at the face of the slope. That is, let us assume that the slope was submerged under water for some time and then subjected to a lowering of the water table, a case known as *drawdown*. Under the initial conditions, total saturation may be assumed, and therefore the slope is not subjected to any excess hydrostatic pore-water pressure. If the drawdown is sudden, the slope remains saturated, at least for a while. In turn, this condition induces changes in hydrostatic pressures on the ground surface and on the failure surface. That is, the removal of the water on the slope face reduces the hydrostatic pressure proportionately to the amount of drawdown, and subsequently thereby decreases the resistance to sliding. In turn, this decreases the total stress on the failure surface, thereby necessitating the mobilization of some additional shear resistance along the surface of failure in order to compensate for a decreasing resisting moment. If this is not available, the stability of the slope may be impaired (38, 41).

As we have seen previously, the shear resistance comprises the effects of both cohesion and friction. Quite apparently a change in the normal forces on the failure surface affects the frictional resistance. That is, a negative change in the pore-water pressure increases the effective normal component and therefore increases the available friction in a c-φ soil. Conversely, an increase in the pore-water pressure decreases the stability by a corresponding decrease in the normal component and

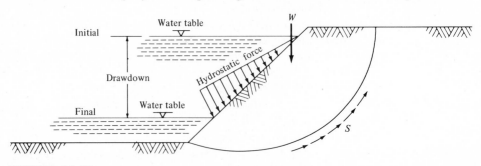

FIGURE 11-20
Slope subjected to drawdown.

subsequent effect on the frictional force. Furthermore one notes that the weight of the sliding mass under total saturation is, at least temporarily, somewhat larger than that for a correspondingly dry mass, thereby further reducing the factor of safety of the slope.

PROBLEMS

11-1. A slope of infinite extent consists of a dry sand which weighs γ kN/m^3. It makes an angle i with the horizontal. At a given point at depth H the normal and shear stresses are 63.6 and 21.7 kN/m^2, respectively. The angle of internal friction is 33°. Determine:
 (a) The maximum shear strength the soil can develop at the given point.
 (b) The safety factor against shear failure.

11-2. A slope of infinite extent consists of a dry clayey silt with $c = 24$ kN/m^2 and $\varphi = 17°$. At a given point along the critical plane the normal and shear stresses are found to be 95.4 and 32.5 kN/m^2, respectively. Determine:
 (a) The maximum shear strength the soil can develop.
 (b) The safety factor against shear failure.
 (c) The critical depth H_c if $\gamma = 18.2$ kN/m^3 and $i = 20°$

11-3. What would be the critical height H_c in Problem 11-2 ($i = 20°$ and $\gamma = 18.2$ kN/m^3) under a condition of seepage and water table at the surface of the slope?

11-4. An infinite slope consisting of sand and having a condition of seepage and water table at the surface is inclined at an angle of 16°. Assuming $\gamma = 18.3$ kN/m^3, determine the value for φ for impending failure.

11-5. An infinite slope constructed of sand is subjected to seepage, with the water at ground surface. If $\varphi = 32°$, and $F = 1.4$ (based on strength) is to be attained, determine the maximum angle of inclination of the slope.

11-6. An infinite slope consists of a uniform layer of silty clay 4 m thick and has an angle of inclination of 23°. A shale ledge runs parallel to the surface (at 4 m below). Assume the following silty clay properties: $\varphi = 16°$, $c = 26$ kN/m^2, and $\gamma = 18.6$ kN/m^3. Determine the safety factor against sliding (shear failure) on the shale ledge:
 (a) If no water exists at the top of the shale.
 (b) If the water level is at the surface of the slope.

11-7. For the slope shown in Fig. P11-7, assume the following data: $W = 3600$ kN, $r = 15$ m, $d = 3.3$ m, $c = 32$ kN/m^2, and $\varphi = 0$. Determine the safety factor against sliding on the circular surface shown.

11-8. What would the developed cohesion have to be for the slope in Problem 11-7 if the safety factor were 2.0?

11-9. For the data of Problem 11-7 and $h = 9$ m, determine the safety factor against sliding if a seismic force of $0.15W$, acting horizontally, is applied at the center of gravity shown.

11-10. For the slope shown in Fig. P11-10 determine the safety factor against sliding by using the method of slices if $\beta_A = 27°$. Divide the wedge into slices approximately $r/10$ wide.

11-11. Rework Problem 11-10 by the "approximate method" of slices. State any assumptions.

11-12. Rework Problem 11-10 via Bishop's method.

11-13. Rework Problem 11-10 by using Cousins's stability charts.

11-14. For the slope in Fig. P11-14 determine the safety factor against sliding by the ordinary method of slices, $\beta_A = 26°$.

11-15. Rework Problem 11-14 by the "approximate method." State any assumptions.

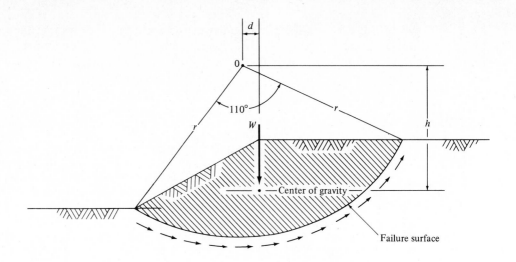

FIGURE P11-7

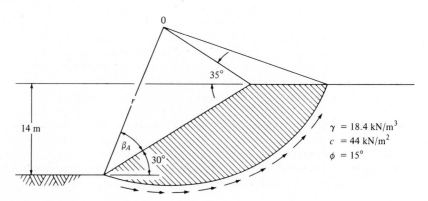

FIGURE P11-10

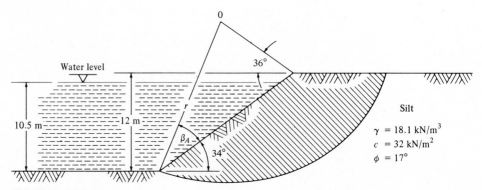

FIGURE P11-14

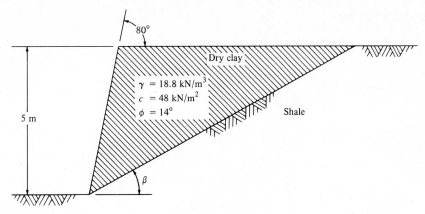

FIGURE P11-21

11-16. Rework Problem 11-14 via Bishop's method.

11-17. Rework Problem 11-14 by using Cousins's stability charts.

11-18. Determine the safety factor against sliding for the slope in Fig. P11-14 by the ordinary method of slices if $\beta_A = 26°$ and the water level drops suddenly from 10.5 to 4 m. State any assumptions and observations.

11-19. Determine the safety factor against sliding for the slope in Fig. P11-14 by the approximate method of slices if $\beta_A = 26°$ and the water level drops suddenly from 10.5 to 0.5 m. State any assumptions and observations.

11-20. Determine the safety factor against sliding for the slope in Fig. P11-14 by the approximate method of slices if $\beta_A = 26°$ and the water level drops suddenly from 10.5 to 0.5 m, and an earthquake force of $0.15W$ is applied horizontally at 7 m from the top. State any assumptions and observations.

11-21. For the cut shown in Fig. P11-21 determine the safety factor on the shale ledge if $\beta_A = 20°$.

11-22. Rework Problem 11-21 if $\beta_A = 25°$.

11-23. Rework Problem 11-21 if $\beta_B = 30°$.

11-24. Determine the maximum depth for the vertical cut shown in Fig. P11-24 at the point of impending failure along a straight line.

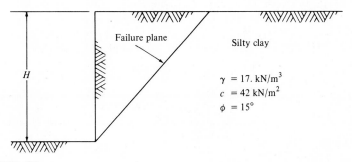

FIGURE P11-24

BIBLIOGRAPHY

1. Bell, J. M., "General Slope Stability Analysis," *ASCE J. Geotech. Eng. Div.*, vol. 94, Nov. 1968.
2. Bell, J. M., "Noncircular Sliding Surfaces," *ASCE J. Geotech. Eng. Div.*, vol. 103, May 1969.
3. Bishop, A. W., "The Use of the Slip Circle in the Stability Analysis of Earth Slopes," *Geotechnique*, vol. 5, 1955.
4. Bishop, A. W., and N. Morgenstern, "Stability Coefficients for Earth Slopes," *Geotechnique*, vol. 10, 1960.
5. Bjerrum, L., "Stability of Natural Slopes in Quick Clay," *Geotechnique*, vol. 5, 1955.
6. Bjerrum, L., and B. Kjaernsli, "Analysis of the Stability of Some Norwegian Natural Clay Slopes," *Geotechnique*, vol. 7, 1957.
7. Bjerrum, L., "Progressive Failure in Slopes of Over-consolidated Plastic Clay and Clay Shales," *Proc. ASCE*, vol. 93, no. SM5 (pt. 1), 1967.
8. Blight, G. W., "Foundation Failures of Four Rockfill Slopes," *ASCE J. Geotech. Eng. Div.*, vol. 95, May 1969.
9. Castro, G., and S. J. Poulos, "Factors Affecting Liquefaction and Cyclic Mobility," *ASCE J. Geotech. Eng. Div.*, vol. 103, June 1977.
10. Catalan, J. M., and C. A. Cornell, "Earth Slope Reliability by a Level-Crossing Method," *ASCE J. Geotech Eng. Div.*, vol. 102, June 1976.
11. Coulomb, C. A., "Essai sur une application des règles de maximia et minimis à quelques problèmes de statique relatifs à l'architecture," *Memoires de la Mathématique et de Physique, presentés a l'Academic Royale des Sciences, par divers savants, et lus dans ces Assemblées*, vol. 7, année 1773, L'Imprimerie Royale, Paris, 1766.
12. Cousins, B. F., "Stability Charts for Simple Earth Slopes," *ASCE J. Geotech. Eng. Div.*, vol. 104, Feb. 1978.
13. Cousins, B. F., "Stability Charts for Simple Earth Slopes Allowing for Tension Cracks," *3rd Australian -N2 Conf. Geomechanics*, Wellington, Australia, vol. 2, May 1980.
14. Crawford, C. B., and W. J. Eden, "Stability of Natural Slopes in Sensitive Clay," *ASCE J. Geotech. Eng. Div.*, vol. 93, 1967.
15. Culmann, K., *Die Graphische Static*, Zurich, 1866.
16. Desai, C. S., "Drawdown Analysis of Slopes by Numerical Method," *ASCE J. Geotech. Eng. Div.*, vol. 103, July 1977.
17. Donovan, N. C., and S. Singh, "Liquefaction Criteria for Trans-Alaska Pipeline," *ASCE J. Geotech. Eng. Div.*, vol. 104, Apr. 1978.
18. Duncan, J. M., and A. L. Bushignani, "Failure of Underwater Slope in San Francisco Bay," *ASCE J. Geotech. Eng. Div.*, vol. 99, Sept. 1973.
19. Fellenius, W., "Calculation of the Stability of Earth Dams," *Trans. 2nd Congr. Large Dams*, Washington, D.C., 1936, vol. 4, U.S. Gov. Printing Office, Washington, D.C., 1938.
20. Finn, W. D. L., P. M. Byrne, and G. R. Martin, "Seismic Response and Liquefaction of Sands," *ASCE J. Geotech. Eng. Div.*, vol. 102, Aug. 1976.
21. Frohlich, O. K., "General Theory of Stability of Slopes," *Geotechnique*, vol. 5, 1955.
22. Goodman, L. E., and C. B. Brown, "Dead Load Stresses and the Instability of Slopes," *ASCE J. Soil Mech. Found. Div.*, vol. 89, no. SM3, 1963.
23. Goodman, R. E., and H. B. Seed, "Earthquake-Induced Displacements in Sand Embankments," *ASCE J. Geotech. Eng. Div.*, vol. 92, 1966.

24. Haldar, A., and W. H. Tang, "Probabilistic Evaluation of Liquefaction Potential," *ASCE J. Geotech. Eng. Div.*, vol. 105, Feb. 1979.

25. Henkel, D. J., and A. W. Skempton, "A Landslide at Jackfield, Shropshire, in a Heavily Overconsolidated Clay," *Geotechnique*, vol. 5, 1955.

26. Horn, J. A., "Computer Analysis of Slope Stability," *ASCE J. Geotech. Eng. Div.*, vol. 86, June 1960.

27. Hovland, H. J., "Three-Dimensional Slope Stability Analysis Method," *ASCE J. Geotech. Eng. Div.*, vol. 103, Sept. 1977.

28. Huang, Y. H., and C. M. Avery, "Stability of Slopes by Logarithmic-Spiral Method," *ASCE J. Geotech. Eng. Div.*, vol. 102, Jan. 1976.

29. Huang, Y. H., "Stability Coefficients for Sidehill Benches," *ASCE J. Geotech. Eng. Div.*, vol. 103, May 1977.

30. Hutchinson, J. N., "A Landslide on a Thin Layer of Quick Clay at Furre, Central Norway," *Geotechnique*, vol. 11, 1961.

31. Janbu, N., "Stability Analysis of Slopes with Dimensional Parameters," Harvard Soil Mech. Series no. 46, Harvard University, Cambridge, Mass., Jan. 1954.

32. Janbu, N., Discussion of J. M. Bell, "Dimensionless Parameters of Homogeneous Earth Slopes," *ASCE J. Soil Mech. Found. Div.*, vol. 93, Nov. 1967.

33. John, K. W., "Graphical Stability Analysis of Slopes in Jointed Rock," *ASCE J. Geotech. Eng. Div.*, vol. 94, Mar. 1968.

34. Lee, K. L., and W. Roth, "Seismic Stability Analysis of Hawkins Hydraulic Fill Dam," *ASCE J. Geotech. Eng. Div.*, vol. 103, June 1977.

35. Liou, C. P., V. L. Streeter, and F. E. Richart, "Numerical Model for Liquefaction," *ASCE J. Geotech. Eng. Div.*, vol. 103, June 1977.

36. Lo, K. Y., "Stability of Slopes in Anisotropic Soils," *ASCE J. Soil Mech. Found. Div.*, vol. 91, SM4, July 1965.

37. Londe, P., G. Vigier, and R. Vormeringer, "Stability of Rock Slopes—Graphical Methods," *ASCE J. Geotech. Eng. Div.*, vol. 96, July 1970.

38. Lowe, J., III, and L. Karafiath, "Stability of Earth Dams upon Drawdown," *1st Pan-Am. Conf. Soil Mech. Found. Eng.*, Mexico, vol. II, 1960.

39. Martin, G. R., W. D. L. Finn, and H. B. Seed, "Fundamentals of Liquefaction under Cyclic Loading," *ASCE J. Geotech. Eng. Div.*, vol. 101, May 1975.

40. May, D. R., and J. H. A. Brahtz, "Proposed Methods of Calculating the Stability of Earth Dams," *2nd Congr. Large Dams*, Washington, D.C., vol. 4, 1936.

41. Morgenstern, N., "Stability Charts for Earth Slopes During Rapid Drawdown," *Geotechnique*, vol. 13, 1963.

42. Morgenstern, N. R., and V. E. Price, "The Analysis of the Stability of General Slip Surfaces," *Geotechnique*, vol. 15, 1965.

43. Newmark, N. M., "Effects of Earthquakes on Dams and Embankments," *Geotechnique*, vol. 15, 1964.

44. Peacock, W. H., and H. B. Seed, "Sand Liquefaction under Cyclic Loading—Simple Shear Conditions," *ASCE J. Geotech. Eng. Div.*, vol. 94, May 1968.

45. Petterson, K. E., "Kajraset i Gotenborg den Ste Mars 1916," *Tek. Tidskr.*, Stockholm, 1916.

46. Petterson, K. E., "The Early History of Circular Sliding Surfaces," *Geotechnique*, vol. 5, 1955.

47. Pyke, R. M., L. A. Knuppil, and K. L. Lee, "Liquefaction Potential of Hydraulic Fills," *ASCE J. Geotech. Eng. Div.*, vol. 104, Nov. 1978.

48. Rankine, W. J. M., "On the Stability of Loose Earth," *Trans. R. Soc.*, London, 1857.

49. Renius, E., "The Stability of Slopes of Earth Dams," *Geotechnique*, vol. 5, 1955.

50. Romani, F., C. W. Lovell, and M. E. Harr, "Influence of Progressive Failure Slope Stability," *ASCE J. Geotech. Eng. Div.*, vol. 98, Nov. 1972.

51. Seed, H. B., and R. E. Goodman, "Earthquake Stability of Slopes of Cohesionless Soils," *ASCE J. Geotech. Eng. Div.*, vol. 90, 1964.

52. Seed, H. B., "A Method for the Earthquake-Resistant Design of Earth Dams," *ASCE J. Geotech. Eng. Div.*, vol. 92, no. SM1, 1966.

53. Seed, H. B., and S. D. Wilson, "The Turnagain Heights Landslide, Anchorage, Alaska," *ASCE J. Geotech. Eng. Div.*, vol. 93, no. SM4, 1967.

54. Seed, H. B., and H. A. Sultan, "Stability Analysis for a Sloping Core Embankment," *ASCE J. Geotech. Eng. Div.*, vol. 93, no. SM4, 1967.

55. Seed, H. B., "The Fourth Terzaghi Lecture: Landslides during Earthquakes due to Liquefaction," *ASCE J. Geotech. Eng. Div.*, vol. 94, Sept. 1968.

56. Seed, H. B., and W. H. Peacock, "Test Procedures for Measuring Soil Liquefaction Characteristics," *ASCE J. Geotech. Eng. Div.*, vol. 97, Aug. 1971.

57. Seed, H. B., K. L. Lee, I. M. Idriss, and F. I. Makdisi, "The Slides in the San Fernando Dams during the Earthquake of February 9, 1971," *ASCE J. Geotech. Eng. Div.*, vol. 101, July 1975.

58. Seed, H. B., I. M. Idriss, K. L. Lee, and F. I. Makdisi, "Dynamic Analysis of the Slide in the Lower San Fernando Dam during the Earthquake of February 9, 1971," *ASCE J. Geotech. Eng. Div.*, vol. 101, Sept. 1975.

59. Seed, H. B., K. Mori, and C. K. Chan, "Influence of Seismic History on Liquefaction of Sands," *ASCE J. Geotech. Eng. Div.*, vol. 103, Apr. 1977.

60. Seed, H. B., F. I. Makdisi, and P. DeAlba, "Performance of Earth Dams during Earthquakes," *ASCE J. Geotech. Eng. Div.*, vol. 104, July 1978.

61. Sevaldson, R. A., "The Slide in Lodalen, October 6th, 1954," *Geotechnique*, vol. 6, 1956.

62. Sherard, J. L., "Earthquake Considerations in Earth Dam Design," *ASCE J. Geotech. Eng. Div.*, vol. 93, 1967.

63. Skempton, A. W., and J. D. Brown, "A Landslip in Boulder Clay at Selset, Yorkshire," *Geotechnique*, vol. 11, 1961.

64. Skempton, A. W., "Long-Term Stability of Clay Slopes," *Geotechnique*, vol. 14, 1964.

65. Spencer, E., "A Method of Analysis of the Stability of Embankments Assuming Parallel Inter-Slice Forces," *Geotechnique*, vol. 17, 1967.

66. Spencer, E., "Circular and Logarithmic Spiral Slip Surfaces," *ASCE J. Geotech. Eng. Div.*, vol. 95, Jan. 1969.

67. Spencer, E., "Earth Slopes Subject to Lateral Acceleration," *ASCE J. Geotech. Eng. Div.*, vol. 104, Dec. 1978.

68. Taylor, D. W., "Stability of Earth Slopes," *J. Boston Soc. Civ. Eng.*, vol. 24, 1937.

69. Taylor, D. W., *Fundamentals of Soil Mechanics*, Wiley, New York, 1948.

70. Vanmarcke, E. K., "Reliability of Earth Slopes," *ASCE J. Geotech. Eng. Div.*, vol. 103, Nov. 1977.

71. Veneziano, D., and J. Antoniano, "Reliability Analysis of Slopes, Frequency-Domain Method," *ASCE J. Geotech. Eng. Div.*, vol. 105, 1979.

72. Whitman, R. V., and P. J. Moore, "Thoughts Concerning the Mechanics of Slope Stability Analysis," *2nd Pan-Am. Conf. Soil Mech. Found. Eng.*, Brazil, vol. 1, 1963.

73. Whitman, R. V., and W. A. Bailey, "Use of Computers for Slope Stability Analysis," *ASCE J. Geotech. Eng. Div.*, vol. 93, no. SM4, 1967.

74. Wolfskill, L. A., and T. W. Lambe, "Slide in the Siburua Dam," *ASCE J. Geotech. Eng. Div.*, vol. 93, no. SM4, 1967.

75. Wu, T. H., W. B. Thayer, and S. S. Lin, "Stability of Embankment on Clay," *ASCE J. Geotech. Eng. Div.*, vol. 101, Sept. 1975.

76. Yen, B. C., "Stability of Slopes Undergoing Creep Deformations," *ASCE J. Geotech. Eng. Div.*, vol. 95, July 1969.

77. Zaruba, Q., and V. Menci, *Landslides and their Control*, Elsevier, London, England, 1968, and Prague.

12
Lateral
Earth Pressure

12-1 INTRODUCTION

The problem associated with *lateral earth pressure* and retaining-wall stability is one of the most common in the civil engineering field, and a segment of soil mechanics that has been receiving widespread attention from engineers for a long time. Historical records indicate that efforts to devise procedures and formulate methodologies for the analysis of and designing for the effects of earth pressure date back to over three centuries—and perhaps much longer. It may be interesting to note that many of the theories developed by some of these early investigators still serve as the basis for present-day analysis of retaining walls.

The typical structures whose primary or secondary purpose is to resist earth pressures may include various types of retaining walls, sheet piling, braced sheeting of pits and trenches, bulkheads or abutments, basement or pit walls, etc. These may be self-supporting (e.g., gravity or cantilever-type concrete walls) or they may be laterally supported by means of bracing or anchored ties. The latter are discussed in some detail in Chapter 13.

The lateral earth pressure depends upon several factors (1, 3): (1) the physical properties of the soil; (2) the time-dependent nature of soil strength; (3) the interaction between the soil and the retaining structure at the interface; (4) the general characteristics of the deformation in the soil–structure composite; and (5) the imposed loading (e.g., height of back-fill, surcharge loads).

Two basic types of soil pressures are evaluated in this chapter, active and passive. If the soil mass pushes against a retaining wall such as to push it away, the soil becomes the actuating element and the pressure resulting thereby is known as *active pressure*. On the other hand, if the wall pushes against the soil (e.g., see Fig. 12-1), the resulting pressure is known as *passive pressure*; in this case the actuating element is the retaining wall itself.

Although much research has been performed and appreciable advancement made during the past two centuries regarding the distribution of earth pressures and on the analysis of a wide range of earth-retaining structures, some of the theories formulated by such persons as Coulomb in 1776 and Rankine in 1857 still remain as the fundamental approaches to the analysis of most earth-supporting structures, particularly for sandy soils. Furthermore, while some research data and experience indicate that assumptions related to pressure distributions on retaining walls, or on the failure surface of back-fills, are not quite those depicted by these early investigators, substantial evidence exists that the analysis and design efforts based on their theories give acceptable results for most cases of cohesionless-type back-fills. The results are significantly less dependable for the more cohesive soils.

12-2 ACTIVE AND PASSIVE EARTH PRESSURES

Figure 12-1 shows some of the forces acting on a typical gravity retaining wall. Frictional forces which may be developed on the front and back faces of the retaining wall are not shown. The lateral force induced by the back-fill pushes against the wall with a resultant pressure P_a. In turn, the retaining wall resists the lateral force of the back-fill, thereby retaining its movement. In this case it is readily apparent that the soil becomes the actuating force. The thrust P_a is the resultant of the active pressure, or simply the *active thrust*. The resistance to the active thrust is provided by the frictional force at the bottom of the wall and by the soil in front of the wall. For the sake of illustration, assume that the wall was pushed to the left by the active thrust P_a. In this case, relative to the soil in front of the wall, the wall becomes the actuating force, with the soil in front of the wall providing the passive resistance to movement. This resistance is known as the *passive earth pressure*, with the resultant of this pressure denoted by P_p.

The magnitude of the lateral force varies considerably as the wall undergoes lateral movement resulting in either tilting or lateral translation, or both. This

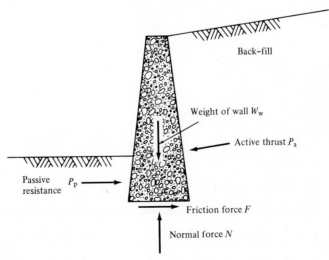

FIGURE 12-1
Forces on gravity retaining wall.

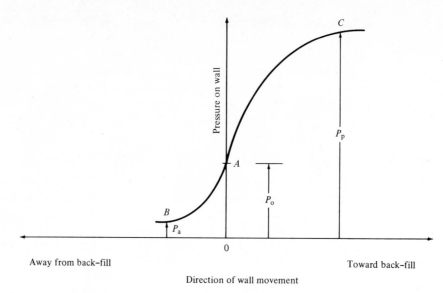

FIGURE 12-2
Relationship between earth pressure and wall movement.

phenomenon was focused upon by Terzaghi (18) through his classic experiments in 1929–1934 and by others (1, 3–5, 20).

Figure 12-2 depicts the relationship between the earth pressure and the wall movement. P_0 represents the magnitude of pressure when no movement of the retaining wall takes place; it is commonly referred to as *earth pressure at rest*. As the wall moves toward the back-fill, the pressure increases, reaching a maximum value of P_p at point C. On the other hand, if the wall moves away from the back-fill, the force decreases, reaching a minimum value of P_a at point B.

The relative magnitude of the active and passive earth pressures may perhaps be better illustrated with the aid of Fig. 12-3. For the sake of simplicity, several assumptions are made:

1. Frictional forces between back-fill and retaining wall are assumed negligible.
2. The wall is vertical, and the surface of the back-fill is horizontal.
3. The backfill is a homogeneous, granular material.
4. The failure surface is assumed to be a plane.

Variations and deviations from these assumptions will be discussed in subsequent sections. Figures 12-3a and 12-3b show the active case, where the wall moves away from the back-fill, together with the corresponding force polygon. Similarly, the case of passive resistance together with its force polygon is shown in Figs. 12-3c and 12-3d.

The magnitudes of the active and passive forces P_a and P_p could be derived from the basic condition of static equilibrium as follows.

Case of Active Pressure.   From Fig. 12-3b,

$$P_a = W \tan(\beta - \varphi) \qquad \text{(a)}$$

From Fig. 12-3a,

$$W = (\tfrac{1}{2}\gamma H)(H \cot \beta) = \tfrac{1}{2}\gamma H^2 \cot \beta \qquad \text{(b)}$$

Substituting Eq. (b) into Eq. (a), we obtain

$$P_a = \tfrac{1}{2}\gamma H^2 \cot \beta \tan(\beta - \varphi) \qquad \text{(c)}$$

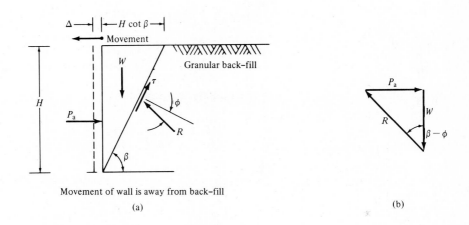

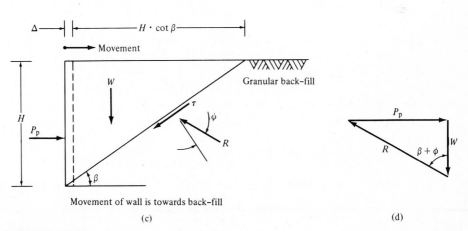

FIGURE 12-3
Relationship between lateral pressure and wall movement for cohesionless soil mass and plane failure surfaces. (a), (b) Development of active pressure P_a. (c), (d) Development of passive pressure P_p.

The maximum P_a may be obtained as follows. Let $\partial P_a / \partial \beta = 0$ and solve for β_{cr}. Then substitute β_{cr} in Eq. (c); hence,

$$\frac{\partial P_a}{\partial \beta} = \tfrac{1}{2}\gamma H^2 [\cot \beta \sec^2(\beta - \varphi) + \tan(\beta - \varphi)(-\csc^2 \beta)] = 0$$

or

$$\frac{\cos \beta \sin \beta - \sin(\beta - \varphi) \cos(\beta - \varphi)}{[\cos(\beta - \varphi) \sin \beta]^2} = 0 \qquad (d)$$

From the trigonometric identities we have

$$\cos \beta \sin \beta = \tfrac{1}{2} \sin 2\beta$$

$$\sin(\beta - \varphi) = \sin \beta \cos \beta - \cos \beta \sin \varphi$$

$$\cos(\beta - \varphi) = \cos \beta \cos \varphi + \sin \beta \sin \varphi$$

Substituting in Eq. (d) and rearranging terms, we have

$$\frac{\cos(2\beta - \varphi) \sin \varphi}{[\cos(\beta - \varphi) \sin \beta]^2} = 0 \qquad (e)$$

Equation (e) is satisfied when $\cos(2\beta - \varphi) = 0$. Thus,

$$2\beta - \varphi = 90$$

or

$$\beta_{cr} = 45 + \frac{\varphi}{2} \qquad (f)$$

Substituting in Eq. (c), we obtain

$$P_a = \tfrac{1}{2}\gamma H^2 \cot\left(45 + \frac{\varphi}{2}\right) \tan\left(45 - \frac{\varphi}{2}\right)$$

But $\cot(45 + \varphi/2) = \tan(45 - \varphi/2)$. Hence we obtain

$$P_a = \tfrac{1}{2}\gamma H^2 \tan^2\left(45 - \frac{\varphi}{2}\right) \qquad (12\text{-}1)$$

Case of Passive Pressure.   From Fig. 12-3d,

$$P_p = W \tan(\beta + \varphi) \qquad (g)$$

From Fig. 12-3c,

$$W = \tfrac{1}{2}\gamma H(H \cot \beta) = \tfrac{1}{2}\gamma H^2 \cot \beta \qquad (h)$$

Thus Eq. (g) becomes

$$P_p = \tfrac{1}{2}\gamma H^2 \cot \beta \tan(\beta + \varphi) \qquad \text{(i)}$$

As previously, letting $\partial P_p/\partial \beta = 0$ and solving for β_{cr}, we obtain

$$\beta_{cr} = 45 - \frac{\varphi}{2} \qquad \text{(j)}$$

and

$$P_p = \tfrac{1}{2}\gamma H^2 \tan^2\left(45 + \frac{\varphi}{2}\right) \qquad \text{(12-2)}$$

Equations (12-1) and (12-2) are frequently written as

$$P_a = \tfrac{1}{2}\gamma H^2 K_a \qquad \text{(12-1a)}$$

and

$$P_p = \tfrac{1}{2}\gamma H^2 K_p \qquad \text{(12-2a)}$$

where

$$K_a = \tan^2\left(45 - \frac{\varphi}{2}\right)$$

and

$$K_p = \tan^2\left(45 + \frac{\varphi}{2}\right)$$

K_a and K_p are generally referred to as *coefficients* for *active* and *passive pressure*, respectively. They are constants for any given soil where $\varphi = $ constant.

It is readily apparent that the coefficient K_p is significantly larger than K_a. However, it is sometimes the practice to assume a value of K_p approximately 10 times K_a. While this is only true for a rather small range of φ values (see Example 12-1), in many instances the value for φ for sand does fall within this range. From the above expression, however, we note a more fundamental relationship:

$$K_a = \frac{1}{K_p}$$

The coefficient of earth pressure at rest K_0 has been shown experimentally to approximate

$$K_0 = 1 - \sin \bar{\varphi}$$

where $\bar{\varphi}$ is the effective angle of internal friction. This expression is acceptable for normally consolidated soils, both cohesionless and cohesive. For overconsolidated clays the value of K_0 is slightly larger than that given by the above expression.

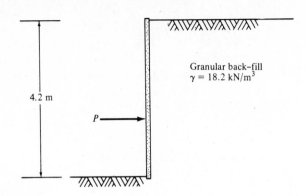

FIGURE 12-4

EXAMPLE 12-1

Given   A retaining wall as shown in Fig. 12-4.

Find   (a) K_a, K_p, and K_p/K_a for $\varphi = 15, 20, 25, 30, 35$, and $40°$.
(b) P_a and P_p for $\varphi = 30°$.

Procedure   (a) $K_a = \tan^2(45 - \varphi/2)$ and $K_p = \tan^2(45 + \varphi/2)$. The coefficients are given in the following table:

	15°	20°	25°	30°	35°	40°
K_a	0.589	0.490	0.406	0.333	0.271	0.217
K_p	1.698	2.040	2.464	3.000	3.690	4.599
K_p/K_a	2.883	4.163	6.069	9.009	13.616	21.193

(b) $P_a = \frac{1}{2}\gamma H^2 K_a = \frac{1}{2}(18.2)(4.2)^2(0.333)$

$P_b = \frac{1}{2}\gamma H^2 K_p = \frac{1}{2}(18.2)(4.2)^2(3.000)$

Answer $P_a = 53.45 \text{ kN}$ and $P_b = 481.57 \text{ kN}$

12-3 RANKINE'S THEORY

Although historical evidence exists which shows that as early as in 1687 Marquis Sebastian de Prestre de Vauban, a French military engineer, formulated some guidelines for designing some earth-retaining structures, it was not until 1776 that Charles Augustin Coulomb published his now famous and fundamental earth pressure theory. Since then a number of investigators, including William John Macquorn Rankine, Jean Victor Poncelot, Karl Culmann, and numerous more current investigators have refined and contributed much toward the solution of problems related to earth pressure. Coulomb's and Rankine's are perhaps the two best known theories and are frequently referred to as the *classical earth pressure theories*.

The theory proposed by Rankine in 1857 is based on the assumption that a conjugate relationship exists between vertical and lateral pressures on vertical planes within a mass of homogeneous isotropic cohesionless material behind a smooth retaining wall. Rankine's theory reflects a simplification of Coulomb's method.

Cohesionless Back-fill and Level Surface

The basic concept behind Rankine's theory can be depicted via Mohr's circle. Consider the element shown in Fig. 12-5a subjected to the geostatic stresses shown. The value for σ_1 could be approximated as the product of the average unit weight times depth, namely, $\sigma_1 \cong \gamma h$. If the wall were to move to the left, thereby creating a case of active stress, the value for σ_1 would become the major principal stress. The corresponding Mohr circle for this case is depicted by circle 1 in Fig. 12-5b. On the other hand, if the wall were to push against the back-fill, a case of passive pressure would be developed. The vertical stress would then become the minor principal stress, and the lateral stress would thus become the major principal stress. The Mohr circle for this condition is depicted by circle 2 in Fig. 12-5b.

For clarity, Fig. 12-6 isolates the corresponding Mohr circles for the active and passive cases depicted in Fig. 12-5b. From Fig. 12-6a we note that

$$\sin \varphi = \frac{(\sigma_1 - \sigma_3)/2}{(\sigma_1 + \sigma_3)/2} = \frac{\sigma_1 - \sigma_3}{\sigma_1 + \sigma_3}$$

or rearranging terms and solving for σ_3,

$$\sigma_3 = \sigma_1 \frac{1 - \sin \varphi}{1 + \sin \varphi} = \gamma h \frac{1 - \sin \varphi}{1 + \sin \varphi} \tag{a}$$

or since $(1 - \sin \varphi)/(1 + \sin \varphi) = \tan^2(45 - \varphi/2)$, Eq. (a) becomes

$$\sigma_3 = \gamma h \tan^2\left(45 - \frac{\varphi}{2}\right) \tag{12-3}$$

For the passive case (Fig. 12-6b) we have

$$\sin \varphi = \frac{(\sigma_3' - \sigma_1)/2}{(\sigma_3' + \sigma_1)/2} = \frac{\sigma_3' - \sigma_1}{\sigma_3' + \sigma_1}$$

Thus after rearranging terms and solving for σ_3',

$$\sigma_3' = \sigma_1 \frac{(1 + \sin \varphi)}{(1 - \sin \varphi)} = \gamma h \frac{1 + \sin \varphi}{1 - \sin \varphi} \tag{b}$$

or

$$\sigma_3' = \gamma h \tan^2\left(45 + \frac{\varphi}{2}\right) \tag{12-4}$$

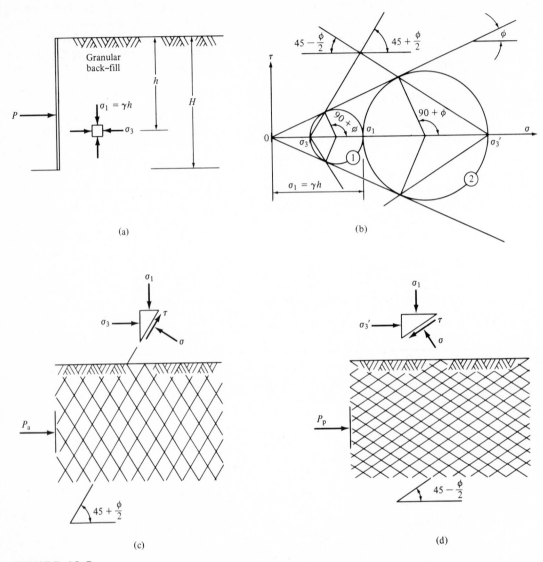

FIGURE 12-5
The orientation of slip planes in granular soil mass with a level surface under active and passive states of stress. (a) Element in granular soil. (b) Mohr's circles for active (1) and passive (2) states of stress in element shown in (a). (c) Slip planes for active case. (d) Slip planes for passive case.

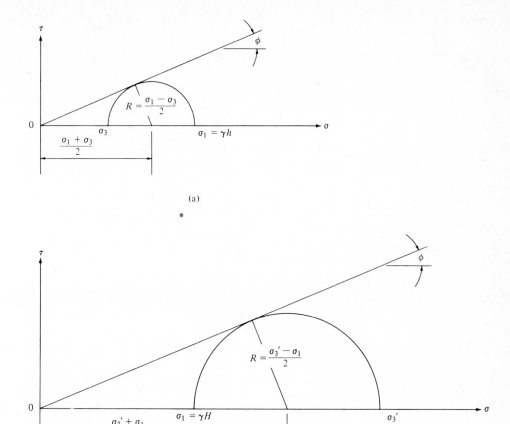

FIGURE 12-6
Mohr's circles for active and passive cases of stress in cohesionless back-fill, level surface, and depth h. (a) Active case. (b) Passive case.

We note that $\tan^2(45 - \varphi/2)$ and $\tan^2(45 + \varphi/2)$ in Eqs. (12-3) and (12-4) are constants for constant values of φ. Hence the corresponding pressures against the retaining wall vary linearly with depth, as indicated by Fig. 12-7. The corresponding resultant pressures, active and passive, can be calculated for a unit length of retaining wall as

$$P_a = \tfrac{1}{2}\gamma H^2 \tan^2\left(45 - \frac{\varphi}{2}\right) = \tfrac{1}{2}\gamma H^2 K_a \tag{12-5}$$

$$P_p = \tfrac{1}{2}\gamma H^2 \tan^2\left(45 + \frac{\varphi}{2}\right) = \tfrac{1}{2}\gamma H^2 K_p \tag{12-6}$$

which correspond to Eqs. (12-1) and (12-2), respectively.

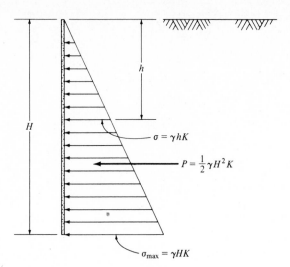

FIGURE 12-7
Pressure distribution.

The corresponding slip planes for the active and passive cases are shown in Figs. 12-5c and d.

Cohesionless Back-fill and Inclined Surface

Let us now consider a cohesionless mass with a sloping surface behind a smooth vertical retaining wall. Assume this condition to be depicted by Fig. 12-8a. The lateral stresses acting on the vertical faces of the element (i.e., the faces parallel to the wall) are parallel to the inclined surface. Thus any such planes experience not only normal but also shear stresses. Needless to say, they are no longer principal planes, as was the case for horizontal surfaces.

The corresponding resultant pressure on the wall could be determined with the aid of Mohr's circle. Figure 12-8c symbolizes an active state of stress. The magnitude of the vertical stress is depicted by the distance $0C$; the lateral stress, acting parallel to the sloped surface, is represented by the distance $0A$. Hence from Fig. 12-8c we have

$$\sigma_h = (0A) \tag{a}$$

But

$$0A = \left(\frac{0B - AB}{0B + AB}\right)(0C) = \left(\frac{0B - AB}{0B + AB}\right)\gamma h \cos i \tag{b}$$

and

$$\left.\begin{aligned}
0B &= (0D)\cos i \\
r &= (0D)\sin \varphi \\
BD &= (0D)\sin i \\
AB &= \sqrt{r^2 - (BD)^2} = \sqrt{(0D \sin \varphi)^2 - (0D \sin i)^2}
\end{aligned}\right\} \tag{c}$$

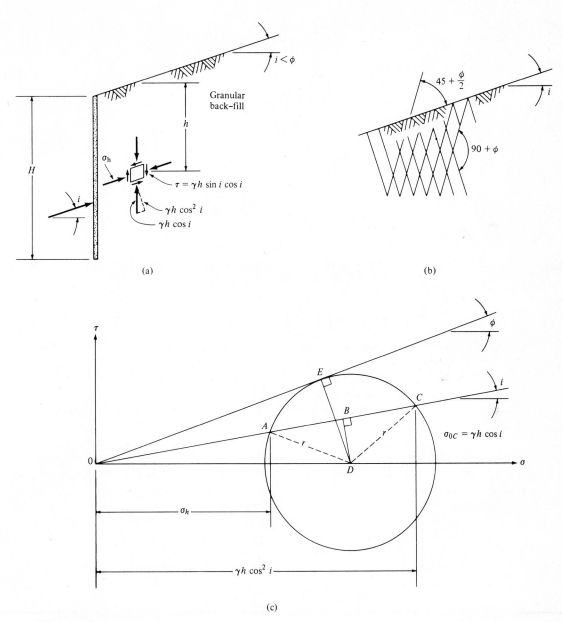

FIGURE 12-8
Lateral pressure and slip planes in granular sloped back-fill under active state of stress. (a) Inclined granular back-fill. (b) Orientation of slip planes. (c) Mohr's circle for active state of stress.

Substituting into Eq. (b), we get

$$0A = \left(\frac{(0D)\cos i - (0D)\sqrt{\sin^2 \varphi - \sin^2 i}}{(0D)\cos i + (0D)\sqrt{\sin^2 \varphi - \sin^2 i}}\right)\gamma h \cos i \tag{d}$$

Factoring and canceling $(0D)$ and substituting $\sin^2 \varphi = 1 - \cos^2 \varphi$ and $\sin^2 i = 1 - \cos^2 i$, Eq. (d) becomes

$$0A = \left(\frac{\cos i - \sqrt{1 - \cos^2 \varphi - 1 + \cos^2 i}}{\cos i + \sqrt{1 - \cos^2 \varphi - 1 + \cos^2 i}}\right)\gamma h \cos i$$

or

$$0A = \left(\frac{\cos i - \sqrt{\cos^2 i - \cos^2 \varphi}}{\cos i + \sqrt{\cos^2 i - \cos^2 \varphi}}\right)\gamma h \cos i \tag{12-7}$$

or

$$\sigma_h = 0A = \gamma h K_a \tag{12-8}$$

and

$$P_a = \tfrac{1}{2}\gamma H^2 K_a = \tfrac{1}{2}\gamma H^2 \left(\frac{\cos i - \sqrt{\cos^2 i - \cos^2 \varphi}}{\cos i + \sqrt{\cos^2 i - \cos^2 \varphi}}\right)\cos i \tag{12-9}$$

For the passive case,

$$P_p = \tfrac{1}{2}\gamma H^2 K_p = \tfrac{1}{2}\gamma H^2 \left(\frac{\cos i + \sqrt{\cos^2 i - \cos^2 \varphi}}{\cos i - \sqrt{\cos^2 i - \cos^2 \varphi}}\right)\cos i \tag{12-10}$$

Equation (12-9) is Rankine's expression for the active lateral pressure at depth H for a back-fill's unit weight γ. Equation (12-10) is Rankine's expression for the passive case.

For a given sloped surface and uniform soil properties K_a becomes a constant. Thus the intensity of load, or stress, varies linearly with depth. Hence, as before, the total resultant active force may be given by Eq. (12-9). Again one notes that the direction of the resultant is parallel to the sloped surface.

For the case of level surface, Eqs. (12-9) and (12-10) reduce to Eqs. (12-5) and (12-6), respectively.

EXAMPLE 12-2

Given   A retaining wall as shown in Fig. 12-9.

Find P_a and P_p.

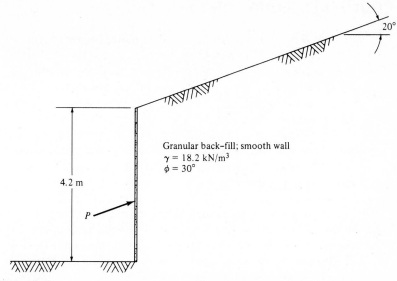

FIGURE 12-9

Procedure   From Eqs. (12-9) and (12-10), respectively

$$P_a = \tfrac{1}{2}\gamma H^2 K_a = \tfrac{1}{2}(18.2)(4.2)^2 \left[\frac{\cos 20° - \sqrt{\cos^2 20 - \cos^2 30}}{\cos 20° + \sqrt{\cos^2 20 - \cos^2 30}} \right] \cos 20°$$

or

$$P_a = 160.52 K_a = 160.52 \left[\frac{0.94 - \sqrt{0.883 - 0.750}}{0.94 + \sqrt{0.883 - 0.750}} \right] \cos 20°$$

$$P_a = 160.52 \frac{0.575}{1.305}(0.939)$$

and

$$P_p = 160.52(0.940)\left(\frac{1.305}{0.575}\right)(0.939)$$

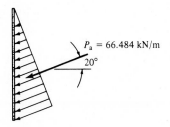

$P_a = 66.484$ kN/m

20°

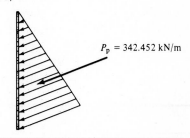

$P_p = 342.452$ kN/m

Answer $P_a = 66.484$ kN/m   and $P_p = 342.452$ kN/m

12-4 COULOMB'S EQUATION

In 1776 Coulomb introduced an expression for determining the active thrust on retaining walls. Although Coulomb was cognizant of the effects of both strength parameters c and φ, and of the likely probability that the sliding surface is not a plane, he elected to base his analysis on the assumption that the sliding surface is a plane and the back-fill is granular ($c = 0$). He did this in order to simplify somewhat the mathematically complex problem introduced when cohesion and nonplane sliding surfaces are considered. He did, however, account for the effects of friction interaction between the soil back-fill and the face of the retaining wall, and he considered the more general case of a sloped backface of the retaining wall. In this respect it is a more general approach than the Rankine cases considered in the previous section.

According to Coulomb's theory, the thrust is induced by the sliding wedge, as shown in Fig. 12-10a. For this reason it is sometimes referred to as the *sliding wedge analysis*. The corresponding force polygon is shown in Fig. 12-10b. The development of Coulomb's equation follows from this basic relationship. From Fig. 12-10b, using the Law of Sines, we have

$$\frac{P_a}{\sin(\alpha - \varphi)} = \frac{W}{\sin(180 - \alpha + \varphi - \beta + \delta)}$$

or

$$P_a = W \frac{\sin(\alpha - \varphi)}{\sin(180 - \alpha + \varphi - \beta + \delta)} \tag{12-11}$$

The weight W of the wedge can be obtained from Fig. 12-11.

$$W = \frac{Lh}{2}\gamma = \frac{(\overline{AD} + \overline{CD})h}{2}\gamma \tag{a}$$

But

$$AB = \frac{H}{\cos(\beta - 90)} = \frac{h}{\sin(\beta - \alpha)}$$

or

$$h = H\left[\frac{\sin(\beta - \alpha)}{\cos(\beta - 90)}\right]$$

Also,

$$\frac{CD}{\sin(90 - \alpha + i)} = \frac{h}{\sin(\alpha - 1)}$$

$$CD = h\frac{\sin(90 - \alpha + i)}{\sin(\alpha - i)}$$

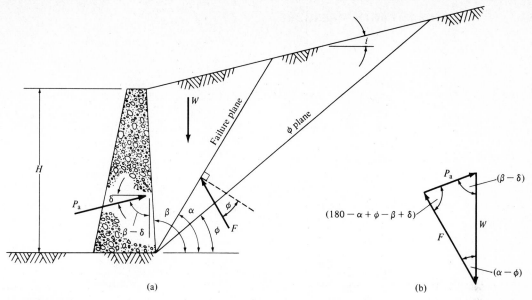

(a) (b)

FIGURE 12-10
Coulomb's theory, general case with cohesionless back-fill, active state. (a) Coulomb's sliding wedge.
(b) Force polygon.

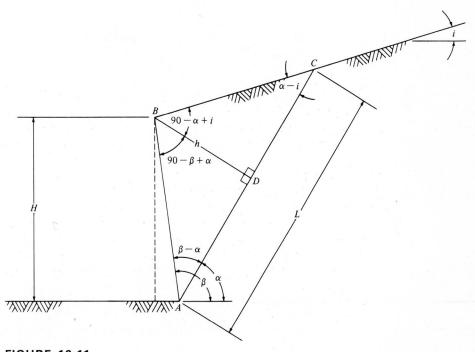

FIGURE 12-11
General cross section of sliding wedge.

and

$$\frac{AD}{\sin(\beta - 90)} = \frac{h}{\sin(\beta - \alpha)}$$

$$AD = h\frac{\sin(\beta - 90)}{\sin(\beta - \alpha)}$$

Substituting into Eq. (a),

$$W = \frac{h^2}{2}\gamma\left[\frac{\sin(90 - \alpha + i)}{\sin(\alpha - i)} + \frac{\sin(\beta - 90)}{\sin(\beta - \alpha)}\right] \qquad \text{(b)}$$

or, in terms of H

$$W = \frac{H^2}{2}\gamma\left[\frac{\sin(\beta - \alpha)}{\cos(\beta - 90)}\right]^2\left[\frac{\sin(90 - \alpha + i)}{\sin(\alpha - i)} + \frac{\sin(\beta - 90)}{\sin(\beta - \alpha)}\right] \qquad \text{(c)}$$

Hence substituting into Eq. (12-11),

$$P_a = \left(\frac{H^2\gamma}{2}\right)\frac{\sin(\alpha - \varphi)}{\sin(180 - \alpha + \varphi - \beta + \delta)}\left[\frac{\sin(\beta - \alpha)}{\cos(\beta - 90)}\right]^2$$

$$\times\left[\frac{\sin(90 - \alpha + i)}{\sin(\alpha - i)} + \frac{\sin(\beta - 90)}{\sin(\beta - \alpha)}\right] \qquad \text{(12-12)}$$

To determine the orientation of the failure plane that produces a maximum P_a, the following condition must be satisfied:

$$\frac{\partial P_a}{\partial \alpha} = 0 \qquad \text{(d)}$$

The solution to this equation is tedious and beyond the scope of this text. Müller-Breslau solved this general problem in 1906 (9). Hence,

$$P_a = \frac{H^2\gamma}{2}\left[\frac{\csc\beta\sin(\beta - \varphi)}{\sqrt{\sin(\beta + \delta)} + \sqrt{\dfrac{\sin(\varphi + \delta)\sin(\varphi - i)}{\sin(\beta - i)}}}\right]^2 = \frac{H^2\gamma}{2}K_a \qquad \text{(12-13)}$$

where

$$K_a = \left[\frac{\csc\beta\sin(\beta - \varphi)}{\sqrt{\sin(\beta + \delta)} + \sqrt{\dfrac{\sin(\delta + \varphi)\sin(\varphi - i)}{\sin(\beta - i)}}}\right]^2$$

The corresponding passive thrust is expressed by Eq. (12-14):

$$P_p = \frac{H^2\gamma}{2}\left[\frac{\csc\beta\sin(\beta + \varphi)}{\sqrt{\sin(\beta - \delta)} - \sqrt{\dfrac{\sin(\varphi + \delta)\sin(\varphi + i)}{\sin(\beta - i)}}}\right]^2 = \frac{H^2\gamma}{2}K_p \qquad \text{(12-14)}$$

where

$$K_p = \left[\frac{\csc \beta \sin(\beta + \varphi)}{\sqrt{\sin(\beta - \delta)} - \sqrt{\dfrac{\sin(\varphi + \delta) \sin(\varphi + i)}{\sin(\beta - i)}}} \right]^2$$

We note that for a vertical smooth backface (that is, $\beta = 90°$ and $\delta = 0$) Eq. (12-13) reduces to Eq. (12-9) derived on the basis of Rankine's theory.

Coulomb arbitrarily placed the resultant thrust P_a at the third point from the bottom. Correspondingly he assumed the pressure distribution to vary linearly with depth. Although this assumption appears to give results acceptable for very rigid walls and granular back-fill, it is not valid for relatively flexible bulkheads, cohesive back-fills, or where the retaining wall rotates about points not close to the bottom.

EXAMPLE 12-3

Given   A retaining wall as shown in Fig. 12-12.

Find P_a and P_p.

Procedure   From Eqs. (12-13) and (12-14),

$$P_a = \frac{H^2 \gamma}{2} \left[\frac{\csc \beta \sin(\beta - \varphi)}{\sqrt{\sin(\beta + \delta)} + \sqrt{\dfrac{\sin(\varphi + \delta) \sin(\varphi - i)}{\sin(\beta - i)}}} \right]^2$$

$$P_a = \frac{(4.2)^2(18.2)}{2} \left[\frac{\csc 98 \sin(98 - 30)}{\sqrt{\sin(93 + 20)} + \sqrt{\dfrac{\sin(30 + 20) \sin(30 - 20)}{\sin(98 - 20)}}} \right]^2$$

Answer

$$\boxed{P_a = 72.94 \text{ kN/m}}$$

and

$$P_p = \frac{H^2 \gamma}{2} \left[\frac{\csc \beta \sin(\beta + \varphi)}{\sqrt{\sin(\beta - \delta)} - \sqrt{\dfrac{\sin(\varphi + \delta) \sin(\varphi + i)}{\sin(\beta - i)}}} \right]^2$$

$$P_p = \frac{(4.2)^2(18.2)}{2} \left[\frac{\csc 98 \sin(98 + 30)}{\sqrt{\sin(98 - 20)} - \sqrt{\dfrac{\sin(30 + 20)\sin(30 + 20)}{\sin(98 - 20)}}} \right]^2$$

Answer

$$\boxed{P_p = 2089 \text{ kN/m}}$$

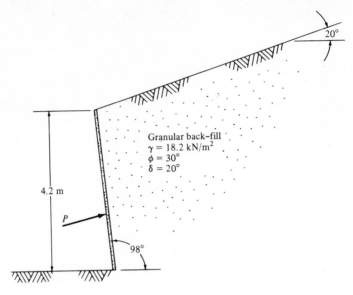

FIGURE 12-12

12-5 LATERAL EARTH PRESSURES IN PARTIALLY COHESIVE SOILS

The Mohr circle may be used to determine the lateral thrust on retaining walls from back-fill soil for which the shear strength may be expressed by $s = c + \sigma \tan \varphi$. In this case the values of c and φ imply effective cohesion and effective angle of internal friction, respectively. As a special case, let us assume a vertical and smooth retaining wall and a $(c\text{-}\varphi)$ soil back-fill with a horizontal surface. At a given depth h on the element shown in Fig. 12-13a, the vertical stress σ_1 is equal to γh. The lateral stress is designated by σ_3. Both of these are principal stresses. The corresponding Mohr circle for this case is shown in Fig. 12-13b.

From Fig. 12-13b we note that

$$\sin \varphi = \frac{(\sigma_1 - \sigma_3)/2}{(\sigma_1 + \sigma_3)/2 + c \cot \varphi} = \frac{\sigma_1 - \sigma_3}{\sigma_1 + \sigma_3 + 2c \cot \varphi}$$

Rearranging terms,

$$\sigma_3 (1 + \sin \varphi) = \sigma_1 (1 - \sin \varphi) - 2c \sin \varphi \cot \varphi$$

or

$$\sigma_3 = \sigma_1 \left(\frac{1 - \sin \varphi}{1 + \sin \varphi}\right) - 2c \left(\frac{\cos \varphi}{1 + \sin \varphi}\right) \tag{a}$$

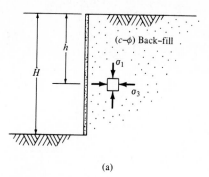

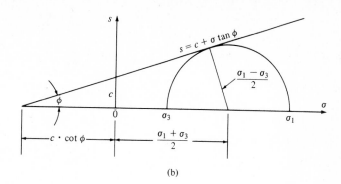

(a)   (b)

FIGURE 12-13
Cohesive (c-ϕ) back-fill, active state.

But

$$\frac{\cos \varphi}{1 + \sin \varphi} = \frac{\sqrt{1 - \sin^2 \varphi}}{1 + \sin \varphi} = \frac{\sqrt{(1 - \sin \varphi)(1 + \sin \varphi)}}{\sqrt{(1 + \sin \varphi)(1 + \sin \varphi)}}$$

or

$$\frac{\cos \varphi}{1 + \sin \varphi} = \sqrt{\frac{1 - \sin \varphi}{1 + \sin \varphi}}$$

Thus substituting into Eq. (a),

$$\sigma_3 = \sigma_1 \frac{1 - \sin \varphi}{1 + \sin \varphi} - 2c \sqrt{\frac{1 - \sin \varphi}{1 + \sin \varphi}} \qquad (b)$$

But $\sigma_1 = \gamma h$ and $(1 - \sin \varphi)/(1 + \sin \varphi) = \tan^2(45 - \varphi/2)$. Hence,

$$\sigma_3 = \gamma h \tan^2\left(45 - \frac{\varphi}{2}\right) - 2c \tan\left(45 - \frac{\varphi}{2}\right) \qquad (c)$$

or

$$\sigma_3 = \gamma h K_a - 2c\sqrt{K_a} \qquad (d)$$

The maximum horizontal stress (or pressure) for the active case occurs when $h = H$. The pressure distribution is shown in Fig. 12-14b. The corresponding resultant P_a is

$$P_a = \tfrac{1}{2}\gamma K_a H^2 - 2c\sqrt{K_a} H \qquad (12\text{-}15)$$

For the passive state the major principal stress would be the horizontal stress, shown as σ_3 in Fig. 12-13a; the minor stress is then σ_1. Thus Fig. 12-13b would show σ_1 and σ_3 in reverse positions. Then

$$\sin \varphi = \frac{\sigma_3 - \sigma_1}{\sigma_3 + \sigma_1 + 2c \cot \varphi}$$

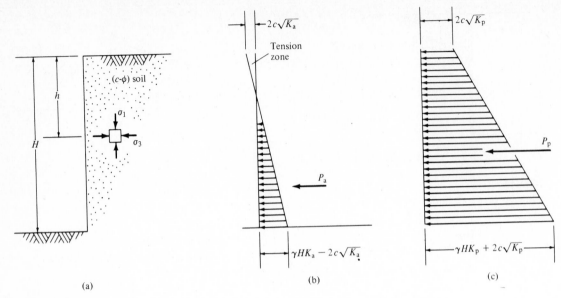

FIGURE 12-14
Pressure distributions in $(c\text{-}\phi)$ soil. (a) $(c\text{-}\phi)$ soil. (b) Active state. (c) Passive state.

Simplifying and solving for σ_3, we get

$$\sigma_3 = \sigma_1 \frac{(1 + \sin \varphi)}{(1 - \sin \varphi)} + 2c \frac{\cos \varphi}{1 - \sin \varphi} \qquad (e)$$

But

$$\frac{\cos \varphi}{1 - \sin \varphi} = \frac{\sqrt{1 - \sin^2 \varphi}}{1 - \sin \varphi} = \sqrt{\frac{1 + \sin \varphi}{1 - \sin \varphi}}$$

Hence,

$$\sigma_3 = \sigma_1 \left(\frac{1 + \sin \varphi}{1 - \sin \varphi}\right) + 2c \sqrt{\frac{1 + \sin \varphi}{1 - \sin \varphi}} \qquad (f)$$

or

$$\sigma_3 = \sigma_1 \tan^2\left(45 + \frac{\varphi}{2}\right) + 2c \tan\left(45 + \frac{\varphi}{2}\right) \qquad (g)$$

or

$$\sigma_3 = \gamma h K_p + 2c \sqrt{K_p} \qquad (h)$$

The stress (pressure) distribution is shown in Fig. 12-14c. From this the pressure resultant P_p is

$$P_p = \tfrac{1}{2}\gamma K_p H^2 + 2c \sqrt{K_p} H \qquad (12\text{-}16)$$

12-6 UNSUPPORTED CUTS IN $(c\text{-}\varphi)$ SOIL

Unsupported excavations would theoretically be possible in $(c\text{-}\varphi)$ soils if the lateral pressure (σ_3 for the active case) would not exceed the strength of the soil. The general expression for the horizontal stress σ_3 in a $(c\text{-}\varphi)$ soil for the active case was given by Eq. (d) in Section 12-5. For convenience it is reproduced here:

$$\sigma_3 = \gamma h K_a - 2c\sqrt{K_a} \tag{d}$$

At ground surface, $h = 0$. Thus,

$$\sigma_3 = -2c\sqrt{K_a} \quad \text{(tension)}$$

This implies the formation of a crack as depicted in Fig. 12-15a. The corresponding pressure distribution based on Eq. (d) is shown in Fig. 12-15b.

The theoretical depth of the crack h_t can be determined by recognizing that, at the bottom of the crack, $\sigma_3 = 0$. Thus from Eq. (d),

$$0 = \gamma h_t K_a - 2c\sqrt{K_a}$$

or

$$h_t = \frac{2c}{\gamma\sqrt{K_a}} \tag{12-17}$$

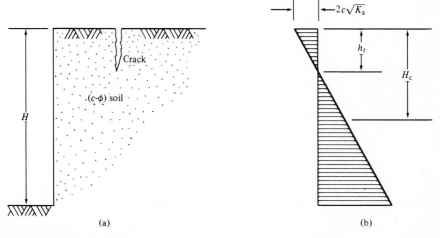

(a)   (b)

FIGURE 12-15
Tension crack in $(c\text{-}\phi)$ soil and corresponding pressure distribution, active state. (a) Tension crack. (b) Pressure distribution.

The theoretical maximum depth H_c of unsupported excavation may be calculated as the point where the tension forces equal the cohesive strength. Hence from Fig. 12-15b, $H_c = 2h_c$. This could also be obtained from Eq. (d) if $\sigma_3 = 2c\sqrt{K_a}$, when $h = H_c$:

$$2c\sqrt{K_a} = \gamma H_c K_a - 2c\sqrt{K_a}$$

or

$$H_c = \frac{4c}{\gamma\sqrt{K_a}} = 2h_t \tag{12-18}$$

While Eq. (12-18) provides a theoretical depth to which an excavation may be made without lateral support, it should be used cautiously. Surface moisture which may enter the crack may induce hydrostatic stresses or may decrease the shear strength of the soil. Hence the unsupported excavation to such depths should be for short duration at best. Even then, judgment reflecting on the potential consequences from unsupported excavation is indeed warranted (10).

In general it is advisable to minimize the use of cohesive back-fill whenever possible. With changes in moisture content the pressure induced by highly cohesive soil may change significantly. For example, as the clay dries up and shrinks, the pressure on the wall is likely to be significantly reduced. On the other hand, a dry soil subjected to moisture may swell significantly, thereby increasing the pressure by a significant amount. Furthermore, repeated shrinking and swelling cycles may result in cumulative movement in the retaining wall itself. Yet to predict forces and subsequent effects with any degree of accuracy may be rather difficult.

EXAMPLE 12-4

Given An unsupported cut as shown in Fig. 12-16.

Find (a) Stress at top and bottom of cut.
(b) Maximum depth of potential tension crack.
(c) Maximum unsupported excavation.

Procedure (a) From Eq. (c), Section 12-5,

$$\sigma_3 = \gamma h \tan^2\left(45 - \frac{\varphi}{2}\right) - 2c \tan\left(45 - \frac{\varphi}{2}\right)$$

at the top, $h = 0$,

$$\sigma_3 = 0 - 2(25) \tan(45 - \tfrac{10}{2})$$

Answer

$$\sigma_3 = -41.95 \text{ kN/m}^2$$

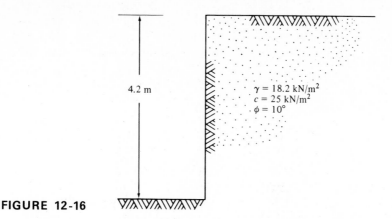

$\gamma = 18.2 \text{ kN/m}^2$
$c = 25 \text{ kN/m}^2$
$\phi = 10°$

4.2 m

FIGURE 12-16

and at the bottom, $h = 4.2$ m,

$$\sigma_3 = 18.2(4.2) \tan^2(45 - \tfrac{10}{2}) - 2(25) \tan(45 - \tfrac{10}{2})$$

$$\sigma_3 = 53.82 - 41.95$$

Answer

$$\sigma_3 = 11.87 \text{ kN/m}^2$$

(b) From Eq. (12-17),

$$h_t = \frac{2c}{\gamma\sqrt{K_a}} = \frac{2(25)}{(18.2)\sqrt{\tan(45 - \tfrac{10}{2})}}$$

Answer

$$h_t = 3.27 \text{ m}$$

(c) From Eq. (12-18),

$$H_c = 2h_t$$

Answer

$$H_c = 6.54 \text{ m}$$

12-7 EFFECTS OF SURCHARGE LOADS

Concentrated and/or uniformly distributed loads, commonly referred to as *surcharge loads*, acting on the surface of the back-fill, will (1) increase the lateral pressure against the retaining wall; and (2) move the point of application of the resultant pressure upward.

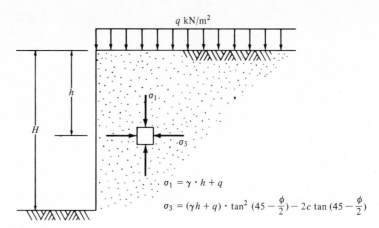

FIGURE 12-17
Uniform surcharge on level back-fill surface.

Figure 12-17 shows a uniformly distributed surcharge q (kN/m²) acting on the surface of the back-fill. The vertical stress σ_1 on an element within the soil back-fill at a depth h is equal to $\gamma h + q$. Correspondingly, the lateral stress σ_3 is, from Eq. (c), Section 12-5, for a $(c$-$\varphi)$ soil,

$$\sigma_3 = (\gamma h + q)\tan^2\left(45 - \frac{\varphi}{2}\right) - 2c\,\tan\left(45 - \frac{\varphi}{2}\right)$$

The intensity of pressure at any given depth can be determined by superimposing the effects of the back-fill and of the surcharge, as indicated in Fig. 12-18. The effect

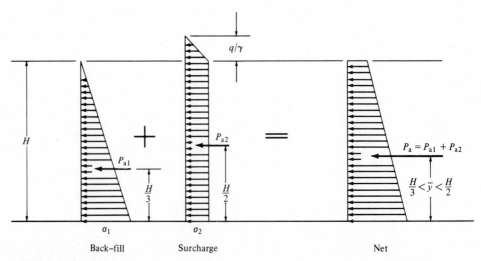

FIGURE 12-18
Stresses induced by back-fill surcharge.

could be envisioned as that of an imaginary equivalent soil layer of thickness q/γ on top of the back-fill. Thus the effect is additive, resulting in the trapezoidal shape shown in Fig. 12-18. Furthermore the resultant thrust of the two superimposed effects acts at a point between the resultants of the two pressure blocks shown in Fig. 12-18. Hence the surcharge increases both the lateral thrust and the overturning moment.

For concentrated surcharge loads Q, such as may be induced by a continuous footing, railroad tracks, etc., running parallel to the wall, it is possible, although rather laborious, to estimate the increased stresses on the wall based on Boussinesq's

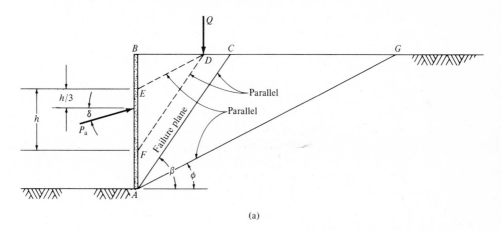

(a)

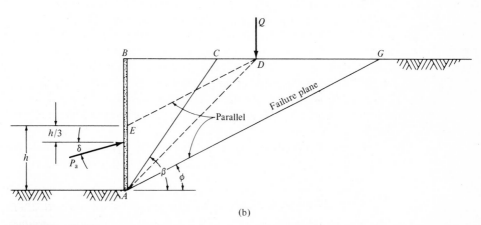

(b)

FIGURE 12-19
Procedure for estimating the line of action of the resultant active thrust P_a, caused by the concentrated surcharge Q on the left and right of the potential slip planes. (a) Location of active thrust induced by concentrated load Q acting between wall and failure plane. (b) Location of active thrust P_a, induced by concentrated load Q acting outside the sliding wedge.

equation, consistent with the theory of elasticity for a semi-infinite homogeneous soil mass. However, graphical methods are more expedient for this purpose (see Examples 12-5 and 12-6).

Experimental data indicate that Boussinesq's formula for lateral stress gives acceptable results where the wall movement is compatible to soil deformations within the back-fill. On the other hand, if the retaining wall is totally rigid, such that the soil deformation is greatly restricted by the rigid boundary, the horizontal stress approaches a value twice that given by Boussinesq's equation. This effect becomes less noticeable as the distance of Q relative to the wall increases.

The line of action of the active thrust induced by the concentrated surcharge loads Q is commonly based on empirical procedures. Such a procedure gives acceptable results for cohesionless back-fill; it may be illustrated with the aid of Fig. 12-19. When the concentrated surcharge load is located to the left of point C, the active thrust P_a may be determined by drawing lines ED and FD parallel to lines AG and AC, respectively. The point of application of P_a is one-third the distance EF from point E, as indicated in Fig. 12-19a. When load Q is located to the right of the failure plane, as indicated in Fig. 12-19b, line ED is parallel to line AG. The point of application of P_a is one-third the distance EA from point E, as shown in Fig. 12-19b. We note that the line of action of the resultant thrust moves up the wall as the load Q approaches the wall. Furthermore, the lateral thrust as well as the overturning effect decrease as the load Q moves away from the wall.

12-8 CULMANN'S METHOD

The following graphical procedure was devised by Karl Culmann over a century ago (1875). It is used to determine the magnitude and the location of the resultant earth pressures, both active and passive, on retaining walls. This method is applicable with acceptable accuracy to cases where the back-fill surface is level or sloped, regular or irregular, and where the back-fill material is uniform or stratified. Also, it considers such variables as wall friction, cohesionless soils, and, with some procedural modifications, cohesive soils, and surcharge loads, both concentrated and uniformly distributed. It does, however, require that the angle of internal friction of the soil be a constant for the total back-fill. The procedure presented here is limited to cohesionless soils.

Reference is made to Fig. 12-20 in describing the procedure for determining the active pressure for a case of cohesionless soil by Culmann's method:

1. Select a convenient scale to show a representative configuration of retaining wall and back-fill. This should include height and slope of the retaining wall, surface configuration of the back-fill, location and magnitude of concentrated (line) surcharge loads, uniformly distributed surcharge, etc.
2. From point A draw line AC which makes an angle of φ with the horizontal.
3. Draw line AD at an angle of ψ from line AC. Figure 12-20 shows the angle ψ to be the angle between the vertical and the resultant active pressure.
4. Draw rays AB_1, AB_2, AB_3, etc., i.e., assumed failure surfaces.

5. Determine the weight of each wedge, accounting for variations in densities if the back-fill is a layered system, for variable moisture content, etc.

6. Select a convenient scale and plot these weights along line AC. For example, the distance from A to W_1 along line AC equals W_1; similarly, the distance from W_1 to W_2 along line AC equals W_2, etc.

7. From each of the points located on line AC, draw lines parallel to line AD to intersect the corresponding assumed failure surfaces; that is, the line from W_1 will intersect line AB_1, the one from W_2 will intersect line AB_2, etc.

8. Connect these points of intersection with a smooth line, *Culmann's curve*.

9. Parallel to line AC draw a tangent to Culmann's curve. In Fig. 12-20 point E represents such a tangent point. More than one tangent is possible if the Culmann line is irregular.

10. From the point of tangency draw line EF parallel to line AD. The magnitude of EF, based on the selected scale, represents the active pressure P_a. If several tangents to the curve are possible, the largest of such values becomes the value of P_a. The failure surface passes through E and A, as shown in Fig. 12-20.

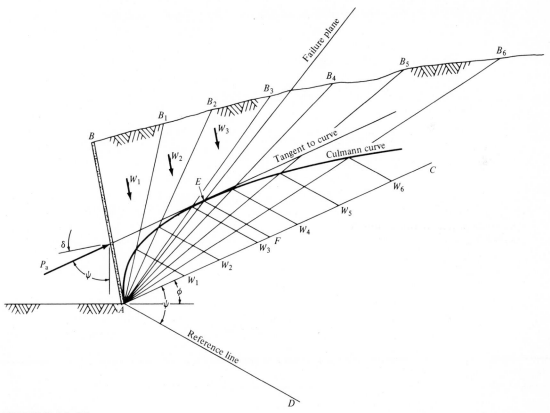

FIGURE 12-20
Culman's active earth pressure for cohesionless soils.

Surcharge loads and their respective effects on the location of the resultant could be accounted for as described in the previous section. Examples 12-5 and 12-6 may further enhance this explanation.

Figure 12-21 illustrates the procedure for determining the *passive* resistance via Culmann's method. The approach is similar to that for the active pressure, with some notable differences: (1) line AC makes an angle of φ degrees below rather than above the horizontal; (2) the reference line makes an angle of ψ with line AC, with ψ measured as indicated in Fig. 12-21. For the assumed sliding wedges, the weights W_1, W_2, etc., are plotted along line AC. From these points lines are drawn parallel to the reference line to intersect the corresponding rays, as shown in Fig. 12-21. The Culmann line represents a smooth curve connecting such points of intersection. A tangent to the Culmann

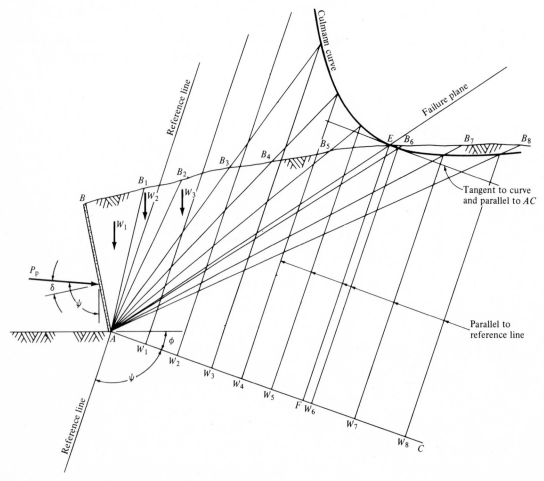

FIGURE 12-21
Culmann's passive earth pressure for cohesionless soils.

curve parallel to line *AC* is drawn, with the resultant earth pressure being the scaled value of line *EF*, as shown in Fig. 12-21.

EXAMPLE 12-5

Given   A retaining wall with back-fill as shown in Fig. 12-22.

Find   The active thrust via Culmann's method.

Procedure   Figure 12-22 shows a 7-m vertical wall supporting a granular back-fill whose φ value equals 30°. The wall is assumed smooth. A line load of 100 kN/m runs parallel to the wall. For an arbitrary scale of 1 cm = 1 m the given data are plotted to scale.

For convenience the bases for all the wedges are the same. Hence the weight of all the wedges equals 127.4 kN, as shown in Fig. 12-23. The corresponding points along *AC* are shown in Fig. 12-22 for an arbitrary scale of 1 cm = 100 kN. From these points lines are drawn parallel to line *AD* so as to intersect rays AB_1, AB_2, AB_3, etc. Note that a similar line is drawn for the line load P_1. By connecting these points of intersection with a smooth curve (Culmann's

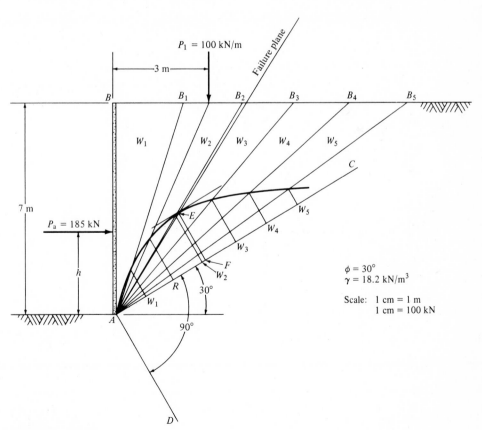

FIGURE 12-22

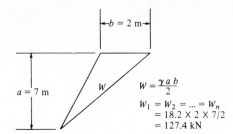

$$W = \frac{\gamma a b}{2}$$

$$W_1 = W_2 = ... = W_n$$
$$= 18.2 \times 2 \times 7/2$$
$$= 127.4 \text{ kN}$$

FIGURE 12-23

curve) and drawing a tangent to this curve parallel to AC, we obtain the value for P_a, which is equal to the corresponding scaled value EF. The scaled value for $EF = 1.85$ cm for 185 kN.

The point where P_a acts is determined as described in the preceding section, and as shown in Fig. 12-24. Line EG is parallel to line AC, and line GF is parallel to the failure plane. P_a therefore acts at one-third distance EF from point E, or a total of 4.25 m above point A.

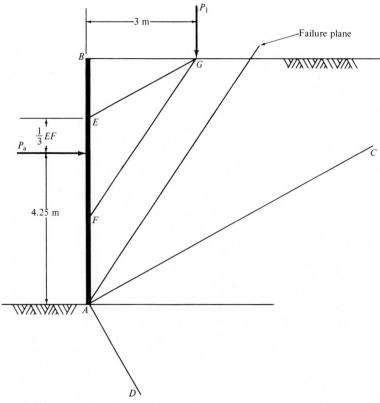

FIGURE 12-24

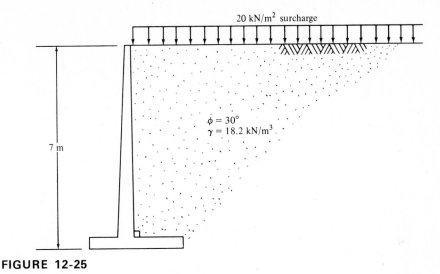

FIGURE 12-25

EXAMPLE 12-6

Given   A surcharge on a back-fill as shown in Fig. 12-25.

Find   The active thrust via Culmann's method.

Procedure   The uniform surcharge shown in Fig. 12-25 is transformed into an equivalent effective weight as shown in Fig. 12-26. From there on the procedure is very similar to that given in Example 12-5. Note, however, that the soil on top of the heel of the retaining wall cannot form a wedge during failure, provided the retaining structure remains intact.

    Hence line AB forms an imaginary rigid surface of the back-fill. Furthermore, the point where P_a acts may be determined by assuming that the pressure distribution and the location of the resultant pressure are as shown in Fig. 12-18. In our case the resultant thrust turns out to be located 2.6 m from the bottom of the wall.

PROBLEMS

12-1.   For the condition outlined in Section 12-2, the expressions for the coefficients for active and passive pressures were determined to be $K_a = \tan^2 (45 - \varphi/z)$ and $K_p = \tan^2 (45 + \varphi/2)$.
    (a) For $5°$ increments, determine the value of K_a and K_p.
    (b) Plot the relationships determined in part (a).
    (c) Is the ratio of K_p/K_a a constant?
    (d) At what value of φ would the passive coefficient approach infinity?

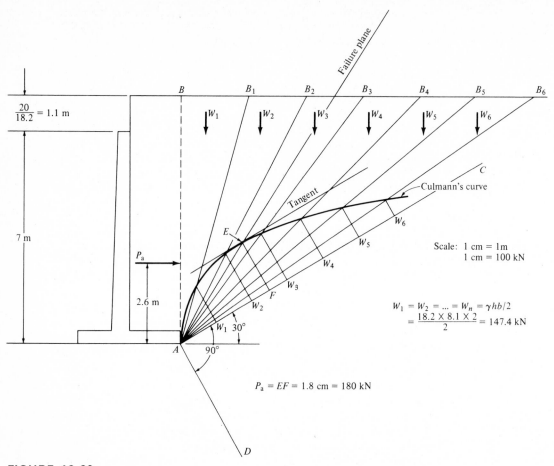

$$\frac{20}{18.2} = 1.1 \text{ m}$$

7 m

P_a

2.6 m

W_1 30°

A 90°

$P_a = EF = 1.8 \text{ cm} = 180 \text{ kN}$

Scale: 1 cm = 1m
        1 cm = 100 kN

$$W_1 = W_2 = ... = W_n = \gamma hb/2$$
$$= \frac{18.2 \times 8.1 \times 2}{2} = 147.4 \text{ kN}$$

FIGURE 12-26

12-2.   A vertical retaining wall retains a granular back-fill whose angle of internal friction and unit weight are 30° and 18.5 kN/m³, respectively. For $H = 5$ m:
   (a)   Determine the active thrust P_a per meter length of wall by Eq. (12-1a).
   (b)   Determine P_a by Rankine's equation (12-9) for $i = 0, 10, 15, 20,$ and 25°.
   (c)   What is the percentage error when calculating P_a by assuming a level back-fill instead of the actual slopes given in part (b)?
   (d)   Plot percentage error vs. i for the results of part (c). Is a trend implied? Explain.

12-3.   A 2-m high, vertical retaining wall pushes against a granular back-fill whose angle of internal friction is 30°. Assume the unit weight of the fill to be 17.6 kN/m³.
   (a)   Determine the total passive resistance P_p via Eq. (12-2a).
   (b)   Determine P_p via Rankine's equation (12-10) for $i = 0, 10, 15, 20,$ and 25°.
   (c)   What is the percentage error when calculating P_p by assuming a level back-fill instead of the actual slopes given?
   (d)   Plot percentage error vs. i for the results of part (c). Is there a distinct trend? Explain.

12-4. A smooth, vertical retaining wall, 5 m high, retains a granular back-fill whose angle of internal friction is 30°. Assume the unit weight of the back-fill to be 18.5 kN/m³.

  (a) Determine the active thrust P_a by Eq. (12-1a).

  (b) Determine P_a by Coulomb's equation (12-13) for $i = 0, 10, 15, 20$, and 25°.

  (c) What is the percentage error when calculating P_a by assuming a level back-fill instead of the actual slopes given in part (b)?

  (d) Plot percentage error vs. i for the results of part (c). What might one conclude? Explain.

12-5. A smooth vertical retaining wall, 2 m high, pushes against a granular back-fill whose angle of internal friction is 30°. Assume the unit of weight of the fill to be 17.6 kN/m³.

  (a) Determine the total passive resistance P_p via Eq. (12-2a).

  (b) Determine P_p via Coulomb's equation (12-14) for $i = 0, 10, 15, 20$, and 25°.

  (c) What is the percentage error when calculating P_p by assuming a level back-fill instead of the actual slopes given in part (b)?

  (d) Plot percentage error vs. i for the results of part (c). What might one conclude? Explain.

12-6. A smooth vertical retaining wall, 5 m high, retains a granular back-fill whose angle of internal friction is equal to 30°. Assume the unit weight of the backfill to be 18.5 kN/m³. For a back-fill slope $i = 0, 10, 15, 20$, and 25°:

  (a) Determine K_a via the Rankine equation and Coulomb equation.

  (b) Determine the corresponding differences in the values of K_a.

  (c) Plot the differences from part (a) vs. i.

12-7. For the retaining wall shown in Fig. P12-7 determine the active thrust via Coulomb's theory for the following given data: $i = 22°$, $\gamma = 17.3$ kN/m³, $\varphi = 32°$, $\delta = 20°$, $\beta = 100°$, and $H = 5$ m.

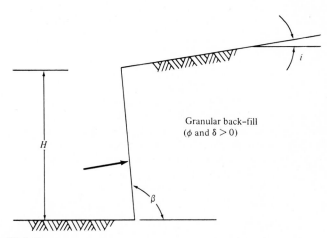

FIGURE P12-7

12-8. Rework Problem 12-7 assuming $\delta = 0$ (a smooth wall). What is the percentage error introduced by assuming a smooth wall when the angle δ actually is 20°?

12-9. Rework Problem 12-7 by assuming that the angle $i = 0°$. What percentage error is introduced by assuming a value of 0° when i actually equals 22°?

12-10. Rework Problem 12-7 assuming the angle $\beta = 90°$. What percentage error is introduced by assuming a value of 90° when it actually equals 100°?

12-11. The depth of the vertical unsupported cut shown in Fig. P12-11 is 3.7 m. Assume $c = 22$ kN/m², $\varphi = 12°$, $\gamma = 17.3$ kN/m³. Determine:
  (a) The stress at the top and bottom of the cut.
  (b) The maximum depth of the potential tension crack.
  (c) The maximum unsupported excavation depth.

12-12. Figure P12-11 shows a vertical unsupported cut of 3.7-m depth. Assume $c = 22$ kN/m², $\gamma = 17.3$ kN/m³, and $\varphi = 6°$. Determine:
  (a) The stress at the top and bottom of the cut.
  (b) The maximum depth of the potential tension crack.
  (c) The maximum unsupported excavation depth.

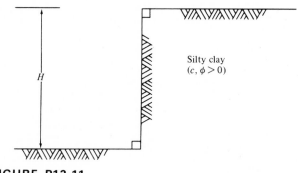

H

Silty clay
$(c, \phi > 0)$

FIGURE P12-11

12-13. Figure P12-11 shows a vertical unsupported cut of 3.7 m. Assume $\gamma = 17.3$ kN/m³ and $\varphi = 12°$. Cohesion is suspected to vary from 20 to 30 kN/m². For the range of 20–30 kN/m², in increments of 2 kN/m², plot c vs. the following quantities:
  (a) The stress at the top and bottom of the cut.
  (b) The maximum depth of the potential tension crack.
  (c) The maximum unsupported excavation depth.

12-14. Rework Problem 12-13 for $\varphi = 0°$.

For Problems 12-15 through 12-20 refer to Fig. P12-15 and determine the active thrust for the respective data given. Assume a frictionless wall.

12-15. $H = 5$ m; $\varphi = 30°$; $\beta = 90°$; $h_w = 0$; $\gamma = 18$ kN/m³; $q = 0$.
12-16. $H = 5$ m; $\varphi = 30°$; $\beta = 90°$; $h_w = 0$; $\gamma = 18$ kN/m³; $q = 250$ kN/m².
12-17. $H = 5$ m; $\varphi = 30°$; $\beta = 90°$; $h_w = 2$ m; $\gamma = 18$ kN/m³; $q = 0$.
12-18. $H = 5$ m; $\varphi = 30°$; $\beta = 90°$; $h_w = 2$ m; $\gamma = 18$ kN/m³; $q = 250$ kN/m².
12-19. $H = 5$ m; $\varphi = 30°$; $\beta = 75°$; $h_w = 0$; $\gamma = 18$ kN/m³; $q = 0$.
12-20. $H = 5$ m; $\varphi = 30°$; $\beta = 95°$; $h_w = 0$; $\gamma = 18$ kN/m³; $q = 0$.

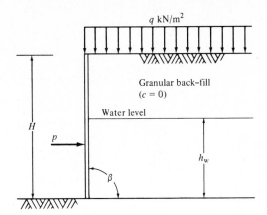

FIGURE P12-15

Referring to Fig. P12-15, solve for P_a via Coulomb's method for the data given in Problems 12-21 through 12-26. Assume γ is a constant for the whole depth and $c = 0$.

12-21. $H = 5$ m; $\varphi = 30°$; $\beta = 100°$; $h_w = 0$; $\gamma = 18$ kN/m³; $q = 0$.
12-22. $H = 5$ m; $\varphi = 30°$; $\beta = 100°$; $h_w = 0$; $\gamma = 18$ kN/m³; $q = 250$ kN/m².
12-23. $H = 5$ m; $\varphi = 30°$; $\beta = 90°$; $h_w = 2$ m; $\gamma = 18$ kN/m³; $q = 0$.
12-24. $H = 5$ m; $\varphi = 30°$; $\beta = 90°$; $h_w = 2$ m; $\gamma = 18$ kN/m³; $q = 250$ kN/m².
12-25. $H = 5$ m; $\varphi = 30°$; $\beta = 95°$; $h_w = 0$; $\gamma = 18$ kN/m³; $q = 0$.
12-26. $H = 5$ m; $\varphi = 30°$; $\beta = 95°$; $h_w = 0$; $\gamma = 18$ kN/m³; $q = 250$ kN/m².
12-27.   Solve for the active thrust via Coulomb's method for the data given in Fig. P12-27, assuming $P_1 = P_2 = 300$ kN/m, and $q = 0$.
12-28.   Referring to Fig. P12-27, for $q = 250$ kN/m² and $P_1 = P_2 = 0$, determine the active thrust on the cantilever retaining wall via Coulomb's method.
12-29.   Referring to Fig. P12-27, if $P_1 = P_2 = 300$ kN/m and $q = 250$ kN/m², determine the active thrust on the retaining wall via Coulomb's method.
12-30.   Referring to Fig. P12-27, for $P_1 = P_2 = 500$ kN/m and $q = 250$ kN/m², determine the active thrust of the retaining wall via Coulomb's method.

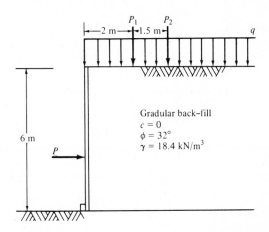

FIGURE P12-27

BIBLIOGRAPHY

1. Andrawes, K. Z., and M. El-Sohby, "Factors Affecting Coefficient of Earth Pressures K_0," *ASCE J. Geotech. Eng. Div.*, vol. 99, July 1973.

2. Brooker, E. W., and H. O. Ireland, "Earth Pressures at Rest Related to Stress History," *Can. Geotech. J.*, vol. 2, no. 1, 1965.

3. Huntington, W. C., *Earth Pressures and Retaining Walls*, Wiley, New York, 1957.

4. Lee, I. K., and J. R. Herington, "Effect of Wall Movement on Active and Passive Pressures," *ASCE J. Geotech. Eng. Div.*, vol. 98, June 1972.

5. Mackey, R. D., and D. P. Kirk, "At Rest, Active and Passive Earth Pressures," *Proc. Southeast Asian Reg. Conf. Soil Eng.*, Bangkok, 1967.

6. Massarsch, K. R., "New Method for Measurement of Lateral Earth Pressure in Cohesive Soils," *Can. Geotech. J.*, vol. 12, Feb. 1975.

7. Massarsch, K. R., and B. B. Broms, "Lateral Earth Pressures at Rest in Soft Clay," *ASCE J. Geotech. Eng. Div.*, vol. 102, Oct. 1976.

8. Moore, P. J., and G. K. Spencer, "Lateral Pressures from Soft Clay," *ASCE J. Geotech. Eng. Div.*, vol. 98, Nov. 1972.

9. Müller-Breslau, H., *Erddruck auf Stutzmauern*, Alfred Kroner Verlag, Stuttgart, Germany, 1906, 1947.

10. Murphy, D. J., G. W. Clough, and R. S. Woolworth, "Temporary Excavation in Varved Clay," *ASCE J. Geotech. Eng. Div.*, vol. 101, Mar. 1975.

11. Peck, R. B., "Earth Pressure Measurements in Open Cuts, Chicago, Illinois, Subway," *Trans. ASCE*, vol. 108, 1943.

12. Peck, R. B., *et al.*, "A Study of Retaining Wall Failures," *2nd Int. Conf. Soil Mech. Found. Eng.*, vol. 3, 1948.

13. Reese, L. C., and R. C. Welch, "Lateral Loading of Deep Foundations in Stiff Clay," *ASCE J. Geotech. Eng. Div.*, vol. 101, July 1975.

14. Rehman, S. E., and B. B. Broms, "Lateral Pressures on Basement Wall: Results from Full-Scale Tests," *5th Eur. Conf. Soil Mech. Found. Eng.*, vol. 1, 1972.

15. Rowe, P. W., and K. Peaker, "Passive Earth Pressure Measurements," *Geotechnique*, vol. 15, Mar. 1965.

16. Shields, D. H., and A. Z. Tolunay, "Passive Pressure Coefficients by Method of Slices," *ASCE J. Soil Mech. Found. Div.*, vol. 99, no. SM12, Dec. 1973.

17. Spangler, M. G., and J. Mickle, "Lateral Pressure on Retaining Walls Due to Backfill Surface Loads," *Highw. Res. Board Bull.* 141, 1956.

18. Terzaghi, K., "Large Retaining Wall Test," *Eng. News Rec.*, Feb. 1, 22; Mar. 8, 20; Apr. 19, 1934.

19. Terzaghi, K., and R. B. Peck, *Soil Mechanics in Engineering Practice*, Wiley, New York, 1967.

20. Tschebotarioff, G. P., *Foundations, Retaining and Earth Structures*, 2nd ed., McGraw-Hill, New York, 1973.

13

Flexible
Retaining
Structures

13-1 INTRODUCTION

It was mentioned in Section 12-1 that the lateral pressure on a retaining structure depends on several factors: (1) the physical properties of the soil; (2) the time-dependent nature of the soil; (3) the imposed loading; (4) the interaction between soil and retaining structure at the interface; and (5) the general characteristics of the deformation in the soil–structure composite.

With flexible earth-retaining structures the last of the above factors assumes particular importance. As a flexible wall deflects, changes occur in the magnitudes and distributions of earth pressures against the wall from those predicted by the methodologies discussed in the previous chapter for rigid walls. Thus magnitudes and locations of pressure resultants P_a and P_p are different from those predicted for rigid walls. Included in the group of flexible retaining structures are sheet-pile installations, braced excavations, and reinforced earth walls. These will be discussed in this chapter.

13-2 CANTILEVER SHEET-PILE WALLS

A cantilever sheet-pile wall is constructed by driving sheet piling to a depth sufficient to develop a cantilever beam-type reaction to resist the active pressures on the wall. That is, the embedment length must be adequate to resist both lateral forces as well as a bending moment.

Lateral deflections in cantilever sheet-pile walls may be rather large for high walls since the piling is relatively flexible and the moment varies as the cube of the height. Also, erosion, scour, or soil compression in front of the wall may further compound this deflection. Hence wall lengths or heights are generally limited to about 5 m. Also, their use is primarily in short-term rather than the long-term, more permanent type of installation.

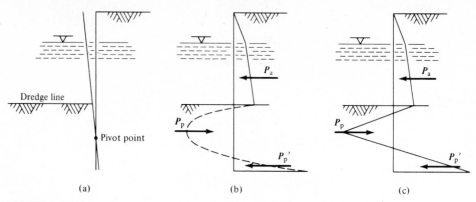

FIGURE 13-1

Sheet-pile deformation and pressures in granular soil. (a) Deflection curve. (b) Likely pressure distribution. (c) Assumed pressure distribution.

The embedment depth varies with different soils, perhaps less than the cantilever length for very dense soils to twice this length for very loose soils. Likewise, the pressure distribution varies with different soils and different water conditions. The actual stress distribution is rather hypothetical, perhaps as depicted in Fig. 13-1b. For computation purposes, however, the distribution is generally simplified as shown in Fig. 13-1c. The lateral deflection is estimated in Fig. 13-1a.

Cantilever Sheet Piling in Granular Soils

The following analysis will be based on a simplified pressure distribution shown in Fig. 13-2, assuming a homogeneous granular soil. For a layered system (density and/or material deviations), the appropriate values for γ and φ should be used in the calculations. Also, if the ground surface is other than level, either the wedge theory or that of Coulomb should be used to approximate the lateral forces. Otherwise, the design concept is identical.

For granular soils it is reasonable to assume the water table to be at the same level on each side of the wall. Thus the pressure distributions (including surcharge effects, etc.) can be established from average values of K_a and K_p. If a safety factor is to be incorporated, either reduce the K_p value (by perhaps 30–50 percent) or increase the depth of penetration (by perhaps 20–40 percent). This will give a safety factor of approximately 1.5–2.0.

With the distribution established, the depth to zero pressure, distance a, can be calculated by similar triangles as

$$a = \frac{p_a}{\gamma_b(K_p - K_a)} \tag{a}$$

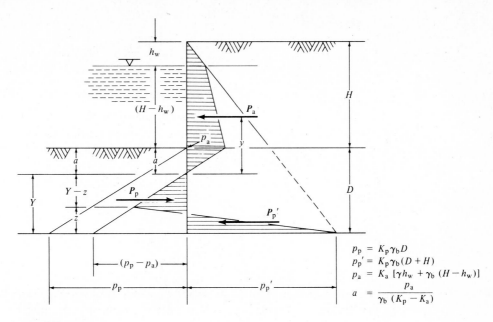

$$p_p = K_p \gamma_b D$$
$$p_p' = K_p \gamma_b (D + H)$$
$$p_a = K_a [\gamma h_w + \gamma_b (H - h_w)]$$
$$a = \frac{p_a}{\gamma_b (K_p - K_a)}$$

FIGURE 13-2
Pressures on cantilever piling in granular soils.

By summing forces in the horizontal direction, we can relate distances z and Y as follows (from Fig. 13-2):

$$\mathbf{P}_a - (p_p - p_p') = 0$$

$$\mathbf{P}_a - \left[p_p \frac{Y}{2} - (p_p + p_p') \frac{z}{2} \right] = 0$$

Solving for z, we get

$$z = \frac{p_p Y - 2\mathbf{P}_a}{p_p + p_p'} \tag{b}$$

Now let us sum moments about the bottom of the piles. This step yields an additional relationship between Y and z. Hence,

$$\mathbf{P}_a(Y + \overline{Y}) - p_p \frac{Y}{3} + p_p' \frac{z}{3} = 0$$

or

$$\mathbf{P}_a(Y + \overline{Y}) - p_p \left(\frac{Y}{2}\right)\left(\frac{Y}{3}\right) + (p_p + p_p')\left(\frac{z}{2}\right)\left(\frac{z}{3}\right) = 0$$

Simplifying, we get

$$6P_a(Y + \bar{Y}) - p_p Y^2 + (p_p + p'_p)z^2 = 0 \qquad (13\text{-}1)$$

The above equation turns out to be a fourth-degree equation of the form

$$Y^4 + C_1 Y^3 + C_2 Y^2 + C_3 Y + C_4 = 0$$

Its solution is usually a trial-and-error one, obtained by first assuming a value for D (see Fig. 13-2), then determining the pressures, distances a and z for the imposed conditions, and subsequently solving. Approximate values for D may be obtained from Table 13-1.

With the final D determined, the maximum bending moment is calculated (at zero shear and prior to increasing the depth), and then a suitable section modulus can be selected.

Cantilever Sheet Piling in Cohesive Soils

The pressures normally assumed to act on a sheet-pile wall embedded in a cohesive soil are shown in Fig. 13-3 for granular and clay back-fill. However, it should be noted that changes such as strength, consolidation, shrinkage, or water in cracks, which are normally time related and rather indeterminate, may result in appreciable changes in the magnitude and location of the pressure resultants acting on the wall.

Immediately after installation it is common practice to calculate pressures assuming that the clay derives all its strength from cohesion and none from internal friction ($c = \frac{1}{2}$ unconfined compressive strength; $\varphi = 0$). $\gamma_e H$ represents the effective pressures at dredge line.

The procedure for the analysis, once the pressure distribution has been established, is basically the same as that previously described. As before, we determine z from the sum of the forces in the x direction, $\sum F_x = 0$. Hence,

$$z = \frac{D(4c - \gamma_e H) - P_a}{4c} \qquad (c)$$

TABLE 13-1
Approximate Values for D

Standard Penetration Resistance N (blows/0.3 m)	Relative Density D_r	Depth* D
0–4	Very loose	$2.0H$
5–10	Loose	$1.5H$
11–30	Medium dense	$1.25H$
31–50	Dense	$1.0H$
Over 50	Very dense	$0.75H$

* See Fig. 13-2

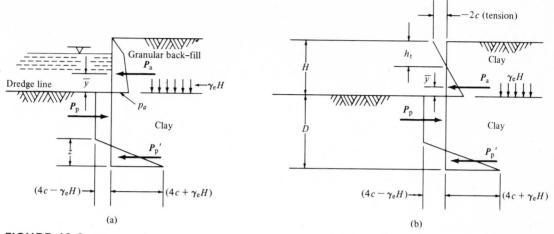

FIGURE 13-3
Pressures on cantilever plies anchored in clay. (a) Pile in clay and granular back-fill. (b) Pile in clay ($\phi = 0$).

and by summing moments about the bottom of the pile, we get

$$\mathbf{P}_a(D + \bar{y}) - (4c - \gamma_e H)D^2 + \left(\frac{4c}{3}\right)z^2 = 0 \tag{13-2}$$

A trial-and-error solution is the normal procedure. For a safety factor greater than 1 one may increase the length by 20–40 percent or reduce the cohesion c by perhaps 20–40 percent.

13-3 ANCHORED BULKHEADS

Anchored sheet-pile walls or *anchored bulkheads* are a type of retaining wall found in waterfront construction which is used to form wharves or piers for loading and unloading ships or barges. Briefly, the construction of such walls is accomplished by first driving the sheet piling into the soil to a designated depth, and then attaching a tie-rod support near or at the upper end of the pile and anchoring it to concrete blocks (*deadmen*), brace piles, sheet piles, or other means of support located at a point safely away from any potential slip surfaces. Frequently the water side of the pile is dredged in order to acquire greater depth, while the back-fill side is built up so as to obtain a level back-fill surface. Depending upon the soil conditions, one level of tie usually suffices for heights of around 10 m. More than one anchor may be necessary for higher walls in order to reduce the tie force, or to decrease the bending moment or deflection of the sheet piles.

Figure 13-4 shows the typical forces acting on an anchored sheet-pile wall. The active thrust is resisted by the tie-rod and by the passive resistance. Hence we need adequate embedment depth in order to develop sufficient passive resistance to

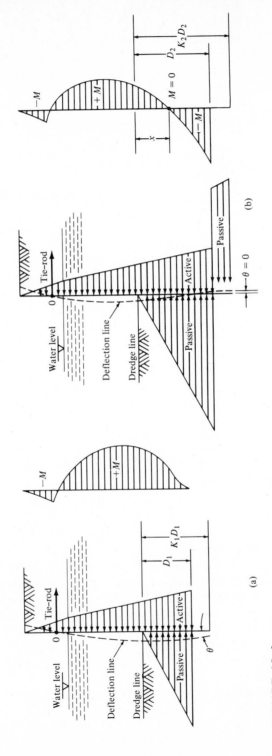

FIGURE 13-4

Effect of depth of penetration on pressure distribution, bending moment, and elastic line. (a) Relatively small depth of penetration; slope of sheet pile at toe is not zero. (b) Greater depth of penetration; slope of sheet pile at toe is approximately zero, similar to that for a fixed-end condition.

resist the active thrust. Likewise, the tie-rod force must be sufficient not only to resist the unbalanced forces, but also to minimize lateral movement of the walls. It is customary to assume that the point where the tie-rod is attached to the wall does not move laterally.

Pressures on and stability of anchored sheet piling are affected by factors such as the depth of piling penetration, the relative stiffness of the piling, the soil properties (e.g., compressibility, cohesive or cohesionless), the amount of anchor yield (e.g., elongation of the rods, yielding in the deadmen or brace-pile support), fluctuating water levels on one or both sides of the pile (e.g., tidal variations), or variable surcharge loads.

The effect of the depth of embedment on the bending moment is illustrated in Fig. 13-4. For relatively short embedment depths, as indicated in Fig. 13-4a, the toe of the pile is not restricted to any significant degree against rotation. The bending moment near the toe, therefore, is negligible. The corresponding moment diagram is indicated to the right of that diagram. The dashed line depicts the shape of the elastic lines for this condition. On the other hand, as the depth of penetration increases, the soil provides some resistance to the toe of the pile in the form of resistance to both translation and rotation. This results in a bending moment as shown in Fig. 13-4b. Subsequent to calculations, the embedment lengths D_1 and D_2 are determined. Thus with the available information on shear, moment, and allowable stresses, a selection is made of the pile size or section modulus. However, because of the many unknowns and variables associated with anchored bulkheads, a safety factor is introduced in the form of added embedment length. That is, the increase in the length of embedment may range anywhere from 20 to 50 percent and more beyond the length calculated.

13-4 FREE-EARTH-SUPPORT METHOD

Granular Soil

Figure 13-5a shows the active and passive forces as well as the supports provided by the tie-rod acting on an anchored sheet-pile wall in sand. The soil pressures may be computed by the Rankine or Coulomb theory. Furthermore, any bending moment induced by the soil on the toe of the pile is assumed negligible for this analysis. Also, point 0, at the level of the tie, is assumed to be stationary.

For convenience, the active thrust is divided into segments designated by A_1, A_2, and A_3 as shown in Fig. 13-5a. Hence employing the equations of equilibrium, we can establish the expression necessary to solve for the sole unknown, the embedment depth D. That is, from $\sum F_x$ the force in the tie-rod is

$$T = \mathbf{P}_a - \mathbf{P}_p$$

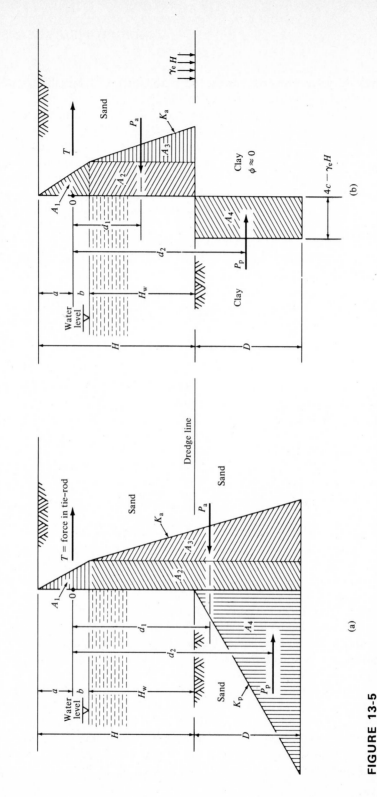

(a)

(b)

FIGURE 13-5

Design of anchored sheet piling by free-earth-support method. (a) Totally granular soil. (b) Granular back-fill and cohesive soil below dredge line. $\gamma_e H$ = weight of back-fill and surcharge. If $P_p \geq 0$, $4c - \gamma_e H \geq 0$, or $c \geq \gamma_e H/4$; unconfined compressive strength $q_u \geq \gamma_e H/2$.

Similarly, the expressions for the active thrust $\mathbf{P_a}$ and the passive resistance $\mathbf{P_p}$ are given by Eqs. (13-3) and (13-4), respectively:

$$\mathbf{P_a} = \frac{\gamma K_a}{2}(a + b)^2 + \gamma K_a(a + b)(H_w + D) + \frac{\gamma_b K_a}{2}(H_w + D)^2 \qquad (13\text{-}3)$$

and

$$\mathbf{P_p} = \frac{\gamma_b K_p}{2} D^2 \qquad (13\text{-}4)$$

By summing moments about point 0 (level of tie), we get

$$d_1 \mathbf{P_a} = -\frac{\gamma K_a}{2}(a + b)^2\left(\frac{a + b}{3} - b\right) + \gamma K_a(a + b)\left(\frac{H_w + D}{2} + b\right)$$

$$+ \frac{\gamma_b K_a}{2}(H_w + D)^2[\tfrac{2}{3}(H_w + D) + b] \qquad (13\text{-}5)$$

and

$$d_2 \mathbf{P_p} = \frac{\gamma_b K_p}{2} D^2(H_w + b + \tfrac{2}{3}D) \qquad (13\text{-}6)$$

From equilibrium we know that the moments about point 0 must equal to zero; thus this is satisfied by

$$d_1 \mathbf{P_a} = d_2 \mathbf{P_p}$$

Substituting the equivalent terms from Eqs. (13-5) and (13-6) one gets an expression of the following form:

$$C_1 D^3 + C_2 D^2 + C_3 D + C_4 = 0 \qquad (13\text{-}7)$$

The solution to Eq. (13-7) is frequently carried out by a process of trial and error. One notes that the only unknown is D. Hence the solution to Eq. (13-7) yields the embedment depth D. For design purposes it is advisable to increase the depth 20–40 or 50 percent to provide a safety factor against possible passive failure. One commonly used factor (or increase) is given by Eq. (13-8):

$$\text{design depth } D_d \approx (\sqrt{2})D \qquad (13\text{-}8)$$

   Example 13-1 gives the sequence of the steps followed in carrying out such a solution.

Cohesive Soils

Figure 13.5b illustrates a case where the sheet piling is embedded in clay and back-filled with a granular material. The active thrust is determined in accordance with the procedure described above, by either Coulomb's or Rankine's theory. The passive resistance, however, is assumed to be a rectangular block with the intensity of pressure as indicated by Fig. 13-5b. The procedure for determining the embedment length D is simply a matter of satisfying equilibrium, very much similar to that described previously.

13-5 ROWE'S MOMENT-REDUCTION METHOD

Because of the flexibility of steel sheet piling, the soil pressure for such installations differs significantly from the classical hydrostatic distribution. Generally the bending moment decreases with increasing flexibility of the piling. Hence the bending moment calculated via the free-earth-support method results in a design usually regarded as too conservative. Rowe (16, 18) proposed a method for reducing the moment in anchored sheet piling, thereby yielding what appears to be a more realistic design.

The significant factors related to Rowe's moment-reduction theory are:

1. The relative density of the soil.
2. The relative flexibility of the piling, expressed in terms of a flexibility number (units are in fps)

$$\rho = H^4/EI$$

where H = total length of piling (ft)
E = modulus of elasticity of section (psi)
I = moment of inertia of section per foot of wall (in^4)

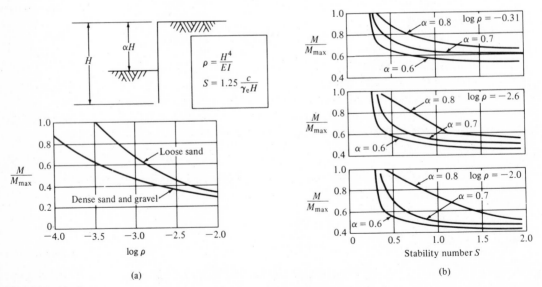

(a)   (b)

FIGURE 13-6
Moment-reduction curves after Rowe (16, 18). (a) Sheet piling in sand. (b) Sheet piling in clay.

3. For cohesion soils a stability number was defined as

$$S = 1.25c/\gamma_e h$$

4. The relative height of piling α.

Figure 13-6 gives a series of curves developed by Rowe which relate the ratio of reduced design moment M to M_{max} determined by the free-earth-support method to $\log \rho$ for sand or stability number S for clays. (Again, note that H here represents the total length of the pile.) An interpolation is to be made for sands which range between loose and dense and for α values that are different from those shown. For known values of H and calculated values of ρ and S, the ratios of M/M_{max} can be determined from the corresponding charts in Fig. 13-6. The actual value of M per foot for a given pile can be determined from $\sigma = M/S$; Table 13-2 provides section properties for some common sheet piling. This value must be equal to or slightly larger than the value obtained via the chart. Example 13-1 may be useful in illustrating the steps.

EXAMPLE 13-1

Given A sheet-pile wall as shown in Fig. 13-7.

Find (a) Total length of the sheet pile.
(b) Force the tie-rod per meter of wall using the free-earth-support method.
(c) Select a suitable piling section based on Rowe's moment-reduction method.

Procedure (a)
$$\mathbf{P}_a = \overbrace{\frac{\gamma K_a}{2}(a + b)^2}^{A_1} + \overbrace{\gamma K_a(a + b)(H_w + D)}^{A_2} + \overbrace{\frac{\gamma_b K_a}{2}(H_w + D)^2}^{A_3}$$

$$\mathbf{P}_a = \frac{19.62 \times 0.3}{2}(3 + 1)^2 + 9.68 \times 0.3(3 + 1)(6 + D) + \frac{9.68 \times 0.3}{2}(6 + D)^2$$

$$\mathbf{P}_a = 47.09 + 23.54(6 + D) + 2.50(6 + D)^2$$

$$\mathbf{P}_p = \overbrace{\tfrac{1}{2}\gamma_b K_p D^2}^{A_4} = \tfrac{1}{2} \times 9.68 \times 3 \times D^2$$

From the sum of the moment about point 0 in Fig. 13-7, $\sum M_0$ (level of tie-rod),

$$\mathbf{P}_a d_1 = -\frac{\gamma K_a}{2}(a + b)^2\left(\frac{a + b}{3} - b\right) + \gamma K_a(a + b)(H_w + D)\left(\frac{H_w + D}{2} + b\right)$$

$$+ \frac{\gamma_b K_a}{2}(H_w + D)^2[\tfrac{2}{3}(H_w + D) + b]$$

$$= -47.09(\tfrac{4}{3} - 1) + 23.54(6 + D)\left(\frac{6 + D}{2} + 1\right) + 2.50(6 + D)^2[\tfrac{2}{3}(6 + D) + 1]$$

TABLE 13-2
Properties of Some Sheet Piling Sections (Manufactured by U.S. Steel Corp.)

Profile	Designation	Driving Distance per Pile (in.)	Weight — Per Foot (lb.)	Weight — Per Square Foot of Wall (lb.)	Web Thickness (in.)	Section Modulus — Per Pile (in.³)	Section Modulus — Per Foot of Wall (in.³)	Area — Per Pile (in.²)	Moment of Inertia — Per Pile (in.⁴)	Moment of Inertia — Per Foot of Wall (in.⁴)
	Interlock with each other									
	PSX32	$16\frac{1}{2}$	44.0	32.0	$\frac{29}{64}$	3.3	2.4	12.94	5.1	3.7
	PS32*	15	40.0	32.0	$\frac{1}{2}$	2.4	1.9	11.77	3.6	2.9
	PS28	15	35.0	28.0	$\frac{3}{8}$	2.4	1.9	10.30	3.5	2.8
	Interlock with each other									
	PSA28*	16	37.3	28.0	$\frac{1}{2}$	3.3	2.5	10.98	6.0	4.5
	PSA23	16	30.7	23.0	$\frac{3}{8}$	3.2	2.4	8.99	5.5	4.1
	PDA27	16	36.0	27.0	$\frac{3}{8}$	14.3	10.7	10.59	53.0	39.8
	PMA22	$19\frac{5}{8}$	36.0	22.0	$\frac{3}{8}$	8.8	5.4	10.59	22.4	13.7

Profile dimensions: 29/64", 1/2", 3/8"; 1 11/32", 1/2"; 1 11/32", 3/8"; 3/8", 5", 3/8"; 3 1/4", 3/8"

Interlock with each other and with PSA23 or PSA28

PZ38	18	57.0	38.0	$\frac{3}{8}$	70.2	46.8	16.77	421.2	280.8
PZ32	21	56.0	32.0	$\frac{3}{8}$	67.0	38.3	16.47	385.7	220.4

Interlocks with itself and with PSA23 or PSA28

PZ27	18	40.5	27.0	$\frac{3}{8}$	45.3	30.2	11.91	276.3	184.2

Suggested allowable design stresses—sheet piling

Steel Brand or Grade	Minimum Yield Point (psi)	Allowable Design Stress* (psi*)
ASTM A328	38,500	25,000
ASTM A572 GR 50 (USS EX-TEN 50)	50,000	32,000
ASTM A690 (USS MARINER GRADE)	50,000	32,000

* Based on 65 percent of minimum yield point. Some increase for temporary overstresses generally permissible.

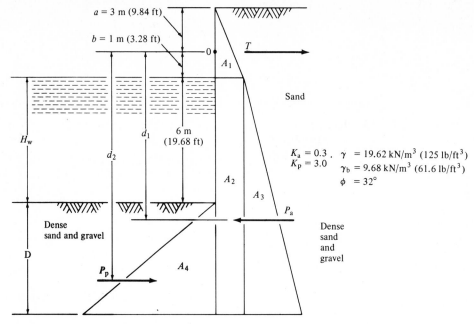

FIGURE 13-7

Simplifying and combining terms,

$$P_a d_1 = 999.26 + 374.78D + 44.27D^2 + 1167D^3$$

$$P_p d_2 = 25.02D^2(H_w + b + \tfrac{2}{3}D) = 25.02D^2(6 + 1 + \tfrac{2}{3}D)$$

$$P_p d_2 = 175.14D^2 + 16.67D^3$$

$$P_a d_1 - P_p d_2 = 0$$

$$15D^3 + 138.87D^2 - 374.78D - 999.26 = 0$$

Solving by trial and error,

$$D = 3.51, \text{ say } 3.5 \text{ m}$$

Design: $D_d \approx D\sqrt{2} \approx 3.5\sqrt{2} \simeq 5$ m

Answer

Hence the total length L of the sheet pile is

$$L = 10 + 5 = 15 \text{ m } (49.2 \text{ ft})$$

(b)   For $D = 3.5$ m, from P_a and P_b above,

$$P_a = 47.09 + 23.54(6 + 3.5) + 2.5(6 + 3.5)^2 = 496.34 \text{ kN}$$

$$P_p = 25.02(3.5)^2 = 306.5 \text{ kN}$$

we find:

$$T = P_a - P_p = 496.34 - 306.5$$

$$T = 189.84 \text{ kN/m (12.98 kips/ft)}$$

(c) Figure 13-8 depicts the forces on the wall for a distance x below the water level, where the maximum moment M_{max} occurs. From equilibrium, $\sum F_x = 0$,

$$0.0093x^2 + 0.492x + 3.23 - 12.99 = 0$$

or

$$9.3x^2 + 492x = 9760$$

Solving for x,

$$x = 15.38 \text{ ft} \qquad \text{(below water table)}$$

Thus,

$$M_{max} = 12.99(15.38 + 3.28) \qquad = 242.4 \text{ kip-ft}$$

$$-3.23\left(15.38 + \frac{13.12}{3}\right) \qquad = -63.8 \text{ kip-ft}$$

$$-0.492(15.38)\left(\frac{15.38}{2}\right) \qquad = -58.2 \text{ kip-ft}$$

$$-0.0093(15.38)^2\left(\frac{15.38}{3}\right) \qquad = -11.3 \text{ kip-ft}$$

$$M_{max} = 109.1 \text{ kip-ft/ft of wall}$$

$$= 1309.2 \text{ kip-in.}$$

The total length of the pile is 15 m = 49.2 ft = H. Then,

$$\rho = \frac{H^4}{EI} = \frac{(49.2)^4}{30 \times 10^6 I} = \frac{0.195}{I}$$

The pile section properties are given in Table 13-3.

From Fig. 13-6a, the moment-reduction curve corresponding to "dense" sand is reproduced in Fig. 13-9. The "dense" classification is part of the data given.

For the three sections selected, the ratio of M/M_{max} was plotted against the corresponding log ρ values. Any of the section plotting above the "dense" curve would be satisfactory regarding stress; any below would not. In our case section $PZ27$ is adequate. Note that the allowable design stress used was 25,000 psi.

Section $PZ27$ is adequate

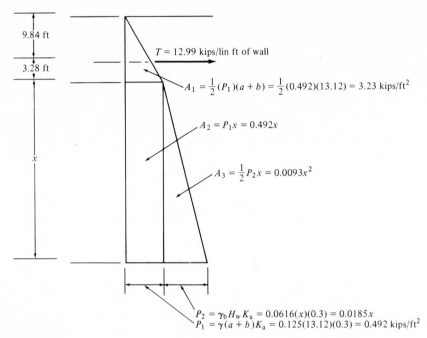

FIGURE 13-8

TABLE 13-3
Pile Section Properties

	Pile Section		
Properties	*PZ38*	*PZ32*	*PZ27*
I (per ft)	280.8	220.4	184.2
$\rho = H^4/EI$	6.94×10^{-4}	8.85×10^{-4}	10.59×10^{-4}
$\log \rho$	-3.16	-3.05	-2.98
Section modulus S	46.8	38.3	30.2
Moment capacity			
$\quad = \sigma S = \frac{25}{12}S$, kip-ft	97.5	79.8	62.9
M/M_{max}	0.89	0.73	0.58

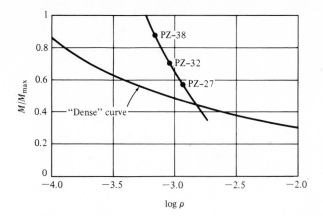

FIGURE 13-9

$\log \rho$

13-6 FIXED-EARTH-SUPPORT METHOD

This method assumes that the toe of the pile is restrained from rotating, as indicated in Fig. 13-10a. The deflection of the sheet piling is indicated by the dashed line. The corresponding moment diagram for this case is shown in Fig. 13-10b. Point C is a point of contraflexure. At this point the pile could be assumed to act as a hinge (zero bending moment). Hence the portion of the piling above point C can be treated as a beam which resists the net earth pressures via the force T and the shear R_C, as indicated in Fig. 13-10c. This is known as the *equivalent-beam method*, first proposed by Blum (3). Blum established a theoretical relationship between the angle of internal friction and the distance from the point of contraflexure to the dredge line X, as shown in Fig. 13-10d.

The lateral pressures, both active and passive, are determined by either the Rankine or the Coulomb formula, as before. For a given angle of internal friction φ the value for X can be calculated as a function of H from Fig. 13-10d. Hence by summing moments about 0 (anchor level), the shear R_C is determined. With R_C known, the summation of moments about point E yields a relationship where the only unknown is D. The depth D determined from this expression is generally increased from 20 to 40 percent or more (K_2 is 1.2–1.4D, etc.). The force P in the tie can likewise be determined by summing moments about the hinge at point C.

With the value of the embedment depth calculated, one may proceed to determine the shear and the moment, as before, and subsequently to select the appropriate size for the sheet piling. Example 13-2 illustrates the sequence of steps summarized above.

EXAMPLE 13-2

Given   The sheet-pile wall of Example 13-1.

Find   Using the fixed-earth-support method, determine: (a) the force T in the tie-rod and (b) the embedment D of sheet piling.

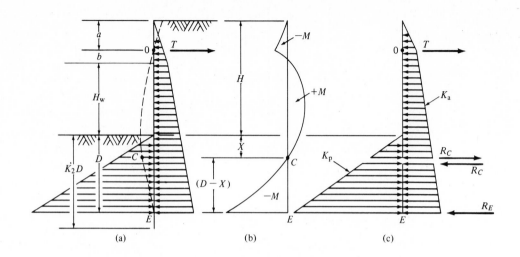

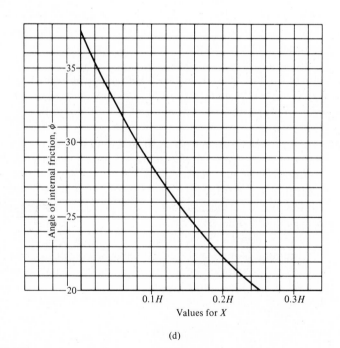

FIGURE 13-10
Fixed-end-support equivalent beam method in sand.

Procedure   (a)   The active and p ssive pressures at the toe of the pile are:

$$\mathbf{P_a} = \gamma K_a(a + b) + \gamma_b K_a(H_w + D)$$

$$\mathbf{P_a} = (19.62)0.3(4) + (9.68)0.3(6 + D)$$

$$\mathbf{P_a} = 41 + 2.9D$$

and

$$p_p = \gamma_b K_p D = 9.68 \times 3 \times D = 29D$$

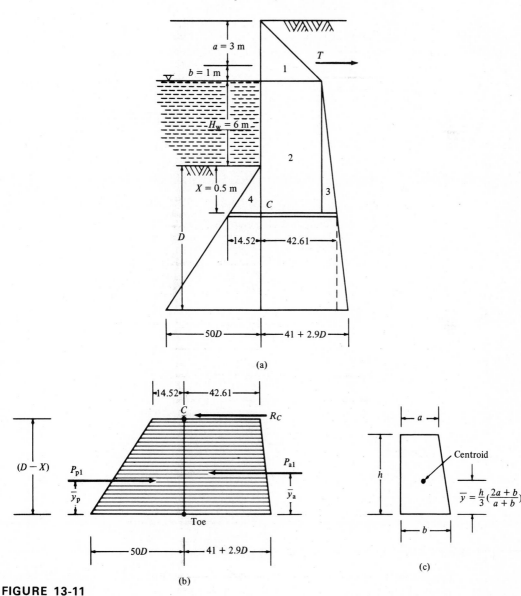

(a)

(b)

(c)

FIGURE 13-11

For $H = 10$ m, $X = 0.05H = 0.5$ m. The active forces above point C, namely, A_1, A_2, A_3, and A_4, are:

$$A_1 = \tfrac{1}{2}\gamma K_a(a + b)^2 = \frac{19.62}{2}(0.3)(4)^2 = 47.09; \ \bar{y} = 6.5 + \tfrac{4}{3} = 7.83$$

$$A_2 = \gamma K_a(a + b)(H_w + X) = 19.62(0.3)(4)(6.5) = 153.04; \ \bar{y} = 3.25$$

$$A_3 = \tfrac{1}{2}\gamma_b K_a(H_w + 6)^2 = \frac{9.68}{2}(0.3)(6.5)^2 = 105.70; \ \bar{y} = 2.17$$

$$A_4 = \tfrac{1}{2}(14.52)(X) = \frac{14.52}{2}(0.5) = 3.63$$

$$\sum M_{C\,(top)}: \ T \times 7.5 + 3.63 \times \frac{0.5}{3} = 47.09(7.83) + 153.04(3.25) + 105.70(2.17)$$

$$7.5T = 368.71 + 497.38 + 229.37 - 0.61$$

Answer

$$T = 146.06 \text{ kN/lin m}$$

(b) $\sum F_X:$ $T + R_C = \sum A = 47.09 + 153.04 + 105.70 - 3.63$

$$146.06 + R_C = 303.21$$

$$R_C = 157.15$$

$$P_{p1} = \left(\frac{50D + 14.52}{2}\right)(D - X); \ \bar{y}_{p1} = \left(\frac{D - X}{3}\right)\left(\frac{2 \times 14.52 + 50D}{50D + 14.52}\right)$$

$$P_{a1} = \left(\frac{41 + 2.9D + 42.61}{2}\right)(D - X); \ \bar{y}_{a1} = \left(\frac{D - X}{3}\right)\left(\frac{2 \times 42.61 + 41 + 2.9D}{41 + 2.9D + 42.61}\right)$$

Thus, $\sum M_{toe}: R_C(D - X) = P_{p1}\bar{y}_p - P_{a1}\bar{y}_a$, or

$$R_C = \frac{D - X}{6}[(29.04 + 50D) - (126.22 + 2.9D)]$$

Substituting, $R_C = 157.15$, $X = 0.5$, and simplifying

$$D^2 - 2.57D - 18.99 = 0$$

Solving for D,

Answer

$$D = 5.83 \text{ m}; D + 20\% = 7 \text{ m}$$

13-7 WALES, TIE-RODS, AND ANCHORAGES

The tie-back support provided for sheet-pile installations is illustrated by the general arrangement shown in Fig. 13-12e. A wale is placed in a horizontal position in front of the piling or attached to the back-fill face of the piling if a flush-front face is required.

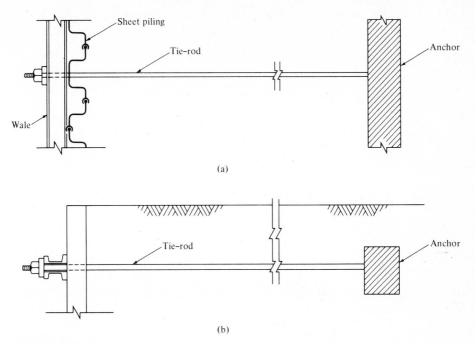

FIGURE 13-12

General arrangement of wale, tie-rod, and anchor. (a) Plan view. (b) Elevation view.

The latter is generally the case for harbor installations. Wales are usually fabricated of two-channel sections placed back to back and separated sufficiently to insert the tie-rod through as shown.

The tie-rod may be a cable or a steel bar threaded to permit vertical alignment and tension adjustments. Paint, asphaltic material, or concerte encasement are materials commonly used to provide corrosion protection to the tie-rod; in Fig. 13-13d the rod is protected by concrete. Vertical piles are frequently used to support tie-rods that rest on soft and compressible material in order to minimize sagging due to settlement. The spacing of the tie-rods depends on the forces assumed by each rod. The design forces, particularly in cohesive soils, are usually assumed to be 20–30 percent larger than those computed via the previously discussed methods, with allowable stresses approaching 80–90 percent of the yield stress of the steel.

The wales may be conservatively designed as simply supported beams with spans equal to the distances between tie-rods or, more realistically, as continuous beams. Splices in channel sections of wales should be staggered to avoid possible weak points. Governing concerns in wale evaluations are bending and web crippling. The latter problem is frequently overcome by web reinforcement of the section.

Figure 13-13 shows a variety of anchor schemes used in the tie-back system. Deadman-type anchors shown in Fig. 13-13a are constructed either by pouring a concrete beam in place, or by embedding a precast beam. Care should be taken to provide reliable lateral resistance by placing the precast beam directly against

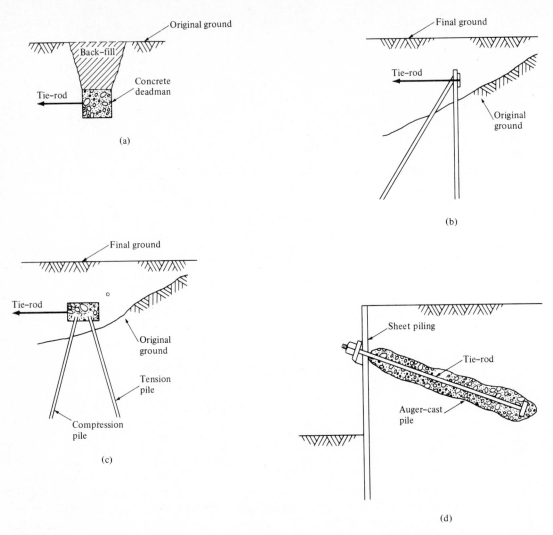

FIGURE 13-13

Some common types of anchors. (a) Deadman anchor, either precast or cast-in-place concrete. (b) Braced piles. (c) Braced piles. (d) Tie-rod anchored in cast-in-place pile (see Chap. 15 for more details).

the original soil and/or to tamp fill material in front of the beam. The deadman is placed at a sufficient distance from the sheet-pile wall to permit full development of the passive resistance. Hence the ultimate capacity per unit length of the deadman is approximately

$$T_u = \mathbf{P}_p - \mathbf{P}_a \qquad (13\text{-}9)$$

where $\mathbf{P}_p$ and $\mathbf{P}_a$ may be regarded as the passive and active resultant forces. Their magnitude is estimated on the assumption that the earth wedge in front of the deadman

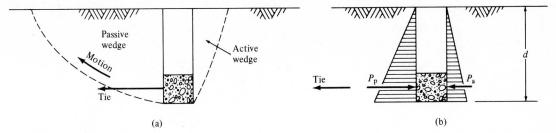

FIGURE 13-14
Forces related to deadman-type anchor. (a) Reaction elements on deadman. (d) Active and passive resultants on deadman.

slides upward and forward, as shown in Fig. 13-14a. An approximate allowable capacity in granular soils per unit length may be obtained from Eq. (13-10):

$$T_a = \frac{1}{F}(A\gamma K_p d^2) \qquad (13\text{-}10)$$

where T_a = allowable capacity per unit length of deadman
F = factor of safety
A = coefficient, usually assumed 0.6
γ = unit weight of soil
K_p = coefficient of passive pressure
d = depth to bottom of deadman

The allowable capacity per unit length in cohesive soils ($\varphi \approx 0$) is sometimes approximated by Eq. (13-11):

$$T_a = \frac{d^2}{F}\left(A\gamma K_p + \frac{c}{2}\right) \qquad (13\text{-}11)$$

where c = soil cohesion.

The anchors shown in Figs. 13-13b, c, and d develop their capacity from the piles Installation and capacity of piles are topics discussed in Chapter 15.

13-8 BRACED CUTS

With the possible exception of excavations in rock formations, steep cuts of any significant depth in soils are laterally supported by either retaining walls or braced sheeting. Generally retaining walls are used as supports for a more permanent construction. On the other hand, braced sheetings are commonly used for temporary supports, perhaps during construction such as for tunneling and subways, deep trenches, deep building foundations, etc. Also the braced supports are appreciably more flexible than most permanent supports such as concrete retaining walls.

Figure 13-15 shows two commonly used techniques for lateral bracing. In Fig. 13-15a steel plates or wood sheeting, placed either horizontally or vertically, are supported directly by cross members or *struts*. This arrangement may be expanded to

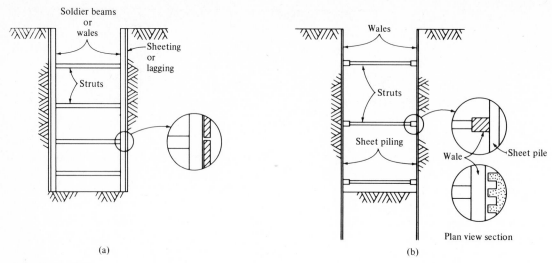

FIGURE 13-15
Common methods of bracing cuts. (a) Lagging. (b) Sheet piles.

include structural members, such as *soldier beams* or *wales*, placed either vertically or horizontally at some designated spacing, as the need dictates. These in turn are also supported by the horizontal struts. This method of bracing is generalized as *lagging*.

Also common is the form of bracing shown in Fig. 13-15b. In this case steel sheet piling is driven into the ground, generally to a depth greater than the excavation in order to develop bottom anchorage. As the excavation progresses, wales are placed horizontally along the length of the excavation at intermittent depths. They in turn are supported by the horizontal struts placed at designated spacings.

The pressure exerted against bracing is different from that usually assumed for retaining walls. The cause for this difference lies in the character of "wall" movements and the subsequent pattern of deformation of the soil associated with the two different types of support. As the top strut is wedged against the wale, it will permit relatively little movement at that point. With advancing excavation below the wale, however, some horizontal movement of the soil will be permitted by the relatively flexible sheeting until the next strut is fixed in position. In other words, the soil in the vicinity of the first strut is rather immobilized, but the lower stratum slips down and toward the excavation until the second strut is fixed in position. Figure 13-16 depicts (although exaggerated) deflections typical of braced excavations. In the process, however, the soil below has a tendency to drag on the soil above it. This continues with progressing excavation until the final strut is in place. The state of stress associated with this type of deformation is closely related to a condition generalized as *arching-active*. Furthermore the distribution of pressure is greatly affected by the deformation conditions, the nature of the soil, and the manner of installing the braces.

At the present time the pressure distribution on braced sheeting is not totally understood. Indeed, almost all of the design for braced sheeting is based upon empirical

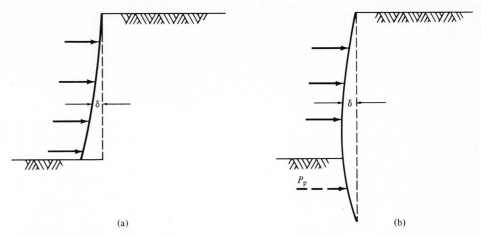

FIGURE 13-16
Deflections of braced excavations. (a) Lagging. (b) Sheet piling.

methods derived as a result of field tests (8, 12, 13, 20). Figure 13-17a provides a qualitative comparison of the type of pressure distribution obtained from field observations and measurements, and the theoretical distribution given by the Rankine theory (dashed line). Field data indicate that the resultant pressure is 10 to 15 percent higher than that predicted by the Rankine distribution. Furthermore, the point at which the resultant force acts is higher than that for the Rankine pressure. It

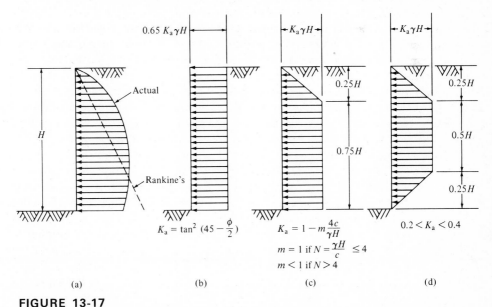

FIGURE 13-17
Pressure distributions on braced excavations. (a) General. (b), (c), and (d) Distributions used for design. [After Terzaghi and Peck, 1967 (21).]

acts near the midpoint of the depth, compared to the third of the height from the bottom for the Rankine distribution.

Figures 13-17b, c, and d show the pressure distribution recommended for the design for various types of soils. Although other distributions have been recommended for this purpose, the ones shown appear to give acceptable results for most design problems. On the other hand, judgment should not be discarded for any individual excavation considered.

EXAMPLE 13-3

Given   An excavation as shown in Fig. 13-18.

Find   Forces in struts per meter depth normal to page.

Procedure   From Fig. 13-17c,

$$N = \frac{\gamma H}{c} = \frac{17.8 \times 7}{25} = 4.98 > 4$$

N is only slightly above the value of 4 recommended. Thus let us assume $m \approx 0.8$. (Values of m measured ranged from 1 in Chicago to 0.4 in Oslo, Norway.) Reliable values of m might better be based on actual measurements of strut loads. Thus,

$$K_a = 1 - m\frac{4c}{\gamma H} = 1 - 0.8\left(\frac{4 \times 25}{17.8 \times 7}\right) = 0.36$$

and

$$K_a \gamma H = 0.36 \times 17.8 \times 7 = 44.86 \text{ kN/m}^2$$

The pressure distribution is as shown in Fig. 13-19. For calculating the strut forces, each strut is assumed to carry the corresponding load based on a simple span.

$$P_1 = 0.5(1.75)(44.86) = 39.25 \text{ kN}$$

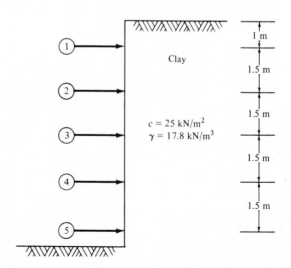

FIGURE 13-18

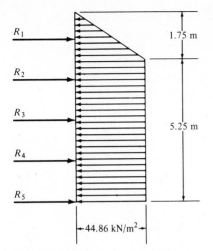
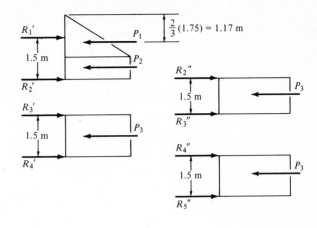

FIGURE 13-19

P_1 acts at 1.17 m from top or 1.33 m from ②

$$P_2 = 0.75(44.86) = 33.64 \text{ kN}$$

P_2 acts at 0.375 m from ②.

$$P_3 = 1.5(44.86) = 67.28 \text{ kN}$$

This force is of the same magnitude for the three rectangles shown.

$\sum M_②:$

$$R_1 = \frac{39.25 \times 1.33 + 33.64 \times 0.375}{1.5} = 43.21 \text{ kN}$$

$\sum F_X:$

$$R_2' = P_2 + P_1 - R_1 = 39.25 + 33.64 - 43.21 = 29.68$$
$$R_2'' = \tfrac{1}{2}P_3 = 33.64$$

Hence,

$$R_2 = R_2' + R_2'' = 63.32 \text{ kN}$$

Since $R_3' = R_3'' = R_4' = \cdots = R_2''$,

$$R_3 = 2R_3' = 2R_2'' = 67.28 \text{ kN}$$
$$R_4 = 67.28 \text{ kN}$$
$$R_5 = \tfrac{1}{2}P_3 = 33.64 \text{ kN}$$

Answer

$$R_1 = 43.21 \text{ kN}$$
$$R_2 = 63.32 \text{ kN}$$
$$R_3 = 67.28 \text{ kN}$$
$$R_4 = 67.28 \text{ kN}$$
$$R_5 = 33.64 \text{ kN}$$

13-9 REINFORCED EARTH

Figure 13-20 depicts the general features of a rather unique earth-retaining structure. It is a relatively new concept which has attracted appreciable interest in research and application since its introduction (2, 5, 8, 14, 15, 23). The earth mass is reinforced with a series of strips which are attached to a face "skin." The skin resists the lateral thrust, induced by the active earth pressure, through the anchorage provided by the strips. The strips derive their tensile capacity from the "bond" developed between the strips and the earth.

The strips may be of any material such as metal, fiber, wood, or plastics, which satisfies the following requirements: (1) has adequate tensile strength; (2) is corrosive resistant; and (3) is suitable for friction or bond development. Also, the strips should provide adequate surface area and display sufficient flexibility to develop the frictional resistance required and they should be compatible with the soil movement and deformation.

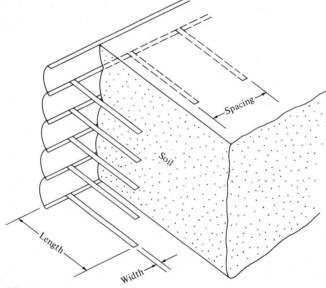

FIGURE 13-20
Reinforced-earth retaining wall.

Granular back-fill is generally desirable. It permits easy drainage and thus creates a state of reduced lateral pressure. Also, it develops better strip bond than do cohesive soils.

The lateral pressure on the wall is assumed to vary linearly with depth, as shown in Fig. 13-21a. The corresponding tensile forces in the strips are therefore computed as indicated in Fig. 13-21b. The length of the strips must be sufficient to extend beyond

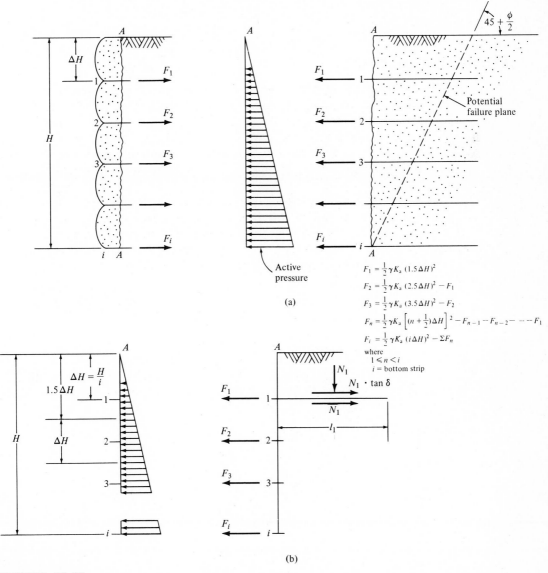

$$F_1 = \tfrac{1}{2}\gamma K_a (1.5\Delta H)^2$$

$$F_2 = \tfrac{1}{2}\gamma K_a (2.5\Delta H)^2 - F_1$$

$$F_3 = \tfrac{1}{2}\gamma K_a (3.5\Delta H)^2 - F_2$$

$$F_n = \tfrac{1}{2}\gamma K_a \left[(n+\tfrac{1}{2})\Delta H\right]^2 - F_{n-1} - F_{n-2} - \cdots - F_1$$

$$F_i = \tfrac{1}{2}\gamma K_a (i\Delta H)^2 - \Sigma F_n$$

where
$1 \leqslant n < i$
i = bottom strip

(a)

(b)

FIGURE 13-21
Forces on reinforced-earth wall. (a) Earth pressure and induced strip tensions. (b) Procedure for estimating strip tensions.

the potential failure plane shown in Fig. 13-21a. The cross section of the strips must be sufficient to satisfy tensile stress limits for the material used.

EXAMPLE 13-4

Given   Granular back-fill; $\varphi = 30°$; $\gamma = 18$ kN/m³; $H = 1.5$ m and $\Delta H = 1.5/5 = 0.3$ m; strip spacing $= 1$ m; $\sigma_s = 14 \times 10^4$ kN/m² (approximately 20,000 psi); $\delta = 20°$ (friction angle for soil strip).

Find   Suitable strip sizes.

Procedure $K_a = 0.333$. Thus,

$$F_1 = \tfrac{1}{2}(18)(0.33)(0.45)^2 = 0.60 \text{ kN/m}$$

$$F_2 = \tfrac{1}{2}(18)(0.33)(0.75)^2 - F_1 = 1.07 \text{ kN/m}$$

$$F_3 = \tfrac{1}{2}(18)(0.33)(1.05)^2 - F_2 - F_1 = 1.06 \text{ kN/m}$$

$$F_4 = \tfrac{1}{2}(18)(0.33)(1.35)^2 - F_3 - F_2 - F_1 = 2.04 \text{kN/m}$$

$$F_5 = \tfrac{1}{2}(18)(0.33)(1.5)^2 - F_4 - F_3 - F_2 = F_1 = 1.37 \text{ kN/m}$$

The cross-sectional area of the strips required is:

$$A = \frac{F_{\max}}{\sigma_s} = \frac{2.04}{14 \times 10^4} = 0.249 \times 10^{-4} \text{ m}^2 \approx 0.15 \text{ cm}^2$$

In order to develop the frictional resistance, a reasonable width must be available if we are to keep the length from becoming excessive. Furthermore, some reasonable thickness must be provided to compensate for some corrosion with time. Hence, somewhat arbitrarily, we select an 8-cm × 0.3-cm section; thus the stress is

$$\sigma_s = \frac{2.04 \text{ kN}}{2.4 \times 10^{-4} \text{ m}^2} = 0.85 \times 10^4 \text{ kN/m}^2 < 14 \times 10^4 \text{ kN/m}^2, \text{ allowable}$$

The length of each strip must extend past the potential failure surface for a sufficient distance to develop the necessary frictional resistance. The distances from the face to the failure surface are:

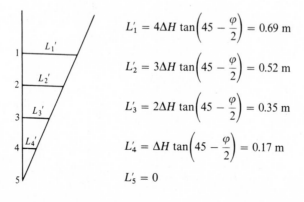

$$L'_1 = 4\Delta H \tan\left(45 - \frac{\varphi}{2}\right) = 0.69 \text{ m}$$

$$L'_2 = 3\Delta H \tan\left(45 - \frac{\varphi}{2}\right) = 0.52 \text{ m}$$

$$L'_3 = 2\Delta H \tan\left(45 - \frac{\varphi}{2}\right) = 0.35 \text{ m}$$

$$L'_4 = \Delta H \tan\left(45 - \frac{\varphi}{2}\right) = 0.17 \text{ m}$$

$$L'_5 = 0$$

The required lengths $L_1, L_2, \ldots, L_i$ are:

$$F_i \times \text{(safety factor)} = \gamma(n\Delta H) \tan \delta(L_n - L_i') \times \text{width} \times 2 \text{ faces}$$

Thus assuming a safety factor of 1.2, we have

$$F_1(1.2) = 18(0.3)(0.36)(L_1 - 0.69)(\tfrac{8}{100})(2)$$

$$0.6(1.2) = 0.35(L_1 - 0.69)$$

and

$$L_1 = 2.75 \text{ m}$$

Also,

$$1.07(1.2) = 0.7(L_2 - 0.52)$$

$$L_2 = 2.35 \text{ m}$$

Similarly,

$$L_3 = 2.85 \text{ m}$$

$$L_4 = 2.91 \text{ m}$$

$$L_5 = 2.40 \text{ m}$$

Answer

Thus assume $L = 3$ m for each strip; the width for each strip is 8 cm and the thickness 0.3 cm.

PROBLEMS

Reference is made to Fig. P13-1 for Problems 13-1 through 13-9. For these problems determine:
(a) The force in the tie-rod per meter of wall.
(b) The total length of the sheet pile by the method stipulated and data given for the corresponding problems.

13-1. $H = 8$ m; $h_w = 4$ m; $a = 2.5$ m; use free-earth-support method.
13-2. $H = 8$ m; $h_w = 2$ m; $a = 2.5$ m; use free-earth-support method.
13-3. $H = 8$ m; $h_w = 0$; $a = 2.5$ m; use free-earth-support method.
13-4. $H = 8$ m; $h_w = 4$ m; $a = 2.5$ m; use fixed-earth-support method.
13-5. $H = 8$ m; $h_w = 2$ m; $a = 2.5$ m; use fixed-earth-support method.
13-6. $H = 8$ m; $h_w = 0$; $a = 2.5$ m; use fixed-earth-support method.
13-7. $H = 10$ m; $h_w = 4$ m; $a = 2.5$ m; use fixed-earth-support method.
13-8. $H = 10$ m; $h_w = 2$ m; $a = 2.5$ m; use fixed-earth-support method.
13-9. $H = 10$ m; $h_w = 0$; $a = 2.5$ m; use fixed-earth-support method.

Reference is made to Fig. P13-10 for Problems 13-10 through 13-12.
(a) The force in the tie-rod per meter of wall.

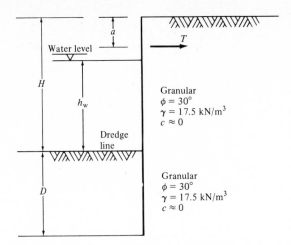

FIGURE P13-1

(b)   The total length of the sheet pile by the method stipulated and for the data corresponding to the following problems.

13-10. $H = 8$ m; $h_w = 4$ m; $a = 2.5$ m; use free-earth-support method.
13-11. $H = 8$ m; $h_w = 2$ m; $a = 2.5$ m; use free-earth-support method.
13-12. $H = 8$ m; $h_w = 0$; $a = 2.5$ m; use free-earth-support method.

For Problems 13-13 through 13-17 determine the forces in the struts shown in Fig. P13-13 for the data given:

13-13. $H = 7$ m; $\gamma = 18$ kN/m³; $\varphi = 32°$; $c = 0$; sand.
13-14. $H = 7$ m; $\gamma = 18$ kN/m³; $\varphi = 0°$; $c = 30$ kN/m²; soft clay.
13-15. $H = 7$ m; $\gamma = 18$ kN/m³; $\varphi = 0°$; $c = 30$ kN/m²; stiff clay.
13-16. $H = 6$ m; $\gamma = 17.5$ kN/m³; $c = 0$; $\varphi = 32°$; sand.
13-17. $H = 6$ m; $\gamma = 17.5$ kN/m³; $c = 25$ kN/m²; $\varphi = 0$; soft clay.

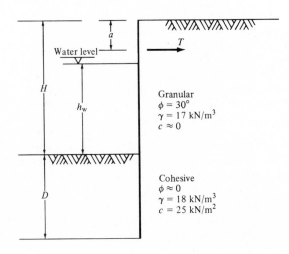

FIGURE P13-10

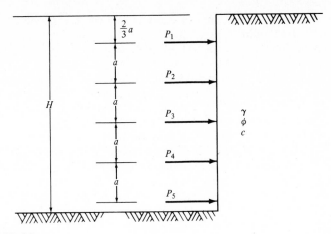

FIGURE P13-13

13-18. For the data of Problem 13-2 select a suitable piling section based on Rowe's moment-reduction method.

13-19. Design a reinforced-earth retaining wall for the following data: height $= 2$ m; granular back-fill; $\varphi = 32°$; $\delta = 22°$; $\gamma = 18.3$ kN/m³; factor of safety $= 1.2$. Assume steel construction.

BIBLIOGRAPHY

1. Al-Hussaini, M., and E. B. Perry, "Field Experiment of Reinforced Earth Wall," *ASCE J. Geotech. Eng. Div.*, vol. 104, Mar. 1978.

2. Binquet, J., and K. L. Lee, "Bearing Capacity Tests on Reinforced Earth Slabs," *ASCE J. Geotech. Eng. Div.*, vol. 101, Dec. 1975.

3. Blum, H., *Einspannungsverhältnisse bei Bohlwerken*, W. Ernst and Sohn, Berlin, Germany, 1931.

4. Broms, B., and H. Stille, "Failure of Anchored Sheet Pile Walls," *ASCE J. Geotech. Eng. Div.*, vol. 102, Mar. 1976.

5. Chang, J. C., and R. A. Forsyth, "Design and Field Behavior of Reinforced Earth Wall," *ASCE J. Geotech. Eng. Div.*, vol. 103, July 1977.

6. Change, J. C., and R. A. Forsyth, "Finite Element Analysis of Reinforced Earth Wall," *ASCE J. Geotech. Eng. Div.*, vol. 103, July 1977.

7. Clough, G. W., and Y. Tsui, "Performance of Tied-Back Walls in Clay," *ASCE J. Geotech. Eng. Div.*, vol. 100, Dec. 1974.

8. Golder, H. Q., *et al.*, "Predicted Performance of Braced Excavation, *ASCE J. Soil Mech. Found. Div.*, vol. 96, May 1970.

9. Haliburton, T. A., "Numerical Analysis of Flexible Retaining Structures," *ASCE J Geotech. Eng. Div.*, vol. 94, Nov. 1968.

10. Hanna, T. H., and I. I. Kurdi, "Studies on Anchored Flexible Retaining Walls in Sand," *ASCE J. Geotech. Eng. Div.*, vol. 100, Oct. 1974.

11. Kay, J. N., and M. I. Qamar. "Evaluation of Tie-Back Anchor Response." *J. Geotech. Eng. Div.*, vol. 104, Jan. 1978.

12. Lambe, T. W., "Braced Excavations," *4th Pan Am. Soil Mech. and Found. Eng. Conf.*, ASCE, 1970.
13. Lambe, T. W., *et al*, "Measured Performance of Braced Excavation," *ASCE J. Soil Mech. Found. Div.*, vol. 96, no. SM3, May 1970.
14. Lee, K. L., B. D. Adams, and J. J. Vagneron, "Reinforced Earth Retaining Walls, *ASCE J. Soil Mech. Found. Div.*, vol. 99, Oct. 1973.
15. Richardson, G. N., and K. L. Lee, "Seismic Design of Reinforced Earth Walls," *ASCE J. Geotech. Eng. Div.*, vol. 101, Feb. 1975.
16. Rowe, P. W., "Anchored Sheet Pile Walls," *Proc. Inst. Civ. Eng.*, vol. 1, pt. 1, 1952.
17. Rowe, P. W., "A Stress–Strain Theory for Cohesionless Soil with Applications to Earth Pressures at Rest and Moving Walls," *Geotechnique*, vol. 4, June 1954.
18. Rowe, P. W., "Sheet Pile Walls in Clay," *Proc. Inst. Civ. Eng.*, vol. 7, July 1957.
19. Rowe, P. W., "Anchored Sheet-Pile Walls," *Proc. Inst. Civ. Eng.*, 1962.
20. Swatek, E. P., Jr., *et al.* "Performance of Bracing for Deep Chicago Excavation," *5th Pan Am. Soil Mech. and Found. Eng. Conf.*, ASCE, vol. 1, pt. 2, 1972.
21. Terzaghi, K., and R. B. Peck, *Soil Mechanics in Engineering Practice*, 2nd ed., Wiley, New York, 1967.
22. Turabi, D. A., and A. Balla, "Distribution of Earth Pressure on Sheet-Pile Walls," *ASCE J. Geotech. Eng. Div.*, vol. 94, Nov. 1968.
23. Vaid, Y. P., and R. A. Campanella, "Time-Dependent Behavior Earth Wall," *ASCE J. Geotech. Eng. Div.*, vol. 103, July 1977.
24. Vidal. H., "The Principle of Reinforced Earth," *Highw. Res. Rec.* 282, 1969.

14
Bearing Capacity- Shallow Foundations

14-1 INTRODUCTION

A safe foundation design provides for a suitable safety factor against (1) shear failure of the soil; and (2) excessive settlement. The ability of a soil to sustain a building load without undergoing excessive settlement or shear failure is a measure of the *bearing capacity* of the soil—a topic to be discussed in this chapter.

Sometimes the foundation loads are transmitted directly to a rock foundation. Except for the possible concerns related to the stability of the rock stratum (e.g., expansive shales, pyrite formations, heavily laminated with clay seams, or highly fractured), rock formations generally have more than ample strength for a safe bearing support. On the other hand, if the stratum directly underneath the building foundation consists of a soil that has very low shear strength and/or is highly compressible, the foundation load may have to be transmitted to a greater depth, to a stiffer stratum or to rock, perhaps by means of piles or caissons. In most instances, however, one may support the foundation loads quite safely by bearing directly on the soil on which the foundation rests. It is this means of support that will be considered in this chapter; piles and caisson-type foundations are evaluated in Chapter 15.

Much investigation on the subject of bearing capacity has been carried out during the past century. Subsequently numerous proposals have been advanced regarding considerations, criteria, and procedures for the evaluation of the bearing capacity in soils (48). Indeed, it may be exhaustive and perhaps counterproductive to attempt to delineate the work of all the many contributors to the subject. Instead, the author elected, not without reluctance, to provide the reader with only a brief coverage of some of the fundamental concepts, plus the more current philosophy associated with the bearing capacity problem.

Essentially the designer must establish: (1) what is the ultimate contact pressure from the foundation loads that will cause a probable or impending failure (or what is the *ultimate bearing capacity*); and (2) what is the safe contact pressure (or what is the *allowable bearing capacity*) of the stratum for which one designs. It is common practice to express the allowable bearing capacity as the ultimate bearing capacity divided by a suitable safety factor. A number of formulas and procedures related to bearing capacity design will be focused upon in this chapter.

Depending upon the importance of the proposed structure, the selection of a safe and economical bearing capacity criterion to be used for design may involve extensive investigations, or it may be a matter of merely complying with the minimum standard of a local building code. Local building codes and handbooks are frequently relied upon as the sole basis for the selection of a bearing capacity figure for ordinary buildings such as homes and other light structures. While this could prove a dangerous practice even for light buildings, it is highly inadvisable as a general practice, particularly for heavy and/or tall structures. The more reliable approach consists of field explorations and testings (e.g., methodical observations via drilling and soil sampling, field-load tests, etc.), laboratory analysis, and subsequent evaluation via analytical tools coupled with judgment on the part of the investigator.

14-2 LOAD–SOIL DEFORMATION RELATIONSHIP

Not unexpectedly, subjected to load, the soil will deform. However, unlike the more homogeneous materials such as steel or other metals, the load–deformation relationship in soil is not nearly as well defined or predictable. Nevertheless a certain degree of generality could be representative even of soils. For example, most fairly dense or stiff soils will depict a reasonable elastic relationship between load and deformation up to a substantial percentage of their ultimate strength. This phase of deformation is primarily attributed to the densification of the stratum as a result of the reduction in voids within a soil mass. Increasing the load results in a further increase in the deformation, but at a more rapid rate. This increased rate of yielding may be due partially to an additional decrease in the void ratio, but also to a combination of high lateral displacement coupled with vertical deformation. Subsequently, with an increase in load, excessive vertical penetration and ultimate shear failure of the soil stratum will take place.

Figure 14-1 depicts a general pattern of soil deformation with increasing load on a typical spread or isolated footing. With loads of magnitudes such that the induced settlement is due primarily to elastic strain (e.g., primarily a reduction in the voids), as shown in Fig. 14-1a, the amount of penetration is relatively small. Commensurate with this range of load, the strain in the soil under and around the footing is likewise rather small. With added load and a corresponding increase in the penetration, the soil under the footing becomes increasingly compressed, while that around the footing has a tendency to bulge laterally and upward, as shown in

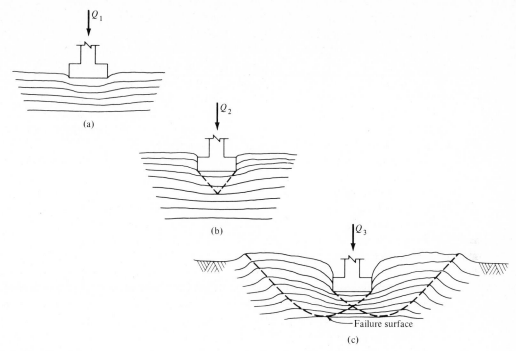

FIGURE 14-1
Soil deformation under and around isolated footing at various stages of increasing load.

Fig. 14-1b. Progressively this becomes more apparent or pronounced as the penetration increases, as indicated in Fig. 14-1c. Under careful scrutiny one may frequently observe from field tests the transition from a reasonable "elastic" condition to one where excessive yielding takes place.

Figure 14-2 depicts a rather typical load–settlement relationship for a case of a spread footing on a hypothetical stratum; that is, it is general in nature and not representative of any particular type of soil, density, or stiffness. In fact, different types of soils, of different densities and degrees of stiffness, display wide variations in load–settlement relationships. This aspect is discussed in Section 14-3. In general, virtually no part of the load–settlement curve is a truly straight line in the case of soils. Furthermore, the yield and the ultimate strength are not well-defined levels. Instead, such points are selected as ranges of load or stress based upon a combination of perhaps numerical guidelines, functional requirements, judgment, and sometimes heavy dependence on field testing and model studies. However, while field tests and model studies may be reflective of the load–settlement characteristics of the soil stratum, and therefore may be useful in predicting the bearing capacity at a given site, the results may be highly misleading if such factors as underlying strata or footing size are not carefully accounted for in the final analysis. Load bearing capacity tests are frequently run in the field on relatively small areas, say $\frac{1}{4}$ to $\frac{1}{2}$ m^2, loaded with

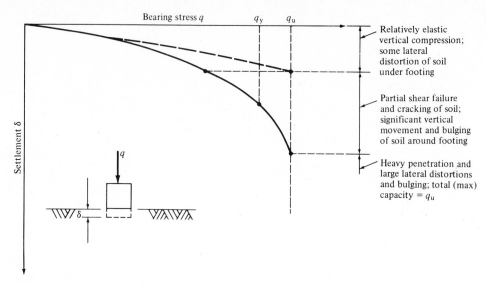

FIGURE 14-2
General load intensity—settlement in soils.

dead loads. Due to the very large loads required for prototype-size loading tests, it is practical to limit the bearing plates to a relatively small size. (Load tests are also undertaken to determine the capacity of piles. However, pile tests are almost always conducted by utilizing the prototype pile, quite frequently one of the piles in the structural pile group.)

Referring back to Fig. 14-2, one notes that as the load is increased, the settlement increases at an increasing rate (e.g., point q_y). A point is reached, q_u, where the slope of the curve becomes almost vertical. This corresponds to the knee portion of the curve and is generally regarded as the *bearing capacity* of the soil. With any further increase in stress, significant shear failure and deep penetration into the soil will occur, perhaps with catastrophic consequences. Consistent with sound engineering practice, the design bearing capacity is but a fraction of q_u.

14-3 BEARING FAILURE PATTERNS

Bearing failure of a foundation usually results because the soil supporting the foundation fails in shear. The modes of shear failure are commonly separated into three categories: (1) *general shear* failure; (2) *local shear* failure; and (3) *punching shear* failure. Based on his personal observations and a comprehensive summary of findings by others, Vesic (51) notes that the mode is linked to several factors. In general it depends on the relative compressibility of the soil in the particular geometrical and loading conditions.

Vesic (51) depicts the general characteristics of these failure modes as shown in Fig. 14-3 for tests conducted in sand. For the case of *general shear* failure, usually

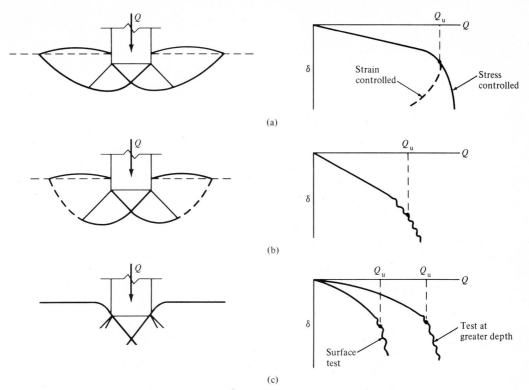

FIGURE 14-3
Modes of bearing failures. (a) General shear. (b) Local shear. (c) Punching shear. [After Vesic (51).]

associated with dense soils of relatively low compressibility, the slip surface is continuous from the edge of the footing to the soil surface, and full shear resistance of the soil is developed along the failure surface. The load–deformation curve has a fairly constant slope for a substantial percentage of the ultimate load. A gradual deviation from a straight line and subsequent yielding are observed as the load intensity approaches Q_u. The slope becomes virtually vertical as Q_u is reached. Under stress-controlled conditions (e.g., dead loads, building loads) an increase in load past Q_u may result in significant continuous and cumulative vertical and/or tilting deformation, with likely sudden and total failure. Under strain-controlled conditions (e.g., constant rate of penetration) deformation is likely even under a reduced level of load. Bulging of the soil near the footing is usually apparent throughout most of the loading cycle, as shown in Fig. 14-3a.

For the *local shear* failure the failure surface extends from the edge of the footing to approximately the boundary of the Rankine passive state. The shear resistance is fully developed over only part of the failure surface. There is a certain degree of bulging on the sides and considerable vertical compression under the footing. This is not usually apparent until significant vertical penetration occurs. As shown

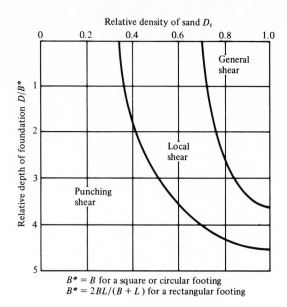

FIGURE 14-4
Modes of failure of model footings in Chattahoochee sand. [After Vesic (51).]

in Fig. 14-3b, the load–deformation curve displays a lesser degree of linearity, a steeper slope, and a smaller Q_u than the load–deformation relationship for general shear.

In the case of *punching shear* failure, a condition common for loose and very compressible soils, the pattern is not easily detected. Generally some vertical shear deformation is visible around the periphery of the footing, for there is little horizontal strain and no apparent bulging of the soil around the footing. Test results for footings at the surface indicate that the ultimate load is significantly lower than for the case where the footing is placed at a greater depth. Furthermore the load–deformation curve has a steeper slope for the surface than it does for the greater depth tests, as shown in Fig. 14-3c. Although deformations are considerable, sudden collapse or tilting failures are not common. Vesic (51) developed Fig. 14-4 which shows the mode of failure that may be expected within the relative density ranges indicated for the various relative depths of penetration. For example, a footing on a very loose sand may fail in punching shear, while the same footing on the surface of a very dense sand would fail in general shear. On the other hand, a footing in a dense sand might fail in general shear near the surface but may fail in local shear or even punching shear if it is located at greater depths.

14-4 PRANDTL'S THEORY FOR ULTIMATE BEARING CAPACITY

Many of the current-day fundamental principles, however limited or incomplete, regarding bearing capacity determination had their substantive beginning with Prandtl's theory (42) of plastic equilibrium. Subsequent to Prandtl's findings in the early 1920s, extensive investigations by numerous individuals have pointed to some deficiencies in Prandtl's theory when used to predict a bearing capacity of soils. These observations led to modifications and improvements in the applicability of Prandtl's theory to soil-bearing-capacity problems. Still, we have not reached the point, and possibly never will, where the bearing capacity of the soil stratum can be predicted with total confidence. Instead, we may be limited to further improving on the work started by Prandtl and amplified and improved by others (see bibliography at the end of this chapter).

Prandtl's theory of plastic equilibrium reflects on the deformation or penetration effects of hard objects into much softer material. Synonymous with these conditions, we make use of this theory to assume a rigid footing penetrating into a relatively soft soil. Contrary to some of Prandtl's assumptions, the typical soil is not isotropic and homogeneous. Furthermore the typical footing assessed in terms of practical design limitations is not infinitely long, not smooth at the interface of footing and soil, and quite probably never applied at the very surface of the material—conditions assumed in Prandtl's work.

Figure 14-5 shows three zones developed within a soil mass as a long footing (say $L/B > 5$) is subjected to increasing load which results in bearing failure within the soil. The load is transmitted through the soil wedge, zone 1, which is assumed to remain in a plastic state, and which remains intact during failure. Zone 2 undergoes considerable plastic flow, with the dashed lines depicting arbitrary failure planes developed throughout this zone. The curved boundaries of this zone are logarithmic spirals with the radius of curvature being represented by $r = r_0 e^{\alpha \tan \varphi}$. Zone 3 represents the passive state as developed by Rankine and discussed in Chapter 12. The failure surface of zone 3 is a straight line, as indicated in Fig. 14-5.

At impending failure, zones 2 and 3 are pushed aside by the penetrating wedge. Subsequently a shear resistance to such movement is developed along the logarithmic spiral and the straight-line segment. Assuming a constant value for the cohesion, the shear strength along the plane of failure could be expressed as $s = c + \sigma \tan \varphi$. On this premise the ultimate bearing capacity of a soil based on Prandtl's theory is given by Eq. (14-1):

$$q_{\mathrm{u}} = c \cot \varphi \left[\tan^2 \left(45 + \frac{\varphi}{2} \right) e^{\pi \tan \varphi} - 1 \right] \qquad (14\text{-}1)$$

An evaluation of Eq. (14-1) reveals that if the cohesion of the soil was 0, the bearing capacity would also equal 0. Quite to the contrary, anyone familiar with the foundation conditions recognizes that a dense granular ($c = 0$) stratum may be a most desirable foundation base when considering not only the bearing capacity but also consolidation and predictable behavior. This and other limitations were recognized and accounted for, to some extent, by Terzaghi and others, as discussed in subsequent sections.

Taylor (47) added a term to Eq. (14-1) to account for the shear strength induced by the overburden pressures. He determined this correction term to be equal to

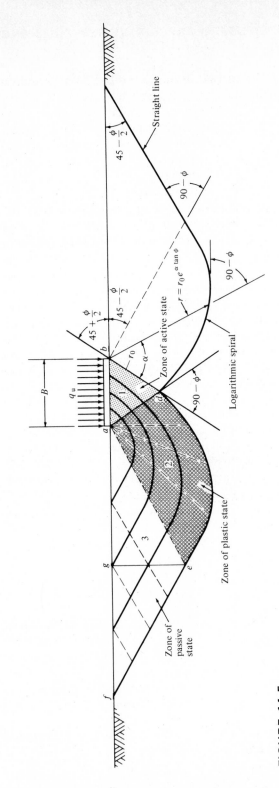

FIGURE 14-5
Prandtl's theory of plastic equilibrium.

$(\gamma B/2)\tan(45 + \varphi/2)$. Hence introducing this term into Eq. (14-1), the expression for the ultimate bearing capacity becomes that given by Eq. (14-2):

$$q_u = \left[c \cot\varphi + \frac{\gamma B}{2}\tan\left(45 + \frac{\varphi}{2}\right) \right]\left[\tan^2\left(45 + \frac{\varphi}{2}\right) e^{\pi\tan\varphi} - 1 \right] \quad (14\text{-}2)$$

14-5 BEARING CAPACITY BASED ON RANKINE WEDGES

Although the following analysis will not yield the final equation employed in the present-day analysis of the bearing capacity in soils, it does provide a dimensionally correct expression for developing the more accurate formulas on which evaluations are currently made. Subsequent sections in this chapter will discuss modifications and alterations to this expression which attempt to account for a number of variables in a way consistent with experimental findings and observations.

Figure 14-6a shows a long narrow (i.e., L/B is very large) footing at a depth D into a $(c\text{-}\varphi)$ soil. Figure 14-6b depicts the Rankine wedges used in this analysis. Wedge I is assumed to be an active Rankine wedge which is pushed down and to the right during the failure sequence; wedge II is assumed to be a passive wedge which is pushed to the right and upward in the process. The horizontal resistances are designated by P and are assumed to act at the interface of the two wedges as shown.

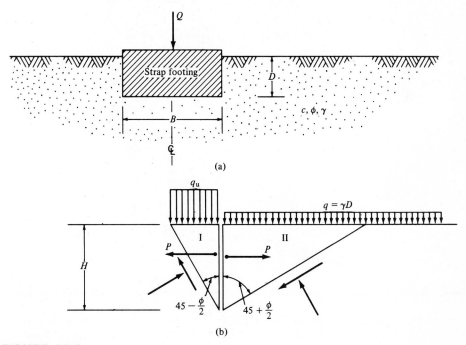

FIGURE 14-6
(a) Footing in $(c\text{-}\phi)$ soil. (b) Rankine wedges.

They have the same magnitude but opposite directions, consistent with requirements for equilibrium for the respective wedges. However, the P associated with wedge I represents the active pressure resultant, while the P for wedge II is the passive thrust. Thus for the active case, wedge I, from Eq. (12-15) we have

$$P = \tfrac{1}{2}\gamma K_a H^2 - 2cH\sqrt{K_a} + q_u K_a H \tag{a}$$

where the last term reflects the influence of the surcharge q_u, and $K_a = \tan^2(45 - \varphi/2)$.

For the passive case, wedge II, from Eq. (12-16) we have

$$P = \tfrac{1}{2}\gamma K_p H^2 + 2cH\sqrt{K_p} + q K_p H \tag{b}$$

where $K_p = \tan^2(45 + \varphi/2)$.

The two resultants are of the same magnitude. Hence

$$\tfrac{1}{2}\gamma K_a H^2 - 2cH\sqrt{K_a} + q_u K_a H = \tfrac{1}{2}\gamma K_p H^2 + 2cH\sqrt{K_p} + q K_p H$$

Solving for q_u, we have

$$q_u = \tfrac{1}{2}\gamma H\left(\frac{1}{K_a}\right)(K_p - K_a) + 2c\left(\sqrt{K_p} + \frac{1}{\sqrt{K_a}}\right) + q K_p^2 \tag{c}$$

But $K_p = 1/K_a$; also, from Fig. 14-6b,

$$H = \frac{B}{2 \tan(45 - \varphi/2)} = \frac{B}{2\sqrt{K_a}}$$

Hence Eq. (c) becomes

$$q_u = \tfrac{1}{4}\gamma B K_p^{3/2}(K_p - K_p^{-1}) + 2c K_p(K_p^{1/2} + K_p^{-1/2}) + q K_p^2$$

or

$$q_u = \tfrac{1}{4}\gamma B(K_p^{5/2} - K_p^{1/2}) + 2c(K_p^{3/2} + K_p^{1/2}) + q K_p^2 \tag{d}$$

Let

$$N_\gamma = \tfrac{1}{2}(K_p^{5/2} - K_p^{1/2})$$
$$N_c = 2(K_p^{3/2} + K_p^{1/2})$$
$$N_q = K_p^2$$

Thus Eq. (d) becomes

$$q_u = cN_c + qN_q + \tfrac{1}{2}\gamma B N_\gamma \tag{14-3}$$

Equation (14-3) is the basic form of the general bearing capacity expression used in the soil mechanics field. Variations in the values for N_c, N_q, and N_γ have been proposed over the years by different investigators, as have been factors to account for footing shapes, depth, inclination, ground, and base variations, etc. These will be discussed in later sections.

The above derivations are based on less than accurate assumptions: (1) the shear at the interface of the two wedges was neglected; and (2) the failure surfaces are not straight lines as was indicated for the two wedges. These and other deficiencies prompted substantial changes in the N values as well as the introduction of a number of dimensionless factors.

14-6 TERZAGHI'S BEARING CAPACITY THEORY

Terzaghi improved on the wedge analysis described in the preceding section by working with trial wedges of the type assumed by Prandtl. However, he expanded on Prandtl's theory to include the effects of the weight of the soil. He assumed the general shapes of the various zones to remain unchanged, as illustrated in Fig. 14-7. Terzaghi (49) assumed the angle that the wedge face forms with the horizontal to be φ rather than the $(45 + \varphi/2)$ assumed in Prandtl's theory.

Figure 14-7 provides the basic elements in the development of Terzaghi's theory. As did Prandtl, Terzaghi assumed a strip footing of infinite extent. Unlike Prandtl, however, Terzaghi assumed a rough instead of a smooth base surface. Furthermore although he neglected the shear resistance of the soil above the base of the footing (above line bf in Fig. 14-7), he did account for the effects of the soil weight by superimposing an equivalent surcharge load $q = \gamma D$. Otherwise the shape of the failure surface is similar to that assumed by Prandtl's theory.

Figure 14-8 shows the penetrating wedge, in equilibrium, where the downward load is resisted by the forces on the inclined faces of the wedge. These forces consist of the cohesion and the resultant of the passive pressure. Thus assuming a unit length of the footing normal to the page, we obtain q_u as follows. From $\sum F_y = 0$, for a unit length, we have, from Fig. 14-8,

$$q_u B = 2P_p + 2(bd)c \sin \varphi$$

But $bd = B/2 \cos \varphi$. Thus,

$$q_u B = 2P_p + Bc \tan \varphi \tag{14-4}$$

Terzaghi represented the value of P_p as the vector sum of three components: (1) that from cohesion; (2) that from surcharge; and (3) that resulting from the weight of the soil ($bdef$ in Fig. 14-7). These are identical in nature (not in values) to those in Eq. (14-1). From this he obtained an expression for the ultimate bearing capacity for general shear conditions and for *long footings*, given by Eq. (14-5):

$$q_u = \tfrac{1}{2}\gamma B N_\gamma + \gamma D N_q + c N_c \tag{14-5}$$

where

$$N_\gamma = \tfrac{1}{2}\tan \varphi \left(\frac{K_p}{\cos^2 \varphi} - 1 \right)$$

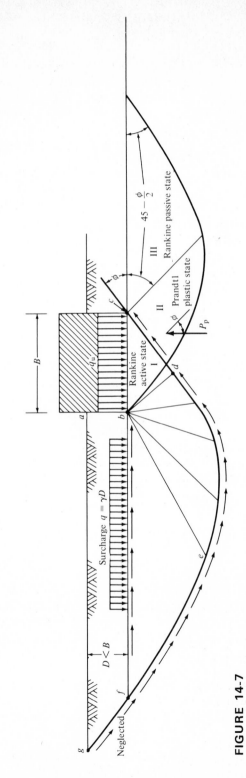

FIGURE 14-7
Terzaghi's bearing capacity theory.

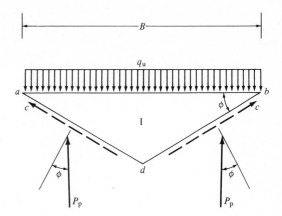

FIGURE 14-8
Forces acting on the wedge.

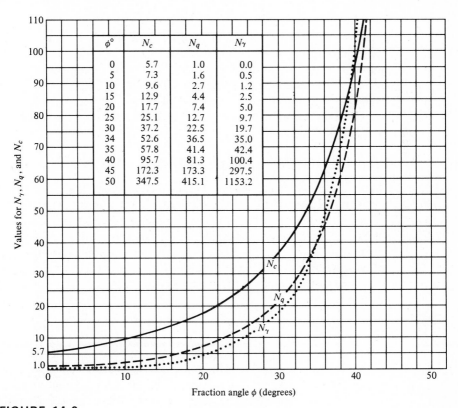

$\phi°$	N_c	N_q	N_γ
0	5.7	1.0	0.0
5	7.3	1.6	0.5
10	9.6	2.7	1.2
15	12.9	4.4	2.5
20	17.7	7.4	5.0
25	25.1	12.7	9.7
30	37.2	22.5	19.7
34	52.6	36.5	35.0
35	57.8	41.4	42.4
40	95.7	81.3	100.4
45	172.3	173.3	297.5
50	347.5	415.1	1153.2

FIGURE 14-9
Bearing capacity factors for footings with rough surface, Eqs. (14-5) to (14-7).

K_p being the coefficient of passive pressure from zones II and III on I,

$$N_q = \frac{a^2}{2 \cos^2(45 + \varphi/2)}$$

with $a = e^{(3\pi/4 - \varphi/2) \tan \varphi}$ and

$$N_c = \cot \varphi \left[\frac{a^2}{2 \cos^2(45 + \varphi/2)} - 1 \right]$$

Also for a general shear condition, Terzaghi's expressions for the ultimate bearing capacity for square and circular footings are given by Eqs. (14-6) and (14-7), respectively. For *square* footings,

$$q_u = 0.4\gamma B N_\gamma + \gamma D N_q + 1.3c N_c \qquad (14\text{-}6)$$

and for *circular* footings,

$$q_u = 0.3\gamma B N_\gamma + \gamma D N_q + 1.3c N_c \qquad (14\text{-}7)$$

Figure 14-9 gives the values for the various bearing capacity factors recommended for Eqs. (14-5) through (14-7).

More recent research data indicate that the angle the face of the penetrating wedge makes with the horizontal is closer to $(45 + \varphi/2)$ than the angle φ that Terzaghi assumed. Furthermore, Terzaghi's equations do not provide for the effects of footer depth, load inclination or eccentricity, soil compressibility, water table, and other factors. With subsequent refinements by others, Terzaghi's equations are not as widely used as before. However, they did and still do serve as a fundamental basis on which modifications and improvements have been and are still being made.

14-7 OTHER INFLUENCES ON BEARING CAPACITY

There are a number of variables and factors other than those mentioned in the preceding sections which are recognized, although not fully understood, to have an effect on the bearing capacity of a soil. Among these factors which may have a direct or indirect influence on the bearing capacity of soils are footing shapes and sizes, depth of footing, inclination and eccentricities of loads, compressibility of the soil, groundwater table, rate of loading, slope of ground adjacent to the footing, roughness of the base, and others. Some of these are discussed here.

Soil Compressibility

All preceding discussions assumed a relatively incompressible soil and a condition associated with *general shear* failure. Thus the bearing capacity equations proposed up to this point would not be applicable for material that experiences relatively large strain prior to ultimate failure and large vertical compression without resulting in general shear failure. For such soils Terzaghi suggested an empirical correction for the soil parameters to be used in the bearing capacity equation: $c' = \frac{2}{3}c$ and $\tan \varphi' = \frac{2}{3} \tan \varphi$.

From his observations of failure loads and small footings on some sands, Vesic (51) suggests that the factor of $\frac{2}{3}$ in the above expression might be replaced by a correction factor which varies with the relative density D_r: $\frac{2}{3} + D_r - \frac{3}{4}D_r^2$, applicable in the range of $0 < D_r < \frac{2}{3}$.

From a practical point of view it is perhaps reasonable to assume that one would not design a foundation for a local shear or a punching shear failure condition, especially for sand. Instead, in all probability, the soil would be compacted to a density state where general shear failure could be assumed.

Water Table

The effective unit weight of a submerged soil is reduced to about half the weight for that same soil above the water table. Furthermore, increasing the moisture content by a rise in the water table may result in swelling of some fine-grain soils, a possible loss of apparent cohesion, a general loosening of the soil, a reduction in the angle of internal friction, a decrease in the shear strength, and subsequently a decrease in the ultimate bearing capacity. Generally, then, all three terms of the bearing capacity equation may become smaller, thereby resulting in considerable reduction in the value of q_u. In this regard it is advisable to assume the highest possible water table when evaluating a bearing capacity. It is customary to neglect the effects when the water table is at a distance B or more below the bottom of the footing. When the water table is at a distance d_w less than B below the footing, the value of γ to be used in the N_γ term of the bearing capacity equation is

$$\gamma = F_w \gamma_{sat}$$

where

$$F_w \approx 0.5 + \frac{d_w}{B \tan(45 + \varphi/2)}$$

Footer Depth

Derivation of the bearing capacity equations neglected the shear resistance of the overburden; that is, referring to Fig. 14-7, one notes that the shear along the footer-overburden interface (a–b) and along bfg was ignored. The shear resistance effects in this regard are thus normally expressed as depth factors to be used in the bearing capacity equations. Such commonly used factors are those proposed by Hanson (21).

Judgment must prevail in using depth factors. Adequate density and minimum compressibility should be evident, thereby ensuring reasonable shear resistance before depth factors are incorporated in the equation. Due to a lack of such assurance their use is not recommended. Indeed, some practitioners advocate the total abandonment of depth factors for shallow footings under all circumstances.

Footing Shapes

Because of mathematical complexity, there are no theoretical analyses for square and rectangular shapes comparable to Eq. (14-5) derived for long strip footings. Hence the correction factors to account for the effect of shapes are empirical. Commonly used shape factors are given in Table 4-1. The three terms fit in the general bearing capacity equations as shown in Example 14-2.

TABLE 14-1
Suggested Factors to Be Used in Eq. (14-8)

(a) Depth Factors*

Footing Depth	d_c	d_q	d_γ
$D \leq B$	$1 + 0.4 \dfrac{D}{B}$	$1 + 2 \tan\varphi (1 - \sin\varphi)^2 \dfrac{D}{B}$	1
$D > B$	$1 + 0.4 \tan^{-1} \dfrac{D}{B}$	$1 + 2 \tan\varphi (1 - \sin\varphi)^2 \tan^{-1} \dfrac{D}{B}$	1

(b) Shape Factors†

Footing Shape	s_c	s_q	s_γ
Rectangle	$1 + \dfrac{B}{L} \dfrac{N_q}{N_c}$	$1 + \dfrac{B}{L} \tan\varphi$	$1 - 0.4 \dfrac{B}{L}$
Circle and square	$1 + \dfrac{N_q}{N_c}$	$1 + \tan\varphi$	0.6

(c) Inclination Factors

i_c	i_q	i_γ
$i_q - \dfrac{1 - i_q}{N_q - 1}$	$\left(1 - \dfrac{H}{V + B'L'c \cot\varphi} \right)^2$	$i_q^{3/2}$

$L' = L - 2e_x$

$B' = B - 2e_y$

* After Hanson (21).

† After Vesic (51).

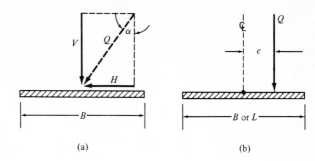

FIGURE 14-10
(a) Inclined loads. (b) Eccentric loads.

(a)   (b)

Load Inclination and Eccentricity

Meyerhof (33) suggested that footing dimensions $B' = B - 2e_y$ and $L' = L - 2e_x$ be used in determining the total allowable load eccentrically applied in the x and y directions, respectively (that is, $Q_u = q_u B'L'$), and in the corresponding terms in the ultimate bearing capacity equations and in the various correction factors for shape and inclination. To account for inclination, he suggested $i_q = i_c = (1 - \alpha/90°)^2$ and $i_\gamma = (1 - \alpha/\varphi)^2$. (Figure 14-10 shows the general notation used herein.) Among other factors are those given in Table 14-1.

14-8 GENERAL BEARING CAPACITY FORMULA

As mentioned in the preceding section, an analytical solution accounting for the many variables associated with the typical foundation-on-soil problems has not been developed. Generally Terzaghi's equations given in Section 14-6 are somewhat outmoded, particularly for his bearing capacity factor N_γ. Subsequent modifications of this basic equation lead to rather improved and somewhat less conservative results. More specifically it is noted that the factors N_c and N_q subsequently proposed by the various investigators lead to rather insignificant differences. However, the value of N_γ is particularly sensitive to the shape of the failure surface, and depending upon the value of the slope which the wedge forms with the horizontal [$\varphi, (45 + \varphi/2)$, or larger], various approaches may lead to substantial differences in N_γ values. From all indications the values for N_c and N_q as proposed by Prandtl (42) and the value for N_γ as proposed by Caquot and Kerisel (7) give acceptable results. The corresponding values are given in Fig. 14-11.

The above discussion inferred a very large, if not infinite, ratio of length to width of footing. However, it is common practice to assume that this discussion applies with acceptable accuracy to footings whose length to width ratio is larger than 5. Accounting for a number of variables discussed in the preceding section, a more *general bearing capacity* expression may be given by Eq. (14-8).

$$q_u = cN_c d_c s_c i_c + \gamma D N_q d_q s_q i_q + \tfrac{1}{2}\gamma B N_\gamma s_\gamma i_\gamma \qquad (14\text{-}8)$$

Table 14-1 gives some of these correction factors to be incorporated in the general bearing capacity equation.

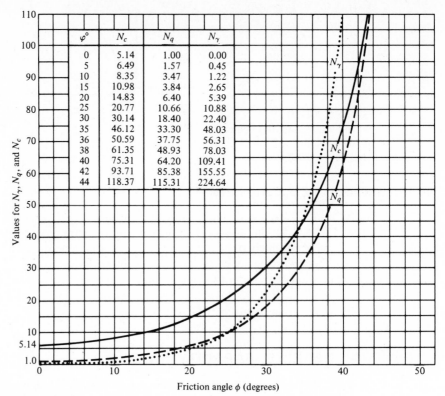

$\varphi°$	N_c	N_q	N_γ
0	5.14	1.00	0.00
5	6.49	1.57	0.45
10	8.35	3.47	1.22
15	10.98	3.84	2.65
20	14.83	6.40	5.39
25	20.77	10.66	10.88
30	30.14	18.40	22.40
35	46.12	33.30	48.03
36	50.59	37.75	56.31
38	61.35	48.93	78.03
40	75.31	64.20	109.41
42	93.71	85.38	155.55
44	118.37	115.31	224.64

FIGURE 14-11
Bearing capacity factors for footings with rough surface, Eq. (14-8).

EXAMPLE 14-1

Given   A footing as shown in Fig. 14-12.

Find   The ultimate bearing capacity via Terzaghi's equation.

Procedure $L/B = 20$. This generally is a long footing if $L/B > 5$. Thus from Eq. (14.5), for general shear,

$$q_u = \tfrac{1}{2}\gamma B N_\gamma + \gamma D N_q + c N_c$$

$$q_u = \tfrac{1}{2}(18.2)(1)N_\gamma + 18.2(1)N_q + 16N_c$$

$$q_u = 9.1N_\gamma + 18.2N_q + 16.0N_c$$

From Fig. 14-9, $N_\gamma = 8$; $N_q = 11.5$; $N_c = 23.5$. Then

$$q_u = 9.1(8) + 18.2(11.5) + 16(23.5)$$

Answer $q_u = 658.10 \text{ kN/m}^2$

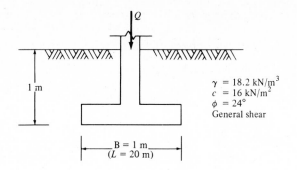

$\gamma = 18.2$ kN/m^3
$c = 16$ kN/m^2
$\phi = 24°$
General shear

B = 1 m
(L = 20 m)

FIGURE 14-12

EXAMPLE 14-2

Given   The data of Example 14-1.

Find   The ultimate bearing capacity via the general bearing capacity formula.

Procedure   From Eq. (14-8),

$$q_u = cN_c d_c s_c i_c + \gamma D N_q d_q s_q i_q + \tfrac{1}{2}\gamma B N_\gamma s_\gamma i_\gamma; \qquad d_\gamma = 1$$

$$d_c = 1 + 0.4\frac{D}{B} = 1 + 0.4(1) = 1.4$$

$$d_q = 1 + 2\tan\varphi(1 - \sin\varphi)^2 \frac{D}{B} = 1 + \tan 24(1 - \sin 24)^2(1) = 1.16$$

From Fig. 14-11, $N_c = 19$; $N_q = 9.5$; $N_\gamma = 9.3$. Also,

$$s_c = 1 + \frac{B}{L}\frac{N_q}{N_c} = 1 + \tfrac{1}{20}\left(\frac{9.5}{19}\right) \approx 1$$

$$s_q = 1 + \frac{B}{L}\tan\varphi = 1 + \tfrac{1}{20}(0.45) \approx 1$$

$$s_\gamma = 1 - 0.4\frac{B}{L} = 1 - 0.4(\tfrac{1}{20}) \approx 1$$

Then,

$$q_u = 16(19)(1.4)(1) + 18.2(1)(9.5)(1.16)(1) + \tfrac{1}{2}(18.2)(1)(9.3)(1)$$

Answer

$$q_u = 709.63 \text{ kN/m}^2$$

From Eq. (14-5) $q_u = 658.1$ via Terzaghi, or 7 percent smaller than from Eq. (14-8).

EXAMPLE 14-3

Given   The data of Example 14-1, except that $L = 2$ m.

Find   The ultimate bearing capacity via Eq. (14-8).

Procedure

$$q_u = cN_c d_c s_c i_c + \gamma D N_q d_q s_q i_q + \tfrac{1}{2}\gamma B N_\gamma s_\gamma i_\gamma; \qquad d_\gamma = 1$$

$$
\left.
\begin{aligned}
d_c &= 1.4 \\
d_q &= 1.16 \\
N_c &= 19 \\
N_q &= 9.5 \\
N_\gamma &= 9.3
\end{aligned}
\right\} \text{see Example 14-2}
$$

$$s_c = 1 + \frac{B}{L}\frac{N_q}{N_c} = 1 + \tfrac{1}{2}\left(\frac{9.5}{19}\right) = 1.25$$

$$s_q = 1 + \tfrac{1}{2}\tan\varphi = 1 + \tfrac{1}{2}(0.45) = 1.225$$

$$s_\gamma = 1 - 0.4(\tfrac{1}{2}) = 1 - 0.2 = 0.8$$

$$q_u = 16(19)(1.4)(1.25) + 18.2(1)(9.5)(1.16)(1.225) + \tfrac{1}{2}(18.2)(1)(9.3)(0.8)$$

Answer

$$q_u = 845.2 \text{ kN/m}^2$$

EXAMPLE 14-4

Given   A footing as shown in Fig. 14-13.

Find   (a) The ultimate bearing capacity via Eq. (14-8).
        (b) Compare result with that from Example 14-3.

Procedure

$$q_u = cN_c d_c s_c i_c + \gamma D N_q d_q s_q i_q + \tfrac{1}{2}\gamma B N_\gamma s_\gamma i_\gamma; \qquad d_\gamma = 1$$

$$d_c = 2.6; \; d_q = 1.64; \text{ other terms as in Example 14-3.}$$

$$q_u = 16(19)(2.6)(1.25) + 18.2(4)(9.5)(1.64)(1.225) + \tfrac{1}{2}(18.2)(1)(9.3)(0.8)$$

Answer

$$q_u = 1963 \text{ kN/m}^2$$

$$q_{u\,D=4}/q_{u\,D=1} = 2.32$$

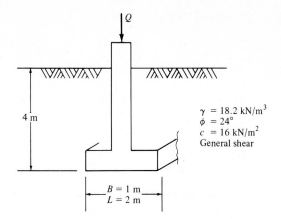

$$\gamma = 18.2 \text{ kN/m}^3$$
$$\phi = 24°$$
$$c = 16 \text{ kN/m}^2$$
General shear

4 m

$B = 1$ m
$L = 2$ m

FIGURE 14-13

14-9 BEARING CAPACITY BASED ON STANDARD PENETRATION TESTS

Cognizant of the relatively tedious and laborious effort that goes into the use of ultimate bearing capacity equations, such as Eq. (14-8), researchers and practitioners made various attempts over the years in obtaining bearing capacity values by more direct approaches. Some such attempts focused upon the use of the data from standard penetration tests (described in Chapter 3—ASTM D-1586-67) toward predicting the allowable bearing capacity for soils (24). Indeed, the allowable bearing capacity as determined from standard penetration test data for sand and gravel deposits is used quite routinely by most practitioners, sometimes as the sole basis for determining the design bearing capacity, especially for more ordinary buildings.

Terzaghi and Peck (50) proposed a series of curves for estimating the allowable soil pressure for footings on sand on the basis of results of standard penetration tests. Teng (48) provided Eq. (14-9) which closely approximates the curves presented by Terzaghi and Peck:

$$q_a = 34.5(N - 3)\left(\frac{0.305B + 1}{0.7B}\right)^2 \text{ kN/m}^2 \qquad (14\text{-}9)$$

where B is given in meters. For B in feet, Eq. (14-9) becomes

$$q_a = 0.72(N - 3)\left(\frac{B + 1}{B}\right)^2 \text{ kips/ft}^2 \qquad (14\text{-}9')$$

Meyerhof (35, 39) presented Eqs. (14-10a) and (14-10b), which give the allowable bearing capacity for a selected range of values for N and B:

$$q_a = 12N \text{ kN/m}^2 \qquad \text{for } B \leq 1.22 \text{ m} \qquad (14\text{-}10a)$$

$$q_a = 8N\left(\frac{0.305B + 1}{0.305B}\right)^2 \text{ kN/m}^2 \qquad \text{for } B > 1.22 \text{ m} \qquad (14\text{-}10b)$$

If B is given in feet, Eqs. (14-10a) and (14-10b) become

$$q_a = \frac{N}{4} \text{ kips/ft}^2 \qquad \text{for } B \leq 4 \text{ ft} \qquad (14\text{-}10a')$$

and

$$q_a = \frac{N}{6} \left(\frac{B+1}{B}\right)^2 \text{ kips/ft}^2 \qquad \text{for } B > 4 \text{ ft} \qquad (14\text{-}10b')$$

where q_a = allowable net increase in soil pressure in 1000 lb/ft² for an estimated maximum settlement of 2.54 cm (1 in)

N = penetration number, or the number of blows per foot adjusted if the penetration test is made at shallow depth

A recommended approach by Bazaraa (2) is that N should be adjusted in accordance with the following:

$$N' = \frac{4N}{1 + 2p_o} \qquad \text{for } p_o \leq 72 \text{ kN/m}^2$$

$$N' = \frac{4N}{3.25 + 0.5p_o} \qquad \text{for } p_o > 72 \text{ kN/m}^2$$

where N = actual blow count

p_o = overburden pressure

B = width of footings, feet

Figure 14-14 is a chart for estimating the allowable soil pressure for footings on sand based on Eqs. (14-10a) and (14-10b) for 1-in settlement. The chart reflects the values presented in Table 14-2.

Some modifications of these equations exist to account for corrections for depth and for different factors of safety. In general it is common practice to increase the allowable bearing capacity with depth from that given by the above equation (perhaps by as much as 50 to 60 percent) on the premise that it is too conservative. Whatever the assertions and the apparent validity for such modifications, it is perhaps worthwhile to reiterate that these are expressions to be used with discreet judgment and common sense. For example, the penetration number in a given clay may vary significantly with seasonal fluctuations in the water table. Furthermore, standard penetration readings may fall drastically short of providing information on the characteristics of the clay (e.g., consolidation, shear strength, shrinkage, or swelling) which are usually relevant to the design of the foundation. The author recalls a situation related to a foundation design which was based on a very high blow count from a standard penetration test obtained during a dry (low water table) season. The architect used this information as the sole basis for determining the allowable bearing capacity and subsequently proceeded to design the foundation for a one-story school building. Gradually, but cumulatively, over a 3-year span from construction, cracks exceeding 5 cm (2 in) developed in some of the masonry walls, and

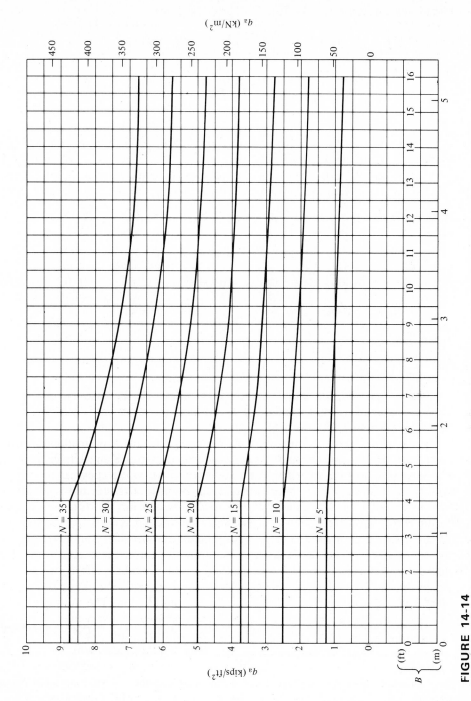

FIGURE 14-14

Allowable bearing capacity for surface-loaded footings for 2.54-cm (1-in) settlement, based on Eq. (14-10).

TABLE 14-2
Allowable Bearing Capacity in kN/m^2 (kips/ft^2) for Surface-Loaded Footings for 2.54-cm (1-in) Settlement, Based on Eq. (14-10)

N	B						
	0–4	6	8	10	12	14	16
5	60	54.3	50.5	48.3	46.83	45.82	45.06
	(1.25)	(1.134)	(1.055)	(1.008)	(0.978)	(0.957)	(0.941)
10	120	108.6	101	96.6	93.66	91.64	90.12
	(2.5)	(2.268)	(2.109)	(2.016)	(1.956)	(1.913)	(1.882)
15	180	162.9	151.5	144.8	140.50	137.46	135.18
	(3.75)	(3.403)	(3.164)	(3.024)	(2.934)	(2.871)	(2.823)
20	240	217.2	202	193.1	187.32	183.28	180.24
	(5)	(4.537)	(4.220)	(4.032)	(3.912)	(3.827)	(3.764)
25	300	271.5	252.5	241.4	234.15	229.10	225.30
	(6.25)	(5.671)	(5.275)	(5.040)	(4.89)	(4.785)	(4.705)
30	360	325.8	303	289.7	281	274.92	270.36
	(7.5)	(6.805)	(6.328)	(6.048)	(5.868)	(5.742)	(5.646)
35	420	378.2	353.5	338	327.8	320.74	315.42
	(8.75)	(7.940)	(7.385)	(7.056)	(6.846)	(6.699)	(6.587)

significant cracking and heaving occurred in the on-grade concrete slabs. A more detailed evaluation revealed the clay stratum to be subject to significant shrinkage and swelling with notable changes in the water content.

Quite apparently the determination of the allowable bearing capacity for the above "clay" case via the standard penetration test proved to be misleading and costly. This is not to imply, however, that granular material is always dependable in this regard. One may obtain totally erroneous results if the split-spoon sampler encounters large gravels, boulders, or rock formation at its tip. Under such circumstances the answers provided by the standard penetration test approach may not only be of dubious value, but downright dangerous, unless treated with utmost scrutiny. Again, therefore, judicial use rather than exclusion of the standard penetration method for determining the allowable bearing capacity is recommended in almost all instances. Relatively speaking, the method is appreciably more dependable in sand and small-gravel deposits than it is for highly cohesive soils or large granular formations.

EXAMPLE 14-5

Given   The blow count from a standard penetration test in a sand and small-gravel stratum was 16 blows per 0.305 m at a depth of 1 m below the surface. The unit weight of the stratum was determined to be 18.1 kN/m^3 for several meters below the surface.

Find   The allowable bearing capacity for a 2.54-cm (1-in) settlement for an anticipated footing size of 2 m × 2 m to be placed at approximately 1-m depth.

Procedure

$$p_o = 1 \, m \, (18.1 \, kN/m^3) = 18.1 \, kN/m^2 < 72 \, kN/m^2$$

In fps units,

$$p_o = 0.38 \, kips/ft^2 \, \text{(from conversion table)}$$

Thus from Eq. (14-10a) the corrected N becomes

$$N = \frac{4N'}{1 + 2p_o} = \frac{4(16)}{1 + 2(0.38)} \cong 37$$

Answer

> From Fig. 14-14,
>
> $$q_a = 8.25 \, kips/ft^2 = 400 \, kN/m^2$$

14-10 SETTLEMENT CONSIDERATIONS

Settlement rather than strength is the criterion that normally governs the allowable bearing capacity used in foundation design. That is, the design load is usually but a small fraction of the ultimate bearing capacity, a requirement dictated by settlement restrictions. Thus in foundation design one does not select an allowable bearing capacity by merely taking the ultimate bearing capacity and dividing it by an arbitrary factor of safety. Instead, the working or design loads must be limited to the magnitudes that induce tolerable settlements.

As we have frequently noted throughout the text, soils are not isotropic, homogeneous, or elastic. Hence accurate settlement predictions of footings on soils are quite difficult to make with any degree of confidence. Estimates of elastic settlements may be made, however, via elastic theories adjusted to fit experimental or in situ findings. Time-related settlements (consolidation type) have been discussed in Chapter 9 and will, therefore, not be repeated here. Within the scope of settlement evaluation, reasonable judgments, assessments, and interpretations become necessary ingredients.

Elastic Settlement under Uniform Loads

Settlement under a corner of a rectangular footing in granular soils, or in fine-grained nonsaturated soils, subjected to a uniform pressure q, can be computed from

$$s = qB \frac{(1 - \mu^2)}{E} I_p \qquad (14\text{-}11)$$

where q = uniform contact pressure
B = footing width ($L/B \geq 1$; L = footing length)
I_p = influence coefficient [see Eq. (14-12) or Fig. 14-15]
E = stress–strain modulus (or modulus of elasticity) of soil
μ = Poisson's ratio of soil

The value of I_p may be calculated from Eq. (14-12) proposed by Steinbrenner (46) for flexible footings (e.g., steel tank resting on ground). For rigid footings (e.g., stiff concrete footings) the I_p values are approximately 7 percent smaller than for the flexible type (44):

$$I_p = \frac{1}{\pi} \left\{ \frac{L}{B} \ln\left[\frac{1 + \sqrt{(L/B)^2 + 1}}{L/B} \right] + \ln\left[\frac{L}{B} + \sqrt{\left(\frac{L}{B}\right)^2 + 1} \right] \right\} \qquad (14\text{-}12)$$

Figure 14-15 gives the relationship between the L/B ratio and I_p as expressed by Eq. (14-12) for a corner of a rectangular area. For any other point I_p can be obtained by dividing the given area into rectangles which have a common corner at the point in question, then superimposing their effects. (A comparable approach was followed when the undersoil pressure was determined at a point from a uniformly loaded surface area; see Section 8-7.)

Table 14-3 provides some approximate average values for E and μ. These may be used in Eq. (14-11) to calculate rough estimates of settlement. Since values of E and μ may vary greatly, such settlement calculations should be viewed accordingly. Experimental verification is indeed advisable for more dependable evaluations and/or more important projects.

Experimental–Empirical Settlement Evaluations

The basic difficulty in calculating elastic settlements in soil by equations such as Eq. (14-11) lies in the heterogeneous nature of virtually all soils and, thereby, in establishing reliable values for E and μ. Estimates of E can be obtained by plotting the stress–strain data from triaxial tests. The initial tangent modulus (for one or more load cycles) is the usual value assumed for E. Also, much evidence exists which

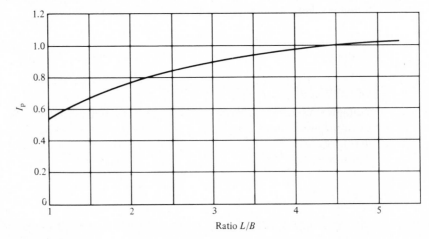

FIGURE 14-15
Influence factor I_p at corner of rectangular footing, based on Eq. (14-12).

TABLE 14-3
Approximate Average Values
for E and μ*

Soil Type	E (MN/m^2)	μ
Clay		
Soft	3	0.4
Medium	7	0.3
Hard	14	0.25
Sandy	36	0.25
Sand		
Loose	15	0.2
Dense	80	0.3
Sand and gravel		
Loose	100	0.2
Dense	150	0.3

* Actual values may vary widely from those shown.

suggests some relationship between E and the N values from Standard Penetration Tests (40). Cone penetration tests provide still another means of estimating E.

Perhaps a more inclusive approach for estimating E and/or settlement, and one that accounts (to some extent) for both values of E and μ, is the *plate-load* test conducted at the building site. Two (or more) plates of different sizes are subjected to loads, say P_1 and P_2. The respective settlements are measured (say s_1 and s_2). If the L/B ratio is the same for both, then I_p is also the same for both. Hence assuming a linear relationship, from Eq. (14-11), (see Example 14-6)

$$\frac{1 - \mu^2}{E} I_p = \frac{s_1 - s_2}{q_1 B_1 - q_2 B_2} \tag{14-13}$$

The quantity $[(1 - \mu^2)/E] I_p$ can be viewed as a slope of the $(s-qB)$ relationship and regarded as a constant for the soil of a given site. Hence this would imply that an estimate could readily be made of a prototype footing from Eq. (14-11) for its value of qB and I_p since the "slope" is known. Size effects, however, require some adjustment in this regard. Terzaghi and Peck (50) proposed a relationship between settlements of a prototype footing and that from a 0.305-m square plate as

$$s = s_1 \left(\frac{6.56B}{3.28B + 1} \right)^2 \tag{14-14}$$

where s = settlement of prototype footing subjected to bearing pressure q
s_1 = settlement of 0.305-m × 0.305-m plate subjected to bearing pressure q
B = width of prototype footing (m)

Tolerable Settlements

Excessive settlements may adversely affect the function and/or aesthetic aspect of a structure. For example, the function of a machine or the supports for a continuous beam may be impaired by large settlements; a masonry wall may experience severe cracking from differential settlement, etc. (See Section 9-12 for a further discussion on tolerable settlements.)

Guidelines or ranges of tolerable settlements, usually based on history of performance, are sometimes found in area building codes or in technical literature. Table 14-4 provides recommended maximum values for differential settlement. Some lower values may be warranted for rather important or critical installations.

TABLE 14-4
Apparent Tolerable Differential Settlements

Structure	Maximums
Single-story masonry buildings	$0.0015L$*
High masonry wall	$0.001L$
Continuous steel frame	$0.002L$
Simple steel frame	$0.005L$
Reinforced concrete building	$0.003L$
Smokestacks and towers	$0.004L$

* L = horizontal distance between two points considered for differential settlement.

EXAMPLE 14-6

Given   A 2-m × 2-m footing supports a 1300-kN column load. Another footing 2.5 m × 2.5 m supports a 2000-kN column load. The two columns are 8 m apart. Plate-load tests were run on two square plates; one plate was 0.305 m × 0.305 m, the other 0.61 m × 0.61 m. The loads on the two plates were equal at 30 kN for each plate. Average corner settlements of the two plates were 2.5 and 1.5 cm, respectively. The Poisson ratio of the soil is approximately 0.35.

Find   (a)   The stress–strain modulus E of the soil.
(b)   The probable elastic settlement of each footing.
(c)   The differential settlement of the footings and the probable effect such settlements may have on the structure if the two columns are part of a system of columns supporting a continuous beam on which rests a masonry wall.
(d)   Compare settlements based on Eqs. (14-11) and (14-14).

Procedure   (a)   Let us refer to the small plate as plate 1 and to the larger one as plate 2. The corresponding soil pressures, settlements, dimensions, etc., will be designated by q_1, s_1, B_1, etc.

$$q_1 = \frac{30 \text{ kN}}{(0.305 \text{ m})^2} = 322.5 \text{ kN/m}^2$$

$$q_2 = \frac{30 \text{ kN}}{(0.61 \text{ m})^2} = 60.6 \text{ kN/m}^2$$

$$q_1 B_1 = 322.5 \times 0.305 = 98.4 \text{ kN/m}$$

$$q_2 B_2 = 60.6 \times 0.61 = 37 \text{ kN/m}$$

$$s_1 = 2.5 \text{ cm} = 0.025 \text{ m}$$

$$s_2 = 1.5 \text{ cm} = 0.015 \text{ m}$$

From Fig. 14-15, for $L/B = 1$,

$$I_p = 0.56$$

which is the same for both. The data may be plotted as shown in Fig. 14-16.

From Eq. (14-13),

$$\frac{1 - \mu^2}{E} I_p = \frac{(2.5 - 1.5) \times 10^{-2}}{98.4 - 37} = \frac{1 \times 10^{-2}}{61.4}$$

Thus,

$$E = (1 - 0.35^2) \times 0.56 \times 61.4 \times 10^2$$

Answer

$$E = 3017 \text{ kN/m}^2 = 3.02 \text{ MN/m}^2$$

(b)   The settlement for the two prototypes can be determined via Eq. (14.14), using the settlement for the 0.305-m $\times$ 0.305-m plate. This was the size on which the equation was suggested. For the prototype, $q \approx 3.25 \text{ kN/m}^2$ each. Thus,

$$s_{2 \times 2} = s_1 \left(\frac{6.56B}{3.28B + 1} \right)^2 = 2.5 \left(\frac{6.56 \times 2}{3.28 \times 2 + 1} \right)^2$$

$$s_{2.5 \times 2.5} = 2.5 \left(\frac{16.40}{9.20} \right)^2$$

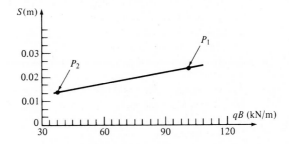

FIGURE 14-16

Answer

$$s_{2 \times 2} = 7.55 \text{ cm (3 in)} \qquad s_{2.5 \times 2.5} = 7.93 \text{ cm (3.12 in)}$$

(c)   The differential settlement = 0.4 cm.
From Table 14-4,

probable tolerable settlement $\approx 0.001L = 0.001(800 \text{ cm}) = 0.8$ cm

Answer

Differential settlement = 0.4 cm

Probable tolerable settlement = 0.8 cm

Thus the wall may be safe against cracking. The total settlement (average $\approx$ 7.75 cm) is, however, rather high and may not be acceptable for reasons other than cracking.

(d) $s'_{2 \times 2} = \left(\dfrac{1300}{B^2}\right) B(\text{slope}) = \left(\dfrac{1300}{2}\right)\left(\dfrac{1}{6140}\right) = 0.11 \text{ m} = 11 \text{ cm}$

$s'_{2.5 \times 2.5} = \left(\dfrac{2000}{2.5}\right)\left(\dfrac{1}{6140}\right) = 0.16 \text{ m} = 16 \text{ cm}$

Answer

$$s_{2 \times 2} = 11 \text{ cm} > 7.55 \text{ cm} \qquad s_{2.5 \times 2.5} = 16 \text{ cm} > 7.93 \text{ cm}$$

PROBLEMS

14.1.   A continuous-wall footing, 1.5 m wide, is supported on a soil whose physical properties are as follows: $\varphi = 18°$; $\gamma = 18.9 \text{ kN/m}^3$; $c = 22 \text{ kN/m}^2$. The water table is 3.2 m below the surface. Determine the ultimate bearing capacity by both Eqs. (14-5) and (14-8), assuming the base of the footing to be:
(a)   At the surface.
(b)   1.3 m below the surface.
(c)   Using the results of Eq. (14-8) as a basis of comparison, what is the percentage difference for case (b) using the two equations.

14-2.   A continuous-wall footing, 2 m wide, is placed on a silty sand stratum whose physical properties are as follows: $c = 19 \text{ kN/m}^2$; $\gamma = 19.3 \text{ kN/m}^3$; $\gamma_b = 9.5 \text{ kN/m}^3$; $\varphi = 15°$. The base of the footing is 1.8 m below the surface, and the water table is 0.8 m below the surface. Calculate the ultimate bearing capacity using Eq. (14-8):
(a)   If the effect of the water table was neglected.
(b)   If the water table was accounted for.

14-3.   Using Terzaghi's bearing capacity formula and the bearing capacity factors given by Fig. 14-9, determine the ultimate bearing capacity for Problem 14-1 when the base of the footing is assumed to be:
(a)   At the surface.
(b)   0.8 m below the surface.

(c)   1.3 m below the surface.

(d)   2.0 m below the surface.

(e)   Plot the relationship between q_u and D for the four cases. Is there a trend?

14-4.   Determine the ultimate bearing capacity using Terzaghi's formula and the corresponding bearing capacity factors for the data given in Problem 14-2 when the water table is:

(a)   At the surface.

(b)   0.8 m below the surface.

(c)   1.6 m below the surface.

(d)   2.4 m below the surface.

(e)   3.2 m below the surface.

(f)   Plot the relationship between q_u and the water table depth by plotting the results of parts (a)–(e). Is there an apparent trend?

14-5.   Assuming that the footing in Problem 14-1 is 10 times as long as it is wide, determine the ultimate bearing capacity for the conditions given in Problem 14-1 using Eq. (14-8) and the corresponding bearing capacity factors given by Fig. 14-11 when the base of the footing is:

(a)   At the surface.

(b)   1.3 m below the surface.

14-6.   Determine the ultimate bearing capacity via Eq. (14-8) for the conditions given in Problem 14-2, assuming that the ratio of length to width of the footing is 10.

14-7.   For the data given in Problem 14-1 determine the ultimate bearing capacity of the soil using Eq. (14-8) for a footing whose ratio $L/B = 10$ if the footing is to be placed at 2.5 m below the surface.

14-8.   Determine the ultimate bearing capacity for the conditions given in Problem 14-2 via Eq. (14-8), assuming a ratio of $L/B = 5$, if the base of the footing is:

(a)   At the surface.

(b)   1 m below the surface.

(c)   2 m below the surface.

(d)   3 m below the surface.

14-9.   Determine the ultimate bearing capacity for the rectangular footing shown in Fig. P14-9. (a) Use Terzaghi's equation and the corresponding bearing capacity factors given by Fig. 14-9. (b) Use the general bearing capacity equation [Eq. (14-8)] for the data shown. $L = 3.5$ m; $B = 2$ m; $D = 2$ m; $D_w = 3$ m; $\gamma = 18.6$ kN/m^3; $c = 12.5$ kN/m^2; $\varphi = 24°$.

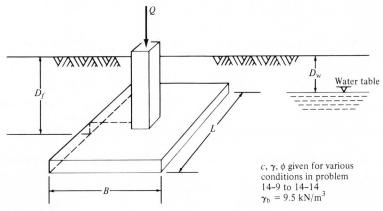

FIGURE P14-9

14-10. Rework Problem 14-9 for the condition where the water table is at the surface.

14-11. Determine the total Q that the footing shown in Fig. 14-9 can support with a safety factor of 5 against ultimate failure for the following conditions: $L = B = 3.0$ m; $D = 2$ m; $D_w = 1$ m; $\gamma = 19.1$ kN/m^3; $c = 14.5$ kN/m^2; $\varphi = 20°$. Use the general bearing capacity formula.

14-12. Rework Problem 14-11 if the water table reaches the ground surface.

14-13. Rework Problem 14-9 if D_f is equal to 4 m.

14-14. Determine the ultimate bearing capacity for the footing shown in Fig. 14-9 for the following data: $L = 3.5$ m; $B = 2$ m; $D = 4$ m; $\gamma = 19.1$ kN/m^3; $c = 0$; $\varphi = 33°$; and the soil is virtually dry.

14-15. A square footing, 3×3 m, rests on a dry sand and gravel stratum with the following physical properties: $\gamma = 20$ kN/m^3; $\varphi = 35°$; $c = 0$. If the base of the footing is at 2.5 m below the surface, determine the total load the footing could support with a safety factor of 4 via the general capacity equation.

14-16. The blow count from a standard penetration test in a relatively dense sand stratum intermixed with relatively small gravel was 20 blows per 30.5 cm, taken at an average depth of 2.4 m. In a moist state the unit weight of the soil was determined to be 18.6 kN/m^3. The water table was approximately 4 m below the surface. A 2-m $\times$ 2.5-m footing is anticipated to be placed at a base depth of 2 m below the surface. Determine:
(a)   The allowable bearing capacity.
(b)   The total load the footing can be expected to support safely for an anticipated 2.5-cm settlement.

14-17. Rework Problem 14-15 if the water table is expected to occasionally reach the surface. In that case a submerged unit weight of the soil of 9.1 kN/m^3 is to be assumed. The blow count from a standard penetration test in a relatively dense sand was 37 blows per 30.5 cm at a depth of 4.8 m. A 3-m $\times$ 3-m footing is anticipated to be placed at a 4.5-m depth. Normally, and at the time of drilling, the water table was determined to be at 4.5 m below the surface. The unit weight of the moist soil was 18.7 kN/m^3. Determine the allowable bearing capacity for an anticipated 2.5-cm maximum settlement.

14-18. Plate-load tests were run on two square plates whose sides were 0.305 and 0.61 m, respectively. The loads were 25 kN for each plate. The observed settlements (average) at the corner of the plates were 1.6 and 0.9 cm, respectively. The Poisson ratio for the soil was estimated at 0.3. Determine:
(a)   The stress–strain modulus of the soil.
(b)   The expected elastic settlement of a 2-m $\times$ 3-m footing subjected to a contact pressure comparable to the small plate.

14-19. For the data of Problem 14-18, what might be the elastic settlement at a corner of the 2-m $\times$ 3-m footing when loaded with a 1500-kN column load, based on Eq. (14-11)?

BIBLIOGRAPHY

1. Balla, A. "Bearing Capacity of Foundations," *ASCE J. Soil Mech. Found. Div.*, vol. 88, no. SM 5, Oct. 1962.
2. Bazaraa, A. R., "Use of the Standard Penetration Test for Estimating Settlements of Shallow Foundations on Sand," Ph.D. thesis, University of Illinois, Urbana, 1967.
3. Bowles, J. E., *Foundation Analysis and Design*, McGraw-Hill, New York, 1977.

4. Caquot, A., and J. Kérisel, *Tables for the Calculation of Passive Pressure, Active Pressure and Bearing Capacity of Foundations* (transl. by M. A. Bec, London), Gauthier-Villars, Paris, France, 1948.

5. Caquot, A., and J. Kérisel, *Traité de Méchanique des Sols*, Gauthier-Villars, Paris, France, 1949.

6. Caquot, A., and J. Kérisel, "Sur le Terme de Surface dans le Calcul des Fondations en Milieu Pulvérulent," *3rd Int. Conf. Soil Mech. Found. Eng.*, Zurich, Switzerland, vol. 1, 1953.

7. Caquot, A., and J. Kérisel, *Traité de Mécanique des Sols*, 3rd ed., Gauthier-Villars, Paris, France, 1956.

8. Chummar, A. V., "Bearing Capacity Theory from Experimental Results," *ASCE J. Soil Mech. Found. Div.*, vol. 98, no. SM12, Dec. 1972.

9. Davis, H. E., and R. J. Woodward, "Some Laboratory Studies of Factors Pertaining to the Bearing Capacity of Soils," *Proc. Highw. Res. Board*, vol. 29, 1949.

10. Davis, H. E., and J. T. Christian, "Bearing Capacity of Anisotropic Cohesive Soil," *ASCE J. Geotech. Eng. Div.*, vol. 97, May 1971.

11. De Beer, E. E., "Bearing Capacity and Settlement of Shallow Foundations on Sand," *Proc. Symp. Soil Mech.*, Duke University, Durham, N.C., 1965.

12. De Beer, E. E., "The Scale Effect on the Phenomenon of Progressive Rupture in Cohesionless Soils," *6th Int. Conf. Soil Mech. Found. Eng.*, Montreal, Canada, vol. II, 1965.

13. De Beer, E. E., and A. Vesic, "Etude Experimentale de la Capacité Portante de Sable sous des Foundations Directes Etablies en Surface," *Ann. Trav. Publiques Belg.*, vol. 59, no. 3, 1958.

14. De Beer, E. E., "Experimental Determination of the Shape Factors and the Bearing Capacity Factors of Sand," *Geotechnique*, vol. 20, no. 4, Dec. 1970.

15. Eden, W., and M. Bozozuk, "Foundation Failure of a Silo on Varved Clay," *Eng. J.*, Montreal, Canada, 1962.

16. Feda, J., "Research on Bearing Capacity of Loose Soil," *5th Int. Conf. Soil Mech. Found. Eng.*, Paris, France, vol. 1, 1961.

17. Hansen, B., "The Bearing Capacity of Sand, Tested by Loading Circular Plates," *5th Int. Conf. Soil Mech. Found. Eng.*, Paris, France, vol. 1, 1961.

18. Hansen, B., "Bearing Capacity of Shallow Strip Footings in Clay," *7th Int. Conf. Soil Mech. Found. Eng.*, Mexico City, Mexico, vol. 2, 1969.

19. Hansen, B., and N. H. Christensen, discussion of A. Larkin, "Theoretical Bearing Capacity of Very Shallow Footings," *ASCE J. Soil Mech. Found. Div.*, vol. 95, no. SM6, Proc. Paper 6258, Nov. 1969.

20. Hansen, J. Brinch, "A General Formula for Bearing Capacity," Dan. Tech. Inst., Copenhagen, Denmark, Bull. no. 11, 1961.

21. Hansen, J. Brinch, "A Revised and Extended Formula for Bearing Capacity," Dan. Geotech. Inst., Copenhagen, Denmark, Bull. no. 28, 1970.

22. Heller, L. W., "Failure Modes of Impact-Loaded Footings on Dense Sand," U.S. Naval Civ. Eng. Lab., Port Hueneme, Calif., Tech. Rep. R-281, 1964.

23. Hough, B. K., "Compressibility as the Basis for Soil Bearing Value," *ASCE J. Geotech. Eng. Div.*, vol. 85, Aug. 1959.

24. Hough, B. K., *Basic Soils*, 2nd ed., Ronald Press, New York, 1969.

25. Hvorslev, M. J., "The Basic Sinkage Equations and Bearing Capacity Theories," U.S. Army Waterways Exp. Stn, Vicksburg, Miss., Tech. Rep. M-70-1, Mar. 1970.

26. Kerisel, J. L., "Vertical and Horizontal Bearing Capacity of Deep Foundations in Clay," *Proc. Symp. Bearing Capacity and Settlement of Foundations*, Duke University, Durham, N.C., 1967.

27. Ko, H. Y., and L. W. Davidson, "Bearing Capacity of Footings in Plane Strain," *ASCE J. Soil Mech. Found. Div.*, vol. 99, no. SM1, Jan. 1973.

28. Krizek, R. J., "Approximation for Terzaghi's Bearing Capacity Factors," *ASCE J. Geotech. Eng. Div.*, vol. 91, Mar. 1965.

29. Lambe, T. W., and R. V. Whitman, *Soil Mechanics*, Wiley, New York, 1969.

30. Lundgren, H., and K. Mortensen, "Determination by the Theory of Plasticity of the Bearing Capacity of Continuous Footings on Sand," *3rd Int. Conf. Soil Mech. Found. Eng.*, Zurich, Switzerland, vol. 1, 1953.

31. Meyerhof, G. G., "An Investigation of the Bearing Capacity of Shallow Footings on Dry Sand," *2nd Int. Conf. Soil Mech. Found. Eng.*, Rotterdam, The Netherlands, vol. 1, 1948.

32. Meyerhof, G. G., "The Ultimate Bearing Capacity of Foundations," *Geotechnique*, vol. 2, 1951.

33. Meyerhof, G. G., "The Bearing Capacity of Foundations under Eccentric and Inclined Loads," *3rd Int. Conf. Soil Mech. Found. Eng.*, Zurich, Switzerland, vol. 1, 1953.

34. Meyerhof, G. G., "Influence of Roughness of Base and Ground Water Conditions on the Ultimate Bearing Capacity of Foundations," *Geotechnique*, vol. 5, no. 3, 1955.

35. Meyerhof, G. G., "Penetration Tests and Bearing Capacity of Cohesionless Soils," *ASCE J. Soil Mech. Found. Div.*, vol. 82, no. SM1, 1956.

36. Meyerhof, G. G., "The Ultimate Bearing Capacity of Wedge-Shaped Foundations," *5th Int. Conf. Soil Mech. Found. Eng.*, Paris, France, vol. 2, 1961.

37. Meyerhof, G. G., "Some Recent Research on the Bearing Capacity of Foundations," *Can. Geotech. J.*, vol. 1, no. 1, 1963.

38. Meyerhof, G. G., "Shallow Foundations," *ASCE J. Soil Mech. Found. Div.*, vol. 91, no. SM2, 1965.

39. Meyerhof, G. G., "Ultimate Bearing Capacity of Footings on Sand Layer Overlying Clay," *Can. Geotech. J.*, vol. 11, no. 2, May 1974.

40. Mitchell, J. K., and W. S. Gardner, "In-Situ Measurements of Volume Change Characteristics," *ASCE J. Soil Mech. Found. Div.*, 1975.

41. Muhs, H., "On the Phenomenon of Progressive Rupture in Connection with the Failure Behavior of Footings in Sand," discussion, *6th Int. Conf. Soil Mech. Found. Eng.*, Montreal, Canada, vol. 3, 1965.

42. Prandtl, L., "Über die Eindringungsfestigkeit plastischer Baustoffe und die Festigkeit von Schneiden," *Z. Angew. Math. Mech.*, Basel, Switzerland, vol. 1, no. 1, 1921.

43. Reddy, A. S., and R. J. Srinivasan, "Bearing Capacity of Footings on Layered Clays," *ASCE J. Soil Med. Found. Div.*, vol. 93, no. SM2, Mar. 1967.

44. Schleicher, F., "Zur Theorie des Baugrundes," *Bauingenieur*, vol. 7, 1926.

45. Skempton, A. W., "The Bearing Capacity of Clays," *Proc. Build. Res. Congr.*, London, England, 1951.

46. Steinbrenner, W., "Tafeln zur Setzungsberechnung," Die Strasse, vol. 1, Oct. 1934.

47. Taylor, D. W., *Soil Mechanics*, Wiley, New York, 1948.

48. Teng, W. C., *Foundation Design*, Prentice-Hall, Englewood Cliffs, N.J., 1962.

49. Terzaghi, K., *Theoretical Soil Mechanics*, Wiley, New York, 1943.

50. Terzaghi, K., and R. B. Peck, *Soil Mechanics in Engineering Practice*, 2nd ed., Wiley, New York, 1967.

51. Vesic, A. S., "Analysis of Ultimate Loads of Shallow Foundations," *ASCE J. Soil Mech. Found. Div.*, vol. 99, Jan. 1973.

15
Pile Foundations

15-1 INTRODUCTION

As was mentioned in the preceding chapter, a large percentage of the building sites display adequate bearing capacity to permit a spread-footing design, especially for light structures. Certain weak strata are sometimes improved by a form of pretreatment. Perhaps this may consist of compacting a loose sand and gravel stratum by means of vibrating compactors, chemical or cement grout, etc., as described in Chapter 16. Then there are times when one may find it necessary and/or possibly more economical to transmit the load from the structure to depths at which adequate support is available by means of deep foundations. Piles are included in this category and will be discussed in this chapter.

The load transfer via piling may be realized from skin friction, from end-bearing, or from a combination of both. If a very large percentage of the total load is resisted by side or skin friction, the pile is known as a *friction pile*. On the other hand, if the pile rests on a very firm stratum (say rock) such that most of the load is transmitted directly to this stratum via point or tip resistance, the pile is known as an *end-bearing pile*. Almost invariably, however, each of the two types derives some of its load from a combination of end-bearing and skin friction. The percentage load assumed by either friction or end-bearing is less discernible in a third type of pile which is installed by either injecting through vibrations or pounding into the soil stratum a column of sand and gravel or concrete. This forms a type of column which acts as a pile and which is sometimes called a *compaction pile*. All three types fall in the general category of bearing piles, and their purpose is in contrast to that of the sheet pile (discussed in Chapter 13), whose primary function usually is to resist lateral forces.

Based on material compositions, the piles may be further classified as *timber*, *concrete*, *steel*, or *composites*. Corrosive properties of the stratum, fluctuations in the water table and dry-rotting effects associated with such fluctuations, ease of installation, length requirement, and availability of material and installation equipment are some of the considerations related to the selection of the type of pile. These are factors which enter into the economic-function phase of the selection effort.

Piles may be driven (hammered) into the stratum, or they may be installed by a cast-in-place process. Furthermore they may be driven vertically or on a batter, tapered or uniform in cross section, of one material or of a combination to form a composite.

Effects on adjacent buildings induced by the vibrations resulting from pile driving, soil composition and stratification (granular or clay, rock seams or boulders, etc.), functional requirements (lateral components of load, uplift anchorage versus only downward support, etc.), and uniform or nonuniform cross section (such as a bell-shaped bottom, a tapered shaft) are some of the considerations related to material selections and installation techniques.

In general a pile foundation is more expensive than an ordinary spread-footing design. On the other hand, it is frequently more economical and, generally, much more reliable than procedures associated with soil stabilization. The evaluation of the subsurface conditions via a detailed program of testing and analysis is a first and most important prerequisite in the total effort of pile selection, economic considera-tion, and subsequent design.

Once the pile type has been selected, the design-construction sequence normally consists of designing the pile by means of empirically established procedures, in-stalling the pile, and subsequently verifying the results by subjecting the pile to a load test. For driven piles the load test is sometimes eliminated if careful observations of the pile's resistance to penetration are made during the driving sequence.

It is common practice to design and test single piles, even when they are part of a pile group or cluster. It is important to note that, depending upon the number of piles, pile spacing, method of installation, soil characteristics, and load distribution, the behavior of the composite may not be totally reflective of that of the individual piles. For example, it is likely that the load capacity of the composite is less than that of the sum of the individual piles; the efficiency of a pile group is less than that of an individual pile; the settlement of a pile group is likely to be greater than that of an individual pile.

15-2 TIMBER PILES

One of the oldest types of piles, the wood pile, is still one of the most common. Fre-quently it is the cheapest type. Essentially wood piles are made from tree trunks with the branches and bark removed. Southern pine and Douglas fir are rather common types, although some hard woods are sometimes used. The choice is frequently dictated by the availability of the material, and this may vary in different parts of the country.

The installation of this type of pile is by driving. Normally the pile is driven with the small end down. Typically the pile has a natural taper, with a top cross section of twice or more that at the bottom end. Depending upon the resistance to driving, and obstacles such as boulders within the soil stratum, wooden piles are some-times tied with steel bands at the bottom end and/or at the top end to avoid splitting.

The size of the pile is frequently dictated by the load capacity expected, by the type of soil, and sometimes by building codes. Generally 20-m maximum lengths are common, although lengths of over 30 m have been used on numerous occasions.

Although the piles could withstand hard handling, their lengths are frequently dictated by transportation restrictions and local availability.

As a general rule, wood piles have a long life provided they are not subject to alternate wetting and drying, or to attacks from marine borers. On the other hand, subjected to repetitive cycles of wetting and drying, the pile may experience rapid decay through dry-rotting within months or perhaps 1 or 2 years; cedar is generally more resistant to such effects. For this reason wood piles are almost always cut off below the permanent groundwater. Furthermore, most frequently wooden piles are treated with preservatives such as creosote oil which is impregnated into the wood. This is an effective means of preventing dry-rotting, particularly for that part of the pile which may undergo cycles of wetting and drying (e.g., marine installations), as well as against damage from most animal and plant attacks. As a general rule, the top end of the pile should be protected, generally by lead paint, zinc coating, or some other treatment, before the concrete cap is poured. Needless to say, such protective treatment, as well as additional costs which may be incurred if excavation, dewatering, and sheeting should prove necessary in connection with cutoffs below the surface, are economic considerations that should be part of the selection process.

Overdriving of timber piles may result in splitting, crushing, and/or shearing of the pile. Buckling of the pile may be still a further effect from overdriving. This is sometimes detected by a sudden reduction in penetration resistance. As a general rule, however, close observations of pile behavior made during the driving operations is the best insurance against such an occurrence. Furthermore designing for a rather limited capacity, perhaps up to 30 tons, may be further assurance against failure from overdriving.

15-3 STEEL PILES

Steel piles are usually rolled-H, fabricated shapes, or pipe piles. Sometimes sections of sheet piles are used to form a pile unit, usually in the form of a box. Because of the relative strength of steel, steel piles withstand driving pressures well and are usually very reliable end-bearing members, although they are found in frequent use as friction piles as well. Pipe piles are normally, although not necessarily, filled with concrete after driving. Prior to driving, the bottom end of the pipe pile usually is capped with either a flat or a cone-shaped point welded to the pipe. Sometimes the pipe is driven open-ended instead of capped, and the soil is removed after the completion of driving either by augering, or by air or water jetting the soil from within the pipe.

Strength, relative ease of splicing, and sometimes economy are some of the advantages cited in the selection of steel piles. Corrosion is a negative aspect of steel piles. Corrosive agents such as salt, acid, moisture, and oxygen are common enemies of steel. Steel piles in disturbed soils appear to corrode more rapidly due to an apparent higher oxygen content in the disturbed soil than in an undisturbed soil. The oxygen content in an undisturbed soil appears to decrease significantly after a few feet from the surface. Furthermore, because of the corrosive effect that salt water has on steel, steel piles have a rather restrictive use for marine installations. Peat deposits, strata containing coal or sulfur, acid clays, and swampy soils are usually of common concern. In general, however, soils with a pH of less than 7 (acidic) should be looked at carefully. Coating such piles with paint or encasing them in a concrete shell may provide some protection against corrosion, although the paint may be damaged during the installation process. For a limited life span of the pile it is sometimes the practice to compensate for the anticipated corrosion by specifying a greater thickness of the steel section. Sometimes cathodic protection is used to reduce such corrosive effects.

15-4 CONCRETE PILES

Under comparable circumstances, concrete piles are much more immune to the detrimental effects of the corrosive elements that rust steel piles or to the effects that cause decay of the wooden piles. Concrete is sometimes used as a protective coating for steel piles. Furthermore, concrete is generally more available than steel in most geographic regions, and in many instances at more economically advantageous terms than wood.

Concrete piles may be of a *precast* or a *cast-in-place* type. Although both types may be plain or reinforced, straight shaft or tapered, they are more conspicuous in their differences rather than in their similarities.

Precast Concrete Piles

These are concrete piles which are formed, cast to specified lengths and shape, and cured before they are driven or jetted into the ground. Reinforced with conventional steel or prestressed, these piles are generally cast with a pointed tip, have square or octagonal cross section, and are usually tapered. They are frequently used in marine installations, or where part of the pile may serve as a column above ground. Although especially designed for each project, the sizes of the precast piles, in both diameter and length, are frequently governed by handling stresses. Usually they are limited to less than 25-m length and generally less than 0.5 m in diameter. The pile capacity is usually limited to about 75 tons.

While cutoffs and splices may be required to adjust for differences in length, these procedures pose a less than desirable situation, particularly for splicing. For example, to splice a new section onto one already driven, the abutting ends of both sections must be chipped so as to expose the reinforcing bars of both. They are subsequently connected by fresh concreting at this junction—a rather time-consuming effort due to both labor and curing time required for the new joint.

Cast-in-Place Piles

The installation process of cast-in-place piles may consist of driving a steel tubing or casing into the ground and then filling it with concrete (*cased pile*); or concrete may be cast into a driven shell which is subsequently extracted as the concrete is poured (*uncased* or *shell-less pile*). An increasingly popular shell-less type is constructed by casting through the auger, under pressure, without the aid of a steel shell.

Depending on wall thickness and general strength, a steel shell or pipe may be driven with or without the aid of a mandrel. That is, for relatively thin shells, a mandrel or core fits into the shell in order to prevent inward collapse and buckling of the shell during the driving operations. Once the shell has been driven to the desired depth, the mandrel is extracted and the shell filled with concrete. Many pipe-type sections and some relatively thin fluted sections, however, may frequently be driven without a mandrel. Sometimes the casing is driven open-ended with subsequent removal of the soil within the pipe, usually via jetting. Subsequently the casing is filled with concrete.

One method of obtaining a shell-less pile consists of extracting the driven shell while the concrete is either hammered down or restricted from moving upward with the pipe by means of a mandrel. Under sufficient force and hammering action, a bulb-shaped mass of concrete could be formed at the lower end of the pile for additional bearing support.

An *auger-cast pile* is a shell-less type formed by using a continuous-flight auger with a hollow stem. The augers range in size from about 25 to 40 cm in diameter, with the flights welded to a hollow stem which may range from around $7\frac{1}{2}$ to 10 cm in diameter. The tip of the auger has a small opening which is plugged up during the downward augering. The hole is augered to a predetermined depth. A concrete grout is then pumped by a positive-displacement pump through high-pressure hoses connected to the stem of the auger. Under pressure the concrete mortar expels the plug from the tip of the auger and permeates into the stratum and into the cavity below the tip of the auger created as the auger is slowly withdrawn. With the pump continuously pumping, the auger is extracted slowly enough to permit maximum grout penetration into the hole and into the adjoining stratum, as well as to prevent choking of the pile by lateral soil pressures, but fast enough to prevent the grout from coming up the shaft. The soil displaced by the auger is brought up by the auger flights during the auger extraction. Reinforcing, in the form of either individual bars, steel sections, or several bars connected by ties or spirals, similar to concrete-column reinforcement, is pushed into the grout at the end of the casting while the grout is still wet.

Particularly advantageous for sand and gravel strata, the auger-cast piles usually provide excellent friction resistance, are vibration free, and usually are more economical than other piles for comparable purposes. Indeed, because it requires neither splicing nor fixed-length dimensions, this type of pile may be injected to varying lengths throughout a site in a most expeditious manner. Furthermore, particularly for granular soils, it compares most favorably with the other types in terms of load capacity. Figure 15-1 shows some common types of cast-in-place piles.

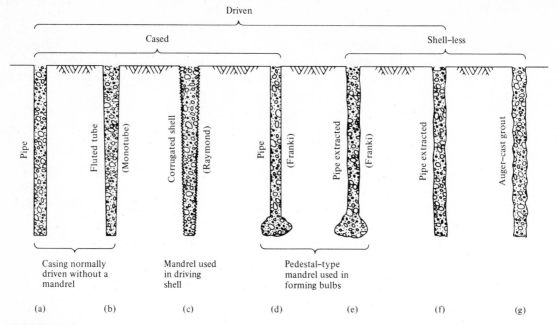

FIGURE 15-1
Some common types of cast-in-place piles. Their lengths and capacities vary for various types and soil conditions.

15-5 ULTIMATE CAPACITY OF A SINGLE PILE

The pile support is realized from two sources: (1) *end-bearing* and (2) *side friction*. As mentioned previously, although the percentage of the pile load assumed by the friction or by the end-bearing will generally vary widely with different situations, static equilibrium dictates that

$$Q_u = Q_p + Q_s \tag{15-1}$$

where Q_u = total pile load
Q_p = point resistance (end-bearing)
Q_s = side-friction resistance

Figure 15-2a shows the forces acting on a pile as described by Eq. (15-1). The magnitude of Q_p can be estimated with acceptable accuracy by using a form of the general bearing capacity formula presented in Chapter 14. The shape and depth correction factors are replaced with compensating terms more representative of the conditions associated with different types of soil, types and lengths of piles, methods of installation, and other factors discussed later in this chapter. However, in general Q_p may be estimated as

$$Q_p = A_p\left(\bar{c}N_c + \gamma L N_q + \frac{\gamma}{2}BN_y\right) \tag{15-2}$$

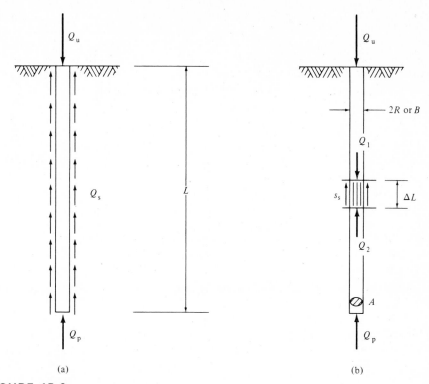

FIGURE 15-2
Components of forces in determining the ultimate bearing capacity of a single pile.

The total support provided by friction Q_s could be estimated as the product of the surface area (product of perimeter p and length L) of the pile in contact with the soil times the average shear resistance per unit area s_s developed between the soil and the pile, that is, $Q_s = pLs_s$. For round piles, $Q_s = 2\pi RLs_s$. However, since the unit shear resistance s_s may vary widely over the length of the pile, an average unit resistance is perhaps difficult to ascertain with any degree of accuracy. In that case it would be more logical to express the total resistance derived from friction as the sum of unit resistances. These are expressed by Eq. (15-3), where ΔL is the increment of pile length. For a layered system and/or varied sections,

$$Q_s = \sum (\Delta L) p_i s_{si} \qquad (15\text{-}3)$$

For uniform soil characteristics (s_s = constant) and constant section (p = constant),

$$Q_s = pLs_s \qquad (15\text{-}3a)$$

Substituting the equivalent quantities from Eqs. (15-2) and (15-3) into Eq. (15-1), one obtains a general ultimate load bearing capacity for piles as given by Eq. (15-4).

That is, for a layered system and/or varying pile sections the ultimate load becomes

$$Q_u = A_p(\bar{c}N_c + \gamma L N_q + \tfrac{1}{2}\gamma B N_\gamma) + \sum \Delta L p_i s_{si} \tag{15-4}$$

As a special case, for uniform soil strength (s_s = constant) or constant pile section (p = constant) and round piles of radius R, we have

$$Q_u = \pi R^2(\bar{c}N_c + \gamma L N_q + \gamma R N_q) + 2\pi R L s_s \tag{15-4a}$$

where

Q_u = ultimate bearing capacity of a single pile
s_{si} = shaft resistance per unit area at any point along pile
B = general dimension for pile width
A_p = cross-sectional area of pile at point (bearing end)
R = radius of pile at a given increment of length
p_i = perimeter of pile in contact with soil at any point
L = total length of embedment of pile
γ = unit weight of soil
$\bar{c}$ = effective cohesion of soil
N_c, N_q, N_γ = bearing capacity factors

15-6 FRICTION PILES IN COHESIONLESS SOILS

When a pile is driven into cohesionless soil, the density of the stratum is increased due to (1) the volume displacement (volume equal to the volume of the imbedded portion of the pile); and (2) the densification resulting from the driving vibrations. Increased density results for a distance of a few pile diameters. In general a significant increase for a distance of 2 diameters and a noticeable increase up to 8 diameters are rather common for cohesionless layers.

The ultimate bearing capacity Q_u is again the sum of the forces realized from skin friction and end-bearing, as expressed by Eq. (15-1). The expansion of this equation into the version given by Eq. (15-4) can be used to determine the capacity of piles in a granular formation with some modification. That is, it is readily apparent that for the cohesionless case ($c = 0$) the first term of Eq. (15-4) drops out. Furthermore the third term (N_γ term) is relatively small when compared with the middle term. Hence a reasonable approximation for the ultimate bearing capacity of piles in cohesionless material could be given by Eq. (15-5):

$$Q_u = A_p \gamma L N_q + \sum \Delta L p_i s_{si} \tag{15-5}$$

The above equation implies that both end-bearing and total skin resistance would increase with increasing depth for a homogeneous stratum. However, experimental evidence and field observations indicate that the end-bearing capacity reaches some upper limit, and does not increase infinitely with depth. It appears that crushing, compressibility, and general failure in the zone near the pile tip as well as other factors impose an upper limit on the ultimate bearing capacity of a given pile. Letting $\bar{\sigma}$

represent the effective overburden pressure at the pile tip, the tip resistance could be represented by

$$Q_p = A_p \bar{\sigma} N_q \leq A_p \sigma_e \qquad (a)$$

where σ_e represents some limiting value for end-bearing for which the depth D equals or is greater than the critical depth (that is, $D \leq D_c$). Although specific values for σ_e still appear questionable at this stage, values of 25 kN/m^2 for loose sand and 100 kN/m^2 for dense sand appear as reasonable approximations for the upper limit.

There is lack of agreement among experts regarding the manner in which to compute the skin resistance which is represented by the last terms in Eq. (15-4). The disagreement lies primarily in the computation of the unit shear resistance s_s developed between pile and soil. In general, however, the value for s_s could be estimated by Eq. (b):

$$s_s = K_s \bar{\sigma} \tan \delta \qquad (b)$$

and therefore the value for Q_s is given by Eq. (c):

$$Q_s = \sum p_i \Delta L K_s \bar{\sigma} \tan \delta \qquad (c)$$

where K_s = average coefficient of earth pressure on the pile shaft (see Table 15-1)
$\bar{\sigma}$ = average effective overburden pressure along pile shaft
δ = angle of skin friction
$\tan \delta$ = coefficient of friction between soil and pile surface

The values of K_s may vary, and are frequently more accurately derived from the results of pile load tests at the building site. These values may range from a value equal to an earth pressure coefficient at rest K_0 to a value equivalent to some passive pressure, approximately 3 to 4 times this value. The lower values are generally associated with bored or jacketed piles into loose granular material, while the upper range applies for driven or pressure-injected piles into relatively dense material. Generally a value of K_s between 1 and 2 appears reasonable, with Table 15-1 providing some approximate values for K_s for some common types of driven piles.

On the basis of the above discussion, the ultimate bearing capacity of a pile in a cohesionless material could be expressed by Eq. (15-6):

$$Q_u = A_p(\bar{\sigma} N_q) + \sum p_i \Delta L K_s \bar{\sigma} \tan \delta \qquad (15\text{-}6)$$

TABLE 15-1
Approximate Values of K_s for Some Driven Piles

Pile Type	Range of K_s
Concrete	1.5 ± 10%
Pipe	1.1 ± 10%
H-section	1.6 ± 10%

It is to be noted that experimental evidence indicates that the value of N_q is influenced by such factors as the friction angle φ, the compressibility of the soil, the method of pile installation, and, to a lesser extent, the type and shape of the pile. Hence the values obtained from Fig. 14-11 are indeed approximate. On the other hand they are deemed acceptable for estimating purposes, particularly in the lower range of values for $\varphi = 25°$ or less.

15-7 BEARING CAPACITY BASED ON STANDARD PENETRATION TESTS OF COHESIONLESS SOILS

A widely used and reasonably reliable method for predicting the load capacity of driven piles into cohesionless soils is an empirical correlation of capacity and blow counts from *standard penetration tests* (SPT). For very coarse material, including large gravel and/or boulders, the blow count on the split-spoon sampler may be exceedingly high and not representative of the true characteristics of the soil to support the driven pile.

The correlation between blow counts and pile capacity has not been adopted as a standard by the profession. Quite to the contrary, wide variations in views exist among practitioners regarding such interpretations. Frequently the engineers will use guidelines based on personal experience for given localities and geographic districts for estimating pile bearing capacities and will verify the design by appropriate tests in the field. Meyerhof (62) suggests a formula that may be used for a *cohesionless* material, particularly in sand deposits, given by Eqs. (a) and (b):

$$q_p = 40N \frac{L}{2R} \le 400N \qquad (kN/m^2) \qquad (a)$$

$$f_s = 2N \qquad (kN/m^2) \qquad (b)$$

where f_s is the average shear resistance per unit surface area and R is the radius of the pile. Hence on the basis of the above expressions, the total capacity of the pile may be approximated by Eq. (15-7), consistent with the limits for the first term of Eq. (a) above. That is,

$$Q_u = A_p q_p + A_s f_s$$

or

$$Q_u = A_p \left(40N \frac{L}{2R} \right) + 2N A_s \qquad (15\text{-}7)$$

where A_s is the surface area that develops the friction = (perimeter) (length).

The end-bearing resistance and average skin support of bored piles in cohesionless material are less than that for driven piles of comparable size. In general the ultimate capacity of a bored pile may be safely estimated as approximately half that of a driven pile. On the other hand, pressure-injected piles installed in cohesionless soil via the auger-cast method (cast under pressure) provide a frictional resistance as well as an

TABLE 15-2

Values of q_p and f_s Used by the Author as a Guide in Estimating Capacities of Auger-Cast Piles Injected under Pressure in Cohesionless Soil*

Average Blow Count Readings N	< 10	10–15	16–20	21–25	26–35	36–50	> 50
q_p (kN/m²)†	0–200	200–300	320–400	420–500	520–700	720–1000	1000
f_s (kN/m²)	0–30	30–45	45–60	60–75	75–105	105–150	175

* For diameters of 25–40 cm (10–16 in) and $L \geq 6$ m long.
† For N readings at tip.

overall ultimate capacity which are favorably comparable to (and in some instances higher than) those of many driven piles. This is generally attributed to (1) a densification of the surrounding stratum induced by some volume displacement and (2) relatively deep lateral penetration of the cement grout (perhaps several centimeters) into the cohesionless stratum. Such penetration creates (1) a shear surface anywhere between 10 and 20 percent larger; and (2) an end-bearing area sometimes 25 to 30 percent larger than for a driven pile of comparable size. Furthermore the friction angle between soil and pile usually approaches φ. In fact, the value of φ is frequently increased around the periphery of the pile due to injection of the mortar grout into the surrounding soils.

Table 15-2 gives a range of values for bearing pressures at the pile tip and side friction f_s with resistance numbers obtained from the standard penetration test. These are perhaps most suitable in sands and small gravels, and less so in cohesive materials; they are least applicable for very coarse gravels and boulders. These results should be verified by testing of a typical pile or piles in accordance with the accepted testing procedures (e.g., ASTM D-1143-74). The essence of such procedures is briefly described in Section 15-10.

15-8 FRICTION PILES IN COHESIVE SOILS

Piles driven into soft saturated clays tend to (1) disturb the clay around the pile; (2) increase the pore-water pressure; (3) increase compressibility in the clay; and (4) remold the clay to varying degrees for a distance of approximately one pile diameter. Temporarily there is an apparent loss of pile capacity. However, the pore-water pressure dissipates rather rapidly, and after some consolidation the shear strength is regained, frequently exceeding the initial value. The increased densities, consolidation effects, and increased horizontal stress after pile driving contribute to an increase in the frictional resistance between soil and pile. This recovery normally takes place within 30 days—ordinarily less than the time in which the total building load is applied to the pile.

The ultimate bearing capacity of a single pile in clay could be estimated by the general expression given by Eq. (15-1). The resistance provided by end-bearing, Q_p, and that from the friction, Q_s, could be determined by use of Eqs. (15-2) and (15-3), respectively. As mentioned previously, the N_γ term is relatively small in comparison with the other two terms, and therefore may be (conservatively) neglected. Hence the total resistance from end-bearing could be expressed by Eq. (15-8):

$$Q_p = A_p(\bar{c}N_c + \gamma L N_q) \tag{15-8}$$

Based on Fig. 14-11, the value of N_c for a value of $\varphi = 0$ is 5.14. Skempton (96) found the value for N_c to be approximately 7.5 for long footings and approximately 9 for circular or square footings when the ratio of footing depth to width exceeded a value of about 10. Values for N_c ranging from 5 for a very sensitive normally consolidated clay to 10 for insensitive, overconsolidated clay are common.

The total resistance from friction Q_s may be estimated from Eq. (15-3). For convenience, this is reproduced as

$$Q_s = \sum (\Delta L) p f_s \tag{15-3}$$

where f_s is the unit skin friction resistance in clay. According to Meyerhof (62), the values for f_s could be approximated as given by Eqs. (a) and (b). For driven piles,

$$f_s = 1.5 c_u \tan \varphi \tag{a}$$

For bored piles,

$$f_s = c_u \tan \varphi \tag{b}$$

where c_u = average cohesion, undrained condition
   φ = angle of internal friction of the clay

Table 15-3 shows a relationship between the unit skin friction resistance f_s and the unconfined compressive strength of clay.

TABLE 15-3
**Values of f_s in kN/m² Based on Un-
confined Compression Tests on Clay
[Tomlinson (106)]**

Unconfined Compressive Strength of Clay (kN/m²)	f_s (kN/m²)	
	Concrete or Timber	*Steel*
0–72	0–34	0–34
72–144	34–48	34–48
144–288	48–62	48–57
288	62	57

Although values for f_s may vary with different sources, particularly for clayey soils, virtually all authorities agree that the values should be used only as guides for estimating the capacity of the pile. As mentioned previously, the actual capacity should be determined via load tests of actual piles in the field.

Based on the above discussion, an expression for estimating the ultimate bearing capacity of a pile installed in a clayey stratum could be given by Eq. (15-9):

$$Q_u = A_p(\bar{c}N_c + \gamma LN_q) + A_s f_s \qquad (15\text{-}9)$$

EXAMPLE 15-1

Given   (1) An auger-cast pile; diameter 0.41 m; $L = 12$ m. (2) Dry sand and gravel stratum; $\gamma = 19.4$ kN/m^3; $\varphi = 32°$; $c \approx 0$.

Find   The ultimate bearing capacity of the pile.

Procedure   From Fig. 14-11, for $\varphi = 32°$, $N_q = 23$ and $N_\gamma = 30$, and

$$s_{s\,avg} = (\tfrac{1}{2})K_s \gamma L \tan \varphi = \tfrac{1}{2}(1.5)(19.6)(12) \tan 32° = 109$$

$K_s \approx 1.5$ from Table 15-1. Furthermore let us assume the shear resistance varies linearly with depth, and $\bar{\varphi} \approx \varphi$. For the typical auger-cast pile, installed under *pressure*, the above assumptions seem reasonable. Hence using Eq. (15-4a),

$$Q_u = \pi\left(\frac{0.41}{2}\right)^2\left[0 + 19.4(12)(23) + 19.4\left(\frac{0.41}{2}\right)(30)\right] + 2\pi\left(\frac{0.41}{2}\right)(12)(109)$$

$$Q_u = 0.132[5354.1 + 119.3] + 1686$$

Answer

$$Q_u = 2408 \text{ kN (541 kips)}$$

EXAMPLE 15-2

Given   (1) An auger-cast pile; diameter 0.41 m; $L = 12$ m. (Note that this is the same size pile as given in Example 15-1.)
   (2) Sandy silt stratum; $\gamma = 18.4$ kN/m^3; $\varphi = 16°$; $c = 13$ kN/m^2; water table at surface.

Find   The ultimate bearing capacity of the pile.

Procedure   The assumptions made in Example 15-1 will appear valid here as well. Hence for $\varphi = 16°$; $N_q = 4$; $N_\gamma = 3$; $N_c = 12$,

$$s_{avg} = 13 + \tfrac{1}{2}(1.5)(18.4 - 9.81)(12) \tan 16° = 35.5, \text{ say } 36 \text{ kN/m}^2$$

Thus,

$$Q_u = \pi\left(\frac{0.41}{2}\right)^2\left[13(12) + 8.59(12)(4) + 8.49\left(\frac{0.41}{2}\right)(3)\right] + 2\pi\left(\frac{0.41}{2}\right)(12)(36)$$

$$Q_u = 0.132[156 + 412.3 + 5.3] + 649$$

$$Q_u = 75.72 + 649$$

Answer

$$Q_u = 725 \text{ kN (163 kips)}$$

15-9 PILE CAPACITY BASED ON DRIVING RESISTANCE

In his book *Pile Foundations* Chellis (19) provides a fascinating short summary of implements man has used to drive piles. The primitive pile-driving equipment consisted of a variety of rams which were worked by hand power or perhaps crudely improvised forms of hammers. During the medieval times the equipment revealed improved features which incorporated some mechanical advantages. With the invention of power engines, the energy for the pile-driving equipment is provided solely by power hammers. There are numerous types and manufacturers of such hammers. In a general category, the following are basic features of several types:

1. *Single-acting hammer.* Steam or compressed air raises the ram, then permits it to drop freely onto the top of the pile.
2. *Double-acting hammer.* Here steam or compressed air not only lifts but also pushes down on the ram.
3. *Diesel hammer.* As the name implies, this hammer is operated by a diesel engine which raises the ram and then permits it to fall freely.

Dynamic Formulas

Numerous empirical formulas have been developed in an attempt to predict the capacity of a driven pile from its resistance to penetration. The basis of most of these formulas is a transfer of the kinetic energy of the hammer to the pile and to the soil. The *Engineering News Record* formula is one formula based on the energy concept. It is one of the simplest and most widely used, although the *modified Engineering News Record* formula appears to give more accurate results.

The energy expended by the hammer weight dropping from a certain height is transferred into work: force × pile penetration plus loss of energy. Symbolically this could be represented by Eq. (a):

$$E = RS + L \tag{a}$$

where E = driving energy = weight × fall of hammer = $W_r h$
  R = resistance to penetration (kN)
  S = pile penetration per blow (mm = 1/1000 m)
  L = loss of energy from impact, soil, pile, cap, etc.

It is common practice to express L as

$$L = RC \tag{b}$$

where C is an arbitrary constant, 2.54 mm for single-acting steam hammers and 25.4 mm for drop hammers.

Adjusting for units, Eq. (a) could be expressed by Eq. (c):

$$E = R(S + C)\frac{1}{1000} \tag{c}$$

Solving for R, we get

$$R = \frac{1000E}{S + C} \tag{d}$$

A safety factor of 6 is common for this equation. Hence on this basis, Eq. (d) would be altered to give expressions (15-10). This is the *Engineering News Record* formula.

For single-acting steam hammers,

$$R = \frac{167E}{S + 2.54} \tag{15-10}$$

and for drop hammers

$$R = \frac{167E}{S + 25.4} \tag{15-10a}$$

Although the above formulas are rather widely used, their accuracy is also greatly questioned. From all indications the safety factor is quite likely appreciably less than 6, perhaps 2 or 3 instead of 6. Hence adjustments were sought. Among the many proposals, the *modified Engineering News Record* formula seems to combine reasonable simplicity and acceptable accuracy. It is given by Eq. (15-11):

$$R = e_h\left(\frac{167E}{S + 2.54}\right)\left(\frac{W_r + n^2 W_p}{W_r + W_p}\right) \tag{15-11}$$

where W_r and W_p are the weights of hammer (or ram) and pile, respectively. Average values for e_h and n are given in Tables 15-4 and 15-5.

TABLE 15-4
Average Values for e_h

Hammer	e_h
Single-acting	0.8
Double-acting	0.85
Drop	0.85

TABLE 15-5
Average Values for n

Material	n
Wood piles	0.25
Wood cushion on steel	0.32
Steel-on-steel anvil	0.5

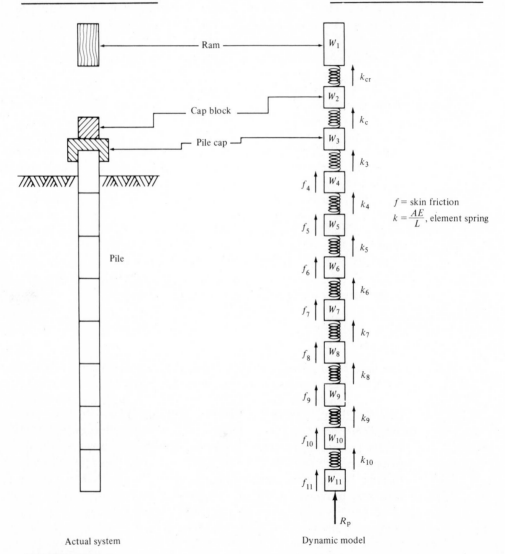

f = skin friction

$k = \dfrac{AE}{L}$, element spring

Actual system   Dynamic model

FIGURE 15-3
Dynamic model used in wave equation analysis.

The Wave Equation

The shortcomings associated with the pile-driving formulas are manifested in that they do not relate to the soil–pile interaction and the time-related behavior of the pile system. These variables were accounted for by Smith (98) as well as by a number of other investigators via an analogy, termed the *wave equation*, of the pile behavior and a mathematical model. For those interested, Bowles (11) gives a more detailed discussion of the approach.

The analogy consists of assuming the pile system to be comprised of a series of small concentrated masses, as shown in Fig. 15-3. The springs simulate the axial resistance of the pile. Elastic deformation, tamping, and friction constitute the surface restraints. The propogation of the elastic wave through the pile is analogous to that induced by a longitudinal impact on a long rod. Thus a partial differential equation can be written to describe the model. This can be solved numerically, with the aid of digital computers.

The wave equation analysis is a useful tool in determining the pile capacity, in selecting appropriate pile-driving equipment (e.g., hammer weights, caps, lengths), and in establishing guidelines (stresses) in the pile-driving effort. With improved computer capabilities the method is finding increasing use in the geotechnical engineering field.

It is a widely accepted consensus that virtually all the dynamic pile-driving formulas are but rough approximations of pile-driving resistance. The lack of accuracy of these formulas, however, does not imply the abandonment of their use. Quite to the contrary, they could be useful if properly supplemented by adequate load tests. Indeed, a common procedure is to observe the penetration resistance for a driven pile, then load-test it, and proceed to relate the load capacity to the resistance to penetration by use of one of these formulas.

15-10 PILE LOAD TEST

The most reliable means of determining the load capacity of a pile is to subject it to a static load test. The test procedure consists of applying static load to the pile in increments up to a designated level of load and recording the vertical deflection of the pile. Occasionally the pile load comes from dead weight (e.g., soil, pig iron) balanced on top of the pile. Most frequently, however, the load is transmitted by means of a hydraulic jack placed between the top of the pile and a beam supported by two or more reaction piles. Figure 15-4 is a schematic view of a typical load test arrangement. In this case four piles are used as reaction piles. Although two reaction piles (one on each end) may provide sufficient reaction resistance, a four-pile reaction setup provides more stability to the physical setup. The vertical deflection of the top of the pile is commonly measured by mechanical gauges attached to a beam which spans over the test pile and which is totally independent (unattached) of the load setup.

The results from pile tests could be the source of most useful information in at least two general aspects: (1) in determining the ultimate bearing capacity of the pile;

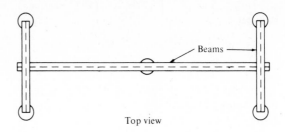

Top view

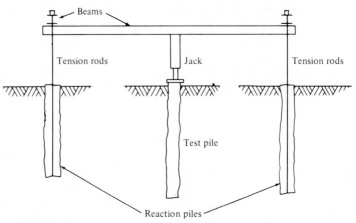

FIGURE 15-4
Schematic arrangement for pile-load test.

and (2) in evaluating the deflection characteristic of the pile. With regard to the ultimate bearing capacity, one may reflect on several additional and related tangible benefits. For example, if the capacity of the pile is different from that desired, the pile length, diameter, and details of installation (e.g., driving criteria, injection pressures, material content) can be adjusted prior to the installation of the rest of the piles. Hence quite apparently, the economic and safety aspects of the design could be more authentically evaluated. Adjustments could likewise be made to account for deflections (settlement), thereby providing a design that would fit within the tolerable deformation requirements. A side, but equally important, benefit from the installation and testing of a test pile prior to the installation of the rest is general information about the site conditions with regard to potential problems (e.g., boulders, water problems) which may or may not be completely reflected upon in prior soil subsurface investigations.

While we have reiterated recommendations that pile capacities based on analytical procedures or pile-driving formulas, etc., should be verified by load tests, one should not assume that load tests are a cure-all, or that they are unquestionably accurate.

Indeed, while tests are considered the most reliable manner of determining the pile capacity, the results from such tests may be misleading if the relevant factors are overlooked:

1. A time lapse should be provided between the time of installation and the time of test loading; this may be anywhere from 3 to 4 days for a granular stratum to perhaps a month for clayey soils. This is the time normally required for the respective soils to regain the strength lost during the driving operation. In the case of concrete piles a minimum time is also required to develop the material strength.
2. The specific location for the installation and subsequent testing of the pile must be representative of the overall site if the test results are to be representative of the rest of the piles. It is common practice on the part of the engineers to select the most unfavorable conditions of the site (one that is expected to be of the least capacity), which thus results in a conservative installation for the rest of the site.
3. The pile characteristics, such as length, diameter, installation method, must closely resemble those of the piles to be installed later. Obviously the results of the test pile would be subject to question if this were not the case.

The manner in which the load is applied to the pile may be designated by local building codes or by the more common ASTM guidelines (e.g., ASTM D-1143-74). Briefly, most codes provide the following guidelines:

1. The load is to be applied in increments of 20 to 25 percent of the design loads. These increments are applied either at specified time intervals, or after a specified rate of settlement has been observed, usually less than half a millimeter per hour.
2. The level of load is generally twice the design load (200 percent of design load). Sometimes the load is increased to failure on the second load cycle, after the pile has successfully resisted the 200 percent load test during the first cycle.
3. The actual load to be used for design purposes is usually 50 percent of the load that results in a settlement of a specified magnitude (usually not to exceed 25 mm).

Figure 15-5 illustrates a typical load–settlement relationship. Figure 15-5a shows the loading, in this case in 20 percent load increments, and the unloading in five steps. Figure 15-5b shows the net settlement for the corresponding loads. For a given point, such as point 1, the net settlement is equal to the gross settlement minus the elastic recovery. Symbolically this represents $(\delta_g - \delta_e)$.

The number of piles tested for any given site is generally left as a judgment factor for the engineer. In turn, the engineer generally will take into account such factors as the overall scope of the project (total cost, foundation cost, number of piles, etc.), the uniformity of the soil characteristics, and perhaps the type of building (height, masonry versus steel, etc.). Inherent in the decision is the general recognition that pile testing is somewhat expensive, and therefore economy is a relevant factor.

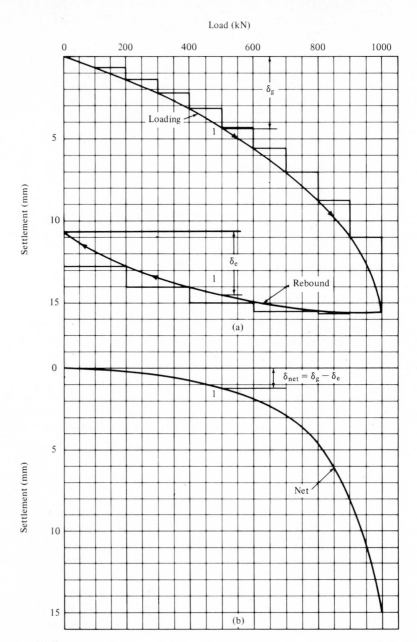

FIGURE 15-5
Typical load–settlement relationship. (a) Gross settlement. (b) Net settlement.

15-11 PILE GROUPS

The behavior of a group of piles is different from that of individual piles in a number of ways:

1. In general the bearing capacity of a group of piles is less than the sum of the individual piles.
2. The settlement of a pile group is larger than that of the individual piles for corresponding levels of load.
3. The efficiency of the pile group is less than that of a single pile.

Yet in spite of these shortcomings, a pile group is a much more common occurrence than a single pile. Single piles lack the overall stability against overturning, a deficiency that is easily overcome by a cluster of piles. Similarly, horizontal thrust in various directions could be more readily resisted by battered piles within the group. Generally, for column supports a minimum of two piles, but more frequently three or more piles, are clustered in a group and connected via a concrete cap to form a unit. For walls, a line of single piles is common.

Load Capacity of Pile Groups

A single pile will transmit its load to the soil in a pressure bulb, as shown in Fig. 15-6a. Depending on the pile spacing, however, the pressure bulbs may overlap. The stresses

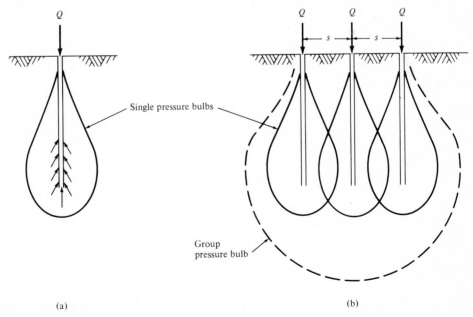

(a)   (b)

FIGURE 15-6
Pressure bulb for friction pile systems. (a) Single pile. (b) Group pile action.

at points of overlap are obviously larger than those for individual piles at the corresponding elevations. That is, the resulting group pressure bulb is the superposition of the individual effects, as depicted by the group pressure bulb in Fig. 15-6b.

The capacity of the group is less than the sum of the individual pile capacities. Hence tests of individual pile capacities do not provide a totally accurate assessment of the group capacity. By increasing the spacing, thereby minimizing overlapping, the effect of superposition is less, or perhaps it is totally eliminated if spacing is large enough. However, large spacings require a correspondingly large (wide) and thick (and thereby more expensive) concrete cap. Most building codes stipulate minimum spacing. Considerations regarding stratum (e.g., clayey, granular), type of pile (e.g., shape, method of installation), settlement, strength criteria, judgment and experience on the part of the engineer are valuable and frequently weighed factors in the determination of spacing. As a guide, however, a center-to-center spacing of 2 diameters ($2D$) is almost always a minimum requirement, with $3D$ to $3.5D$ being quite common.

The stresses within the soil stratum in which a pile group is located are difficult to estimate for a number of reasons such as:

1. The distribution of friction stresses along the pile length is rather indeterminate; and so is the superposition of stresses resulting from overlapping of pressure bulbs.
2. Consolidation, fluctuation in the water table, and other time-related variables are not easily assessed.
3. The interaction of a pile cap with the soil is difficult to evaluate with any degree of accuracy.

Although analytical solutions for such stresses were proposed by a number of investigators [e.g., Mindlin (64), Geddes (38), Poulos and Davis (72), and others], a widely used empirical method consists of treating the pile group as an "equivalent" footing whose length and width are equal to the length and width of the pile group. The support is provided by the end-bearing of this "footing" and the friction along the sides of this long (deep) footing. This is shown in Fig. 15-7.

The general bearing capacity formula for deep foundations may be used to provide an approximate value for the group capacity. The approach for estimating soil stresses from pile groups is described in the following section.

Settlement of Pile Groups

The settlement of pile groups may be viewed as the result of four separate causes:

1. The axial deformation of the pile
2. The deformation of the soil at the pile–soil interface
3. The compressive deformation of the soil between the piles
4. The compression deformation in the stratum below the tips of the piles

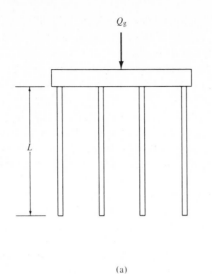
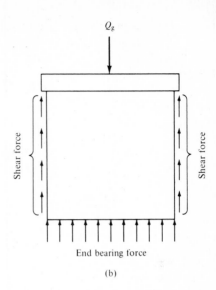

FIGURE 15-7
Estimate of bearing capacity of pile groups via equivalent deep footing. (a) Pile group. (b) Equivalent deep footing. Q_g = capacity of pile group.

The settlements reflected by items (1) and (2) are relatively small and are, therefore, usually neglected. The settlement described by item (3) is difficult to evaluate with reliable accuracy, and therefore it is also ignored. The settlement reflected by item (4), however, can be estimated by treating the pile group as an equivalent footing (Fig. 15-7) and computing the stresses as described in Chapter 8, for a square rectangular or a circular shape, as the case may be. When the piles are end-bearing, the base of the equivalent footing is at the pile tip, with the corresponding stress isobars as illustrated in Fig. 15-8a. When the piles are essentially friction type, the base of the equivalent footing is assumed to act at a depth of approximately two-thirds of the pile length, as shown in Fig. 15-8b. The settlement in both instances, however, is computed only for any compressible stratum below the pile tips.

Pile Group Efficiency

The efficiency of a pile group is the ratio of the capacity of the group to the sum of the capacities of the individual piles. Symbolically this may be expressed by

$$E_g = \frac{Q_g}{nQ_i} \times 100 \text{ percent}$$

where E_g = efficiency of pile group
Q_g = capacity of group
Q_i = capacity of individual pile
n = number of piles in group

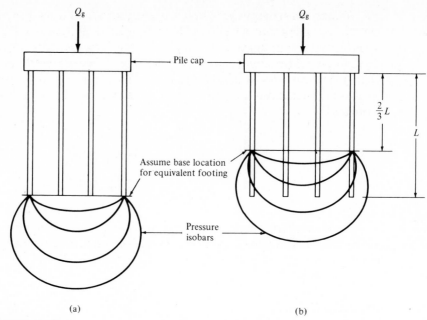

FIGURE 15-8
Method for determining pressures under pile groups. (a) Group of end-bearing piles in dense sands or sand–gravel deposit. (b) Group of friction piles.

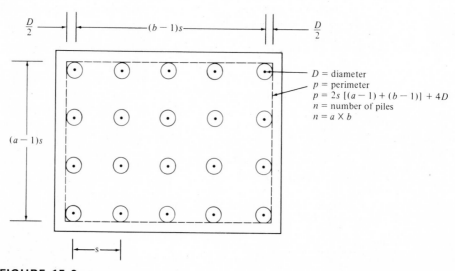

FIGURE 15-9
Group of piles, equal spacing s.

If we assumed the pile group to be an equivalent footing, as shown in Fig. 15-7, the capacity of a pile group of n piles could be estimated as follows: Let Fig. 15-9 represent an arbitrary pile arrangement with a rows and b columns and an equal spacing s. The capacity of the group Q_g is equal to (group perimeter) × (pile length) × (friction factor). The capacity of a single pile Q_i is equal to (single-pile perimeter) × (pile length) × (friction factor). Symbolically, therefore, the efficiency expression could be written as:

$$E_g = \frac{Q_g}{nQ_i} = \frac{[2s(a + b - 2) + 4D]Lf_s}{n(\pi D)Lf}$$

$$E_g = \frac{2s(a + b - 2) + 4D}{\pi abD} \times 100 \qquad (15\text{-}11)$$

Equation (15-12) indicates that as the spacing is increased, the efficiency of the group increases. As mentioned previously, however, practical considerations (e.g., footing size, costs) generally limit the spacing to approximately $3D$.

EXAMPLE 15-3

Given   A pile group with characteristics as shown in Fig. 15-10a and b.

Find   (a) σ_z at middepth of the clay layer.
             (b)   Efficiency of the pile group.

Procedure   (a)   The depth at the midsection of the clay layer is 8.5 m below the hypothetical bottom of the footing, as shown in Fig. 15-10c. An estimate of σ_z can be made using Boussinesq's values (Section 8-7) or by a $30°$ (or 2:1) "slope" method (Section 8-10). Via Boussinesq, Table 8-4, for $z = 8.5$ (see Fig. 15-10c) and $m = 1.65/8.5$ and $n = 1.15/8.5$, we have:

$$m = \frac{1.65}{8.5} = 0.195$$

$$n = \frac{1.15}{8.5} = 0.135$$

$$f(m, n) = 0.0122$$

$$q = \frac{30,000 \text{ kN}}{2.3 \times 3.3} = 3952 \text{ kN/m}^2$$

Then,

$$\sigma_z = 3952(4)(0.0122)$$

Answer

$$\sigma_z = 193 \text{ kN/m}^2 \ (4 \text{ kips/ft}^2)$$

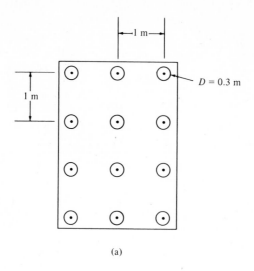

(a)

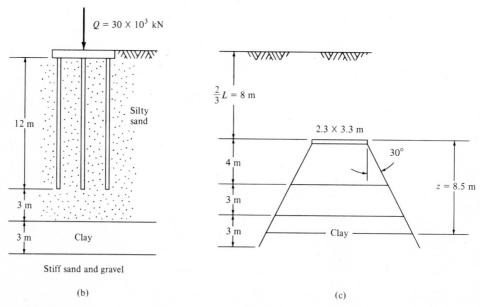

(b)

(c)

FIGURE 15-10
(a) Top view of pile group. (b) Soil profile and piles. (c) σ_z at middepth of clay.

Based on the 30° slope,

$$q_z = \frac{30{,}000}{(2.3 + 1.16 \times 8.5)(3.3 + 1.16 \times 8.5)}$$

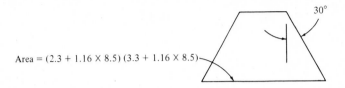

Area = $(2.3 + 1.16 \times 8.5)\,(3.3 + 1.16 \times 8.5)$

Answer

$$q_z = 188 \text{ kN/m}^2$$

This compares well with the 193 kN/m² via Boussinesq.

(b)   The efficiency of a pile group is

$$E_g = \frac{2s(a + b - 2) + 4D}{\pi a b D} \times 100$$

or

$$E_g = \frac{2(1)(3 + 4 - 2) + 1.2}{\pi(12)(0.3)} \times 100$$

Answer

$$E_g = 94 \text{ percent}$$

PROBLEMS

15-1.   State some of the (relative) advantages and disadvantages of: (a) Wood piles. (b) Steel piles. (c) Precast concrete piles.

15-2.   Compare features of cast-in-place piles:
(a)   Constructed by a driven steel shell with concrete.
(b)   Auger-cast type.

15-3.   A cast-in-place concrete pile, 10 m long, with a uniform diameter of 0.4 m, is cast in a steel shell driven in a sandy stratum whose properties are $\gamma = 17.6 \text{ kN/m}^3$; $\varphi = 28°$; $c = 0$. Determine the ultimate bearing capacity of a single pile.

15-4.   Determine the ultimate bearing capacity of the pile in Problem 15-3 if the water table rises close to the surface and $\gamma_b = 8.6 \text{ kN/m}^3$.

15-5.   Determine the ultimate capacity of a concrete-filled pipe pile, 10 m long and 0.3 m in diameter, installed in a sandy stratum whose properties are $\gamma = 18.1 \text{ kN/m}^3$; $\varphi = 28°$; $c = 0$.

15-6.   An auger-cast concrete pile, 13 m long and 0.4 m in diameter, is constructed in moist sandy stratum. $\gamma = 18.5 \text{ kN/m}^3$; $\varphi = 30°$; $c = 0$; average SPT number ≈ 17 blows/0.3 m.

Determine the bearing capacity of a single pile:
(a) Based on the general bearing capacity formula.
(b) Estimated based on N.

15-7. A 0.3-m-diameter pipe was driven into a clay stratum to a depth of 15 m, then filled with concrete. The clay properties are $\gamma = 19 \text{ kN/m}^3$; $\varphi = 18°$; $c = 24 \text{ kN/m}^2$. Determine the ultimate bearing capacity of a single pile.

15-8. A single acting steam hammer weighs 18.8 kN and falls 0.6 m. Estimate the capacity of the pile, via the modified *Engineering News Record* formula, when the resistance to penetration is approximately 17 mm per hammer blow. The pile is yellow pine, 10 m long and 0.4 m diameter. State any assumptions.

15-9. Determine the capacity of a steel pile 14 m long when the blow resistance is 10 mm per hammer blow. The pile is HP 30 × 1.49 (cm × kN/m); $A = 194 \text{ cm}^2$; depth = 31.97 cm; $I_{xx} = 33,800 \text{ cm}^4$; $I_{yy} = 11,500 \text{ cm}^4$. State any assumptions.

15-10. A pile group of nine piles is arranged as shown in Fig. P15-10a and embedded as shown in Fig. P15-10b. Determine:
(a) The efficiency of the group.
(b) The settlement of the group (neglect the compressibility of the sand).

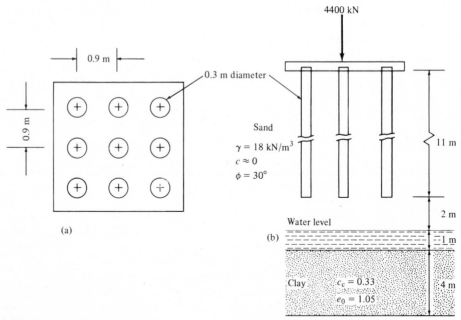

FIGURE P15-10

15-11. Estimate the ultimate bearing capacity of the group if the group was treated as an "equivalent" deep footing.

BIBLIOGRAPHY

1. Alizadeh, M., and M. T. Savisson, "Lateral Load Tests on Piles—Arkansas River Project," *ASCE J. Geotech. Eng. Div.*, Sept. 1970.

2. American Railway Engineering Association Committee, "Steel and Timber Pile Tests—West Atchafalaya Floodway," *Area Bull.* 489, 1950.

3. Ashton, W. D., and P. H. Schwartz, "H-Bearing Piles in Limestone and Clay Shales," *ASCE J. Geotech. Eng. Div.*, July 1974.

4. ASTM, "Standard Method of Testing Piles under Axial Compressive Load," *Annual Book of ASTM Standards*, p. 19, Designation D-1143-74, 1974.

5. Baligh, M. M., V. Vivatrat, and H. Figi, "Downdrag on Bitumen-Coated Piles," *ASCE J. Geotech. Eng. Div.*, Nov. 1978.

6. Barden, L., and M. F. Monckton, "Tests on Model Pile Groups in Soft and Stiff Clay," *Geotechnique*, vol. 20, no. 1, Mar. 1970.

7. Barker, W. R., and L. C. Reese, "Load Carrying Characteristics of Drilled Shafts Constructed with the Aid of Drilling Fluids," Cent. Highw. Res., University of Texas at Austin, Austin, Tex., Res. Rep. 89-9, Aug. 1970.

8. Begemann, H. K., "The Friction Jacket Cone as an Aid in Determining the Soil Profile," *6th Int. Conf. Soil Mech. Found. Eng.*, Montreal, Canada, vol. 1, 1965.

9. Berezantzev, V. G., *et al.* "Load Bearing Capacity and Deformation of Piled Foundations," *5th Int. Conf. Soil Mech. Found. Eng.*, Paris, France, vol. 2, 1961.

10. Bjerrum, L., I. J. Johannessen, and O. Eide, "Reduction of Negative Skin Friction on Steel Piles to Rock," *7th Int. Conf. Soil Mech. Found. Eng.*, Mexico, vol. 2, 1969.

11. Bowles, J. E., *Foundation Analysis and Design*, 2nd ed., McGraw-Hill, New York, 1977.

12. Bozozuk, J., "Downdrag Measurements on a 160-ft Floating Pipe Test Pile in Marine Clay," *Can. Geotech. J.*, vol. 9, 1972.

13. Broms, B. B., "Design of Laterally Loaded Piles," *ASCE J. Geotech. Eng. Div.*, May 1965.

14. Broms, B. B., and L. Hellman, "End Bearing and Skin Friction Resistance of Piles," *ASCE J. Geotech. Eng. Div.*, Mar. 1968.

15. Broms, B. B. "Settlements of Pile Groups," *5th Pan American Soil Mech. Conf.*, ASCE, vol. 3, 1972.

16. Broms, B. B., and P. Boman, "Lime Columns—A New Foundation Method," *ASCE J. Geotech. Eng. Div.*, Apr. 1979.

17. Butterfield, R., and P. K. Banerjee, "The Problem of Pile Group–Pile Cap Interaction," *Geotechnique*, vol. 21, no. 2, June 1971.

18. Caquot, A., and J. Kérisel, *Traité de Mécanique des Sols*, Gauthier-Villars, Paris, France, 1966.

19. Chellis, R. D., *Pile Foundations*, 2nd ed., McGraw-Hill, New York, 1961.

20. Chin, F. K., "Estimation of the Ultimate Load of Piles Not Carried to Failure," *2nd Southeast Asian Conf. Soil Eng.*, 1970.

21. Clark, J. C., and G. G. Meyerhof, "The Behavior of Piles Driven in Clay—Investigation of the Bearing Capacity Using Total and Effective Strength Parameters," *Can. Geotech. J.*, vol. 10, 1973.

22. Clemence, S. P., and W. F. Brumond, "Large-Scale Model Test of Drilled Pier in Sand," *ASCE J. Geotech. Eng. Div.*, June 1975.

23. Coyle, H. M., and L. C. Reese, "Load Transfer for Axially Loaded Piles in Clay," *ASCE J. Soil Mech. Found. Div.*, Mar. 1966.

24. Cummings, A. E., G. O. Kerkhoff, and R. B. Peck, "Effect of Driving Piles into Soft Clay," *Trans. ASCE*, vol. 115, 1950.

25. D'Appolonia, E., and J. P. Romualdi, "Load Transfer in End-Bearing Steel H-Piles," *ASCE J. Geotech. Eng. Div.*, Mar. 1963.

26. D'Appolonia, E., and J. A. Hribar, "Load Transfer in a Step-Taper Pile," *ASCE J. Soil Mech. Found. Div.*, vol. 89, no. SM6, Nov. 1963.

27. Davisson, M. T., and J. R. Salley, "Model Study of Laterally Loaded Piles," *ASCE J. Geotech. Eng. Div.*, Sept. 1970.

28. De Beer, E. E., "Some Considerations Concerning the Point Bearing Capacity of Bored Piles," *Proc. Symp. Bearing Capacity of Piles*, Roorkee, India, 1964.

29. Desai, C. S., "Numerical Design Analysis for Piles in Sands," *ASCE J. Geotech. Eng. Div.*, June 1974.

30. Dos Santos, M. P., and N. A. Gomes, "Experiences with Piled Foundations of Maritime Structures in Portuguese East Africa," *4th Int. Conf. Soil Mech. Found. Eng.*, London, England, vol. 2, 1957.

31. Eide, O., J. N. Hutchinson, and A. Landva, "Short- and Long-Term Test Loading of a Friction Pile in Clay," *5th Int. Conf. Soil Mech. Found. Eng.*, Paris, France, vol. 2, 1961.

32. Ellison, R. D., E. D'Appolonia, and G. R. Thiers, "Load-Deformation for Bored Piles," *ASCE J. Soil Mech. Found. Div.*, vol. 97, no. SM4, Proc. Paper 8052, Apr. 1971.

33. Endo, M., *et al.*, "Negative Skin Friction Acting on Steel Pipe-Piles in Clay," *7th Int. Conf. Soil Mech. Found. Eng.*, Mexico, vol. 2, 1969.

34. Fellinius, B. H., "Down-Drag on Piles in Clay Due to Negative Skin Friction," *Can. Geotech. J.*, vol. 9, Nov. 1972.

35. Fellinius, G. H., "Test Loading of Piles and New Proof Testing Procedure," *ASCE J. Geotech. Eng. Div.*, vol. 101, no. GT9, Proc. Paper 11551, Sept. 1975.

36. Fox, D. A., G. F. Parker, and V. J. R. Sutton, "Pile Driving in the North Sea Boulder Clays," *2nd Conf. on Offshore Technol.*, Houston, Tex, vol. 1, 1970.

37. Fuller, R. M., and H. E. Hoy, "Pile Load Tests Including Quick-Load Test Method, Conventional Methods and Interpretations," *Pile Foundations*, Highw. Res. Board no. 333, Washington, D.C., 1970.

38. Geddes, J. D., "Stresses in Foundation Soils Due to Vertical Subsurface Loading," *Geotechnique*, Sept. 1966.

39. Hagerty, D. J., and R. B. Peck, "Heave and Lateral Movements Due to Pile Driving," *ASCE J. Geotech. Eng. Div.*, vol. 97, Nov. 1971.

40. Housel, W. S., "Pile Load Capacity, Estimates and Test Results," *ASCE J. Soil Mech. Found. Div.*, vol. 92, no. SM 4, July 1966.

41. Ismael, N. F., and T. W. Klym, "Uplift and Bearing Capacity of Short Piers in Sand," *ASCE J. Geotech. Eng. Div.*, vol. 105, May 1979.

42. Kay, J. N., "Safety Factor Evaluation for Single Piles in Sand," *ASCE J. Geotech. Eng. Div.*, vol. 102, Oct. 1976.

43. Kerisel, J., "Deep Foundations—Basic Experimental Facts," *Proc. Deep Foundations Conf.*, Sociedad Mexicana de Mecania de Suelos, Mexico City, Mexico, Dec. 1964.

44. Kim, J. B., R. J. Brungraber, "Full-Scale Lateral Load Tests of Pile Groups," *ASCE J. Geotech. Eng. Div.*, vol. 102, Jan. 1976.

45. Kim, J. B., L. P. Singh, and R. J. Brungraber, "Pile Cap Soil Interaction from Full-Scale Lateral Load Tests," *ASCE J. Geotech. Eng. Div.*, vol. 105, May 1979.

46. Koerner, R. M., and A. Partos, "Settlement of Building on Pile Foundation in Sand," *ASCE J. Geotech. Eng. Div.*, vol. 100, no. GT3, Mar. 1974.

47. Kuhlemeyer, R. L., "Static and Dynamic Laterally Loaded Floating Piles," *ASCE J. Geotech. Eng. Div.*, vol. 105, Feb. 1979.

48. Kuhlemeyer, R. L., "Vertical Vibration of Piles," *ASCE J. Geotech. Eng. Div.*, vol. 105, Feb. 1979.

49. Kulhawy, F. H., D. E. Kozera, and J. L. Withiam, "Uplift Testing of Model Drilled Shafts in Sand," *ASCE J. Geotech. Eng. Div.*, vol. 105, Jan. 1979.

50. Lambe, T. W., and H. M. Horn, "The Influence on an Adjacent Building of Pile Driving for the M.I.T. Materials Center," *6th Int. Conf. Soil Mech. Found. Eng.*, Montreal, Canada, vol. 2, 1965.

51. Lee, K. L., "Buckling of Partially Embedded Piles in Sand," *ASCE J. Geotech. Eng. Div.*, vol. 94, Jan. 1968.

52. Liao, S., and D. A. Sangrey, "Use of Piles as Isolation Barriers, *ASCE J. Geotech. Eng. Div.*, vol. 104, Sept. 1978.

53. Mansur, C. I., and J. A. Focht, "Pile Loading Tests, Morganza Floodway Control Structure," *ASCE J. Geotech. Eng. Div.*, vol. 79, paper no. 324, 1953.

54. Mansur, C. I., and A. H. Hunter, "Pile Tests—Arkansas River Project," *ASCE J. Soil Mech. Found. Div.*, vol. 96, no. SM5, 1970.

55. Mattes, N. S., and H. G. Poulos, "Settlement of a Single Compressible Pile," *ASCE J. Soil Mech. Found. Div.*, vol. 95, no. SM1, 1969.

56. McClelland, B., J. A. Focht, Jr., and W. J. Emrich, "Problems in Design and Installation of Heavily Loaded Pipe Piles," *Proc. ASCE Spec. Conf. on Civ. Eng., in the Oceans*, 1967.

57. McClelland, B., "Design of Deep Penetration Piles for Ocean Structures," *ASCE J. Geotech. Eng. Div.*, vol. 100, July 1974.

58. Meigh, A. C., "Some Driving and Loading Tests on Piles in Gravel and Chalk," *Proc. Conf. Behavior of Piles*, London, England, 1971.

59. Meyerhof, G. G., and L. J. Murdock, "An Investigation of the Bearing Capacity of Some Bored and Driven Piles in London Clay," *Geotechnique*, vol. 3, no. 7, 1953.

60. Meyerhof, G. G., "Penetration Tests and Bearing Capacity of Cohesionless Soils," *ASCE J. Soil Mech. Found. Div.*, vol. 82, no. SM1, Jan. 1956.

61. Meyerhof, G. G., "Compaction of Sands and Bearing Capacity of Piles," *ASCE J. Soil Mech. Found. Div.*, vol. 85, no. SM6, Proc. Paper 2292, Dec. 1959.

62. Meyerhof, G. G., "Bearing Capacity and Settlement of Pile Foundations," *ASCE J. Geotech. Eng. Div.*, vol. 102, no. GT3, Proc. Paper 11962, Mar. 1976.

63. Michigan State Highway Commission, "A Performance Investigation of Pile Driving Hammers and Piles," Lansing, Mich., 1965.

64. Mindlin, R. D., "Force at a Point in the Interior of a Semi-Infinite Soild," *J. Am. Inst. Phys.*, May 1936.

65. Moorhouse, D. C., and J. V. Sheehan, "Predicting Safe Capacity of Pile Groups," *ASCE J. Civ. Eng.*, vol. 38, no. 10, Oct. 1968.

66. Nordlund, R. L., "Bearing Capacity of Piles in Cohesionless Soils," *ASCE J. Soil Mech. Found. Div.*, vol. 89, no. SM 3, May 1963.

67. Novak, M., and J. F. Howell, "Torsional Vibration of Pile Foundations," *ASCE J. Geotech. Eng. Div.*, vol. 103, Apr. 1977.

68. Novak, M., and J. F. Howell, "Dynamic Response of Pile Foundations in Torsion," *ASCE J. Geotech. Eng. Div.*, vol. 104, May 1978.

69. Nunes, A. J. C., and M. Vargas, "Computed Bearing Capacity of Piles in Residual Soil Compared with Laboratory and Load Tests," *3rd Int. Conf. Soil Mech. Found. Eng.*, Zurich, Switzerland, vol. 2, 1953.

70. Olsen, R. E., and K. S. Flaate, "Pile Driving Formulas for Friction Piles in Sand," *ASCE J. Soil Mech. Found. Div.*, vol. 93, no. SM6, Nov. 1967.

71. O'Neil, M. W.. and L. C. Reese, "Behavior of Bored Piles in Beaumont Clay," *ASCE J. Soil Mech. Found. Div.*, vol. 98, no. SM2, 1972.

72. Poulos, H. G., and E. H. Davis, "The Settlement Behavior of Single Axially Loaded Incompressible Piles and Piers," *Geotechnique*, vol. 18, no. 3, Sept. 1968.

73. Poulos, H. G., "The Influence of a Rigid Pile Cap on the Settlement Behavior of an Axially Loaded Pile," *Civ. Eng. Trans., Inst. Eng. Aust.*, vol. CE 10, 1968.

74. Poulos, H. G., "Analysis of the Settlement of Pile Groups," *Geotechnique*, vol. 18, 1968.

75. Poulos, H. G., and N. S. Mattes, "The Analysis of Downdrag in End-Bearing Piles Due to Negative Friction," *7th Int. Conf. Soil Mech. Found. Eng.*, vol. 2, 1969.

76. Poulos, H. G., and N. S. Mattes, "The Behavior of Axially Loaded End-Bearing Piles," *Geotechnique*, vol. 19, 1969.

77. Poulos, H. G., "The Behavior of Laterally Loaded Piles: I. Single Piles," *ASCE J. Soil Mech. Found. Div.*, vol. 97, no. SM5, 1971.

78. Poulos, H. G., "The Behavior of Laterally Loaded Piles: II. Pile Groups," *ASCE J. Soil Mech. Found Div.*, vol. 97, 1971.

79. Poulos, H. G., "Lateral Load-Deflection Prediction for Pile Groups," *ASCE J. Geotech. Eng. Div.*, vol. 100, Jan. 1975.

80. Poulos, H. G., and E. H. Davis, "Prediction of Downdrag Forces in End-Bearing Piles," *ASCE J. Geotech. Eng. Div.*, vol. 101, Feb. 1975.

81. Poulos, H. G., "Settlement of Single Piles in Nonhomogeneous Soil," *ASCE J. Geotech. Eng. Div.*, vol. 105, May 1979.

82. Ramiah, B. K., and L. S. Chickanagappa, "Stress Distribution around Batter Piles," *ASCE J. Geotech. Eng. Div.*, vol. 104, Feb. 1978.

83. Randolph, M. F., and C. P. Wroth, "Analysis of Deformation of Vertically Loading Piles," *ASCE J. Geotech. Eng. Div.*, vol. 104, Dec. 1978.

84. Reese, L. C., *et al.* "Generalized Analysis of Pile Foundations," *ASCE J. Soil Mech. Found. Div.*, vol. 96, no. SM1, Jan. 1970.

85. Reese, L. C., M. W. O'Neill, and F. T. Touma, "Bored Piles Installed by Slurry Displacement," *8th Int. Conf. Soil Mech. Found. Eng.*, Moscow, USSR, vol. 3, 1973.

86. Reese, L. C., F. T. Touma, and M. W. O'Neill, "Behavior of Drilled Piers under Axial Loading," *ASCE J. Geotech. Eng. Div.*, vol. 102, no. GT 5, Proc. Paper 12135, May 1976.

87. Reese, L. C., "Design and Construction of Drilled Shaft," *ASCE J. Geotech. Eng. Div.*, vol. 104, Jan. 1978.

88. Robinsky, E. I., and C. F. Morrison, "Sand Displacement and Compaction around Model Friction Piles," *Can. Geotech. J.*, vol. 1, 1964.

89. Saul, W. E., "Static and Dynamic Analysis of Pile Foundations," *ASCE J. Soil Mech. Found. Div.*, vol. 94, no. ST5, May 1968.

90. Scalan, R. H., and J. J. Tomko, "Dynamic Prediction of Pile Static Bearing Capacity," *ASCE J. Geotech. Eng. Div.*, vol. 95, Mar. 1969.

91. Schlitt, H. G., "Group Pile Loads in Plastic Soils," *Proc. Highw. Res. Board*, Washington, D.C., vol. 31, 1951.

92. Scott, R. F., "Freezing of Slurry around Piles in Permafrost," *ASCE J. Geotech. Eng. Div.*, vol. 85, Aug. 1959.

93. Seed, H. B., and L. C. Reese, "The Action of Soft Clay along Friction Piles," *Trans. ASCE*, vol. 22, 1957.

94. Sherman, W. C., "Instrumental Pile Tests in a Stiff Clay," *7th Int. Conf. Soil Mech. Found. Eng.*, Mexico, vol. 2, 1969.

95. Singh, J. P., C. Donovann, and A. C. Jobsis, "Design of Machine Foundations on Piles," *ASCE J. Geotech. Eng. Div.*, vol. 103, Aug. 1977.

96. Skempton, A. W., "The Bearing Capacity of Clays," *Proc. Build. Res. Cong.*, div. 1, p. III, London, England, 1951.

97. Skempton, A. W., "Large Bored Piles—Summing Up," *Symp. Large Bored Piles*, Inst. of Civ. Eng. and Reinf. Conc. Assoc., London, England, Feb. 1966.

98. Smith, J. E., "Pile Driving by the Wave Equation," *ASCE J. Soil Mech. Found. Div.*, no. SM4, 1960.

99. Stermac, A. G., K. G. Selby, and M. Devata, "Behavior of Various Piles in Stiff Clay," *7th Int. Conf. Soil Mech. Found. Eng.*, Mexico, vol. 2, 1969.

100. Swedish Pile Commission, "Recommendations for Pile Driving Test and Routine Test Loading of Piles," Swed. Acad. Eng. Sci., Comm. on Pile Res., Rep. 11, Stockholm, Sweden, 1970.

101. Tavenas, F. A., "Load Tests Results on Friction Piles in Sand," *Can. Geotech. J.*, vol. 8, no. 1, Feb. 1971.

102. Thorburn, S., and R. S. MacVicar, "Pile Load Tests to Failure in the Clyde Alluvium," *Proc. ICE Conf. Behavior of Piles*, London, England, 1971.

103. Thurman, A. G., "Discussion: Bearing Capacity of Piles in Cohesionless Soils," *ASCE J. Soil Mech. Found. Div.*, vol. 90, no. SM1, Jan. 1964.

104. Tomlinson, M. J., "The Adhesion of Piles Driven in Clay Soils," *4th Int. Conf. Soil Mech. Found. Eng.*, London, England, vol. 2, 1957.

105. Tomlinson, J. J., "Adhesion of Piles in Stiff Clays," Constr. Ind. Res. and Inform. Assoc., London, England, Rep. 26, Nov. 1970.

106. Tomlinson, M. J., "Some Effects of Pile Driving on Skin Friction," *Proc. ICE Conf. Behavior of Piles*, London, England, 1971.

107. Touma, F. T., and L. C. Reese, "Behavior of Bored Piles in Sand, *ASCE J. Geotech. Eng. Div.*, vol. 100, July 1974.

108. Tsinker, G. P., "Performance of Jetted Anchor Piles with Widening," *ASCE J. Geotech. Eng. Div.*, vol. 103, Mar. 1977.

109. Van Weele, A. F., "Negative Skin Friction on Pile Foundations in Holland," *Proc. Symp. Bearing Capacity of Piles*, Roorkee, India, 1964.

110. Vesic, A. S., "Tests on Instrumental Piles, Ogeechee River Site," *ASCE J. Soil Mech. Found. Div.*, vol. 96, no. SM2, Proc. Paper 7170, Mar. 1970.

111. Vesic, A. S., "Load Transfer in Pile–Soil Systems," *Proc. Conf. Design Installation of Pile Foundations*, Lehigh University, Bethlehem, Pa., 1970.

112. Watt, W. G., P. J. Kurfurst, and Z. P. Zeman, "Comparison of Pile Load-Test Skin Friction Values and Laboratory Strength Tests," *Can. Geotech. J.*, vol. 6, 1969.

113. Wess, J. A., and R. S. Chamberlin, "Khazzan Dubai No. 1: Pile Design and Installation," *ASCE J. Geotech. Eng. Div.*, vol. 97, Oct. 1971.

114. Whitaker, T., "Experiments with Model Piles in Groups," *Geotechnique*, vol. 7, 1957.

115. Whitaker, T., "Some Experiments on Model Piled Foundations in Clay," *Proc. Symp. Pile Foundations*, Stockholm, Sweden, 1960.

116. Whitaker, T., and R. W. Cooke, "An Investigation of the Shaft and Base Resistance of Large Bored Piles in London Clay," *Proc. Symp. Large Bored Piles*, Inst. Civ. Eng., London, England, 1966.

117. Woodward, R. J., R. Lundgren, and J. D. Boitano, "Pile Loading Tests in Stiff Clay," *5th Int. Conf. Soil Mech. Found. Eng.*, Paris, France, vol. 2, 1961.

118. Woodward, R. J., W. S. Gardner, and D. M. Greer, *Drilled Pier Foundations*, McGraw-Hill, New York, 1972.

119. Wooley, J. A., and L. C. Reese, "Behavior of an Axially Loaded Drilled Shaft under Sustained Loading," Cent. Highw. Res., University of Texas at Austin, Austin, Tex., Res. Rep. 176-2, May 1974.

120. York, D. L., V. G. Miller, and N. F. Ismael, "Long-Term Load Transfer in End Bearing Pipe Piles," *Can. Geotech. J.*, vol. 12, 1975.

121. Zeevaert, L., "Compensated Friction—Pile Foundation to Reduce the Settlement of Buildings on Highly Compressible Volcanic Clay of Mexico City," *4th Int. Conf. Soil Mech. Found. Eng.*, London, England, vol. 2, 1957.

16

Site Improvement

16-1 INTRODUCTION

It is seldom, if ever, that soil properties and functional requirements at a given site are totally compatible. In general the functional requirements are too diverse for this to happen. For example, it is readily apparent that the requirement regarding soil properties for a dam project may be greatly different from those for a multistory building. In one case a primary consideration may be permeability, while bearing capacity may be of particular importance regarding the other. Perhaps too frequently the in situ soil characteristics of a building site are different from those desired and, almost always, far from ideal for a designated need.

As the more "suitable" sites are built upon, the problems associated with the poorer sites appear to become more prevalent. Thus the engineer may frequently be faced with a choice of one of the following:

1. Adapt the design to be compatible with the soil conditions (e.g., use piles, increase footing dimensions to compensate for poor bearing capacity).
2. Abandon the site in favor of one with more favorable soil characteristics.
3. Alter the soil properties toward a designated goal (e.g., increase strength, reduce permeability, reduce compressibility).

The process, natural or artificial, of altering the soil properties of a given site is generally referred to as *soil stabilization*.

Encompassed by the term soil stabilization are a number of techniques: (1) densification of soil via compaction, precompression, drainage, vibrations, or a combination of these; (2) mixing and/or impregnation of the soil formation with chemicals or grouting; (3) replacement of an undesirable soil with a suitable one under controlled conditions.

Generally not all methods and techniques are practical for all sites and for all types of soils. For example, the impregnation of a clay stratum with grout or chemicals is likely to be appreciably more difficult than the injection of these materials into a granular layer; the permeability of the stratum, in this case, is of obvious relevance.

451

Similarly, the details (e.g., type of equipment, field control) are generally different for the compaction of a clay layer than for the compaction of a granular type soil; also, the same corrective treatment for two different soil conditions is quite likely to yield two different end products. Furthermore, economic considerations are usually an indispensable part of the decision-making process. Indeed, the stabilization procedures are quite expensive and should not be used without a careful assessment of the benefit–cost relationship.

16-2 COMPACTION

One of the most widely used methods of soil densification is a process whereby a soil mass is reduced in volume by tamping, rolling, vibrations, or by the addition of temporary surcharge loads. This process is commonly referred to as *compaction*. It is a procedure employed frequently to densify in situ soils and almost always when new fill is added to a building site. Benefits from compaction stabilization include:

1. Increase in soil strength and bearing capacity
2. Reduction in void ratio and thereby reduction of settlement and permeability
3. Reduced shrinkage

In this section we shall limit the discussion to *surface* compaction of relatively shallow depths, say less than 3 m; compaction associated with deeper formations will be discussed in Section 16-4.

The compaction of layers via the application of compacting equipment at the surface of a stratum is generally referred to as *field compaction,* and in situ formations may be compacted to varying degrees for some depths with surface-type vibrators. For example, some noticeable increase in density in sands has been realized in projects with which the author was involved to depths of about 1.5 m, although depths of 2 m and more were reported by others (8, 20). The effect from this method of vibration compaction appears to be practically nil, however, beyond about 2 m. Interestingly the maximum density from surface compaction occurs not at the surface, but at approximately 0.5 m below. Compaction of granular fills has been accomplished via repeated droppings of large weights (15, 17), with densification realized to depths exceeding 6 m.

The typical new fill is placed in uniform layers, which generally vary in thickness from 0.2 to as much as 0.5 m, depending on building requirements, type of equipment, and type and gradation of the material. The layers are compacted to a specified density; generally this is a percentage of optimum based on a specified test (e.g., standard Proctor, ASTM D-698-78 or AASHTO T-99; modified Proctor, ASTM D-1557-78 or AASHTO T-180). In general each layer is tested and approved by a qualified technician before subsequent layers are placed. If the compaction falls below the desired level, it is likely that (1) the material is too dry or too wet; or (2) additional passes with the roller must be made. In other words, the technician in the field not only takes density tests, but also makes tests for moisture content, observes the soil classification, and guides the compaction contractor to obtain a specified result at minimum cost and delay. With proper control, the controlled fill is frequently better than the soil on which it is placed.

Density–Water Content Test

The degree of compaction in the field is commonly measured relative to a *soil density–water content test*, generally run in the laboratory. The basis for such a test is usually attributed to R. Proctor, who pioneered this effort in connection with his construction of dams in the late 1920s, and to subsequent applications of several articles in this regard (23). Subsequent efforts induced a standardization for some widely used compaction standards. Proctor's work was influential in showing that subjected to a given compaction effort, the soil increases in density with an increase in moisture content up to a certain point; beyond this point there is a decrease in density with further increase of moisture content. Figure 16-1 is a graphical depiction of this statement. Although various modifications of methods and procedures have been developed for determining the density–water content relationship, the basic test, essentially described by and named after Proctor, remains perhaps by far the most widely used test for this purpose.

The details for the performance of density–water content tests can be obtained from a number of laboratory manuals dealing with soil testing (3, 13, ASTM or AASHTO specifications, etc.). Table 16-1 gives the adopted standards for some of the more common density–water content tests. Briefly, the test consists of compacting a soil sample which is placed into a cylindrical mold of specified size in a designated number of layers. Each layer is tamped by a freely falling hammer of designated mass, imparting a specific number of evenly distributed loads over the surface of each layer. When all the layers are compacted, the soil in the mold is weighed and the water content measured. This is repeated several times for the same soil, but at a slight increase in moisture content with each complete test. Each test result is plotted as a point on a unit weight–water content test curve, as shown in Fig. 16-1.

Figure 16-1 depicts the results from both the standard as well as the modified Proctor tests for the same soil. They are frequently referred to as *compaction-control charts*. Typical of such test results, the modified Proctor test (1) yields a maximum density at a somewhat lower optimum water content; and (2) yields a maximum density which, as might be expected, is larger than that from the standard Proctor test.

One notes that the energy expended via the modified Proctor test is approximately four times that for the standard test. The increase in maximum density, however, is not commensurate with the much larger proportion of expended energy. In fact, the increase in density from the modified Proctor test is generally only a few percent, varying somewhat with different types of soil and gradation, but it seldom exceeds 10 percent that of the standard tests.

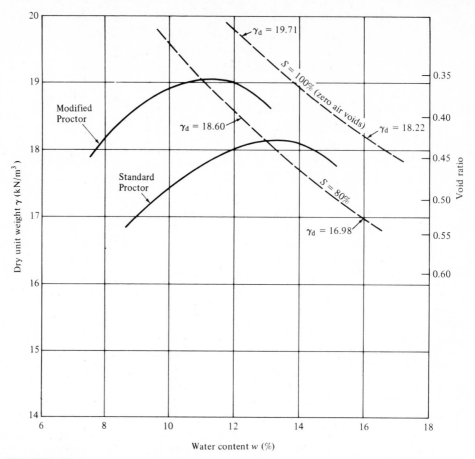

FIGURE 16-1
Standard and modified Proctor compaction tests, run-of-bank mixture (gravelly, silty sand, G = 2.65).

The dry density* could be expressed as a function of the void ratio, as given by Eq. (16-1):

$$\gamma_{dry} = \frac{G\gamma_w}{1 + e} \tag{16-1}$$

Figure 16-1 relates the dry density–void ratio for the particular soil tested. Similarly, the dry density could be related to the water content and degree of saturation via Eq. (16-2):

$$\gamma_{dry} = \frac{G\gamma_w}{1 + wG/S} \tag{16-2}$$

---

* It is common to discuss compaction in terms of dry density as a substitute for dry unit weight.

TABLE 16-1
Adopted Standards Related to Optimum Density—Water Content Test

Test	Reference		Hammer		Mold Volume	Layers	Blows per Layer	Energy* per Test
	ASTM	AASHTO	Weight	Fall				
Standard Proctor	D-698–78	T-99	2.49 kg (5.5 lb)	305 mm (12 in)	9.44×10^{-6} m^3 ($\frac{1}{30}$ ft^3)	3	25	592 kJ/m^3 (12,375 ft-lb/ft^3)
Modified Proctor	D-1557–78	T-180	4.54 kg (10 lb)	457 mm (18 in)	9.44×10^{-6} m^3 ($\frac{1}{30}$ ft^3)	5	25	2695 kJ/m^3 (56,250 ft-lb/ft^3)

*1 kJ = 1 kN-m.

455

Figure 16-1 shows dry density calculations at a water content of 12 and 16 percent for the corresponding values of $S = 100$ percent and $S = 80$ percent. [For the derivation of Eqs. (16-1) and (16-2), see Example 4-1(c) and (c').] Furthermore, the dry density and wet density could be expressed by Eq. (16-3):

$$\gamma_{dry} = \frac{\gamma_{wet}}{1 + w} \tag{16-3}$$

The curve representing 100 percent saturation is a theoretical limit, something never reached in practice since this implies zero air voids; that is, some air voids will always remain within the mass.

Degree of Compaction

Typical specification requirements for compacted fills (frequently referred to as controlled fills) will specify the percentage of compaction based on maximum density. Sometimes both the density and the optimum moisture content are designated. For example, a specification of 95 percent of maximum density is frequently designated for the interior of buildings. Correspondingly a range of water content from optimum could be tolerated without negative effects. That is, referring to Fig. 16-2, a range on the dry or wet side of the optimum can be tolerated without undue energy requirements. On the other hand, if the material is too dry or too wet, optimum

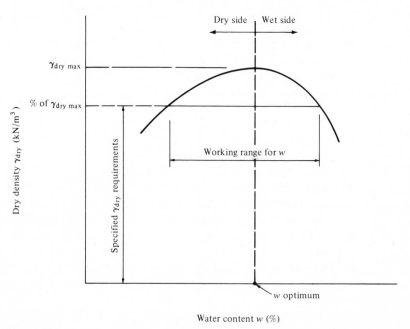

FIGURE 16-2
Variations in γ_d with w near peak density.

compaction is difficult, if not impossible, to obtain. A soil which is too dry is likely to form lumps, which must be crushed with additional energy for an increase in density. On the other hand, a slight increase in the water content may not only reduce the strength of the lumps, but it may also lubricate the particles during the compaction shifting. Of course if excessive water is permitted within the mass, pore pressures are developed from the applied energy, while the mass is merely shifted and not packed. Needless to say, the range of water content for maximum benefit could be obtained by drying (e.g., scarifying) the soil, if possible, if too wet; or by adding additional moisture (e.g., via spray-bar trucks) in order to maintain a most favorable moisture condition, if too dry.

Soil Types

As might be expected, a granular material will behave differently than a cohesive material under a given compaction effort and a given moisture content. Furthermore the selection of equipment and the subsequent results from the use of this equipment will yield different results for the two types of material. For example, clays compacted dry of optimum display an arrangement somewhat independent of the type of compaction (25). On the other hand, when subjected to compaction under moisture conditions higher than the optimum (i.e., wet of optimum), the particle orientation is significantly affected, as are the strength, permeability, and compressibility of the clay. This is further discussed in Section 16-3.

In general, while the shear strength varies with soil types, samples compacted dry of optimum appear stronger and more stable than those compacted wet of optimum. Similarly while increasing the compactive effort reduces the permeability via a reduction of voids, the permeability is also increased with an increase in the water content within the range on the dry side of optimum. However, a slight decrease in permeability is experienced if the water content is increased on the wet side of optimum.

Clay behaves more uniquely. The compressibility of two saturated clay samples at the same density is affected by both stress and the water content at the time of compaction. That is, the sample compacted on the wet side is more compressible than the one compacted on the dry side at low stresses, while the sample compacted on the dry side is more compressible than that compacted on the wet side at high stresses. Likewise, clays compacted on the dry side tend to shrink less upon drying and swell more when subjected to moisture than those compacted on the wet side. Hence it is quite apparent that the engineer must weigh the effect of loading or stresses (e.g., from the addition of building loads, surcharge) and changes in water content and pore-water pressures as well as the subsequent effects that these changes in load and/or moisture conditions may have on the intended function of the compacted stratum.

The engineering properties of cohesionless soils are significantly affected by the relative density of the soil, and not as much by the many variables cited in connection with the compaction of cohesive soils. Generally an increase in density increases the shear strength of the soil and reduces its compressibility. On the other hand, for a given compacting effort the density also increases with an increase in water content, up to a point; then it decreases with a further increase in water content. Hence usually density is the only specified criterion for the compaction of cohesionless soils; the degree of moisture is not a specified criterion, as may be frequently the case for cohesive soils. For optimum density the material selected is usually a well-graded material, which may range in size from as much as 15 cm (6 in) in diameter to clay (e.g., run of bank is a common selection for this type of fill). Many specifications will dictate either a relative density or a percentage of maximum density as the criterion for the degree of compaction.

Equipment

A variety of equipment for compaction purposes is available, with the choice for the proper equipment usually left to the engineer. Usually it is advisable to use vibration-type equipment (e.g., vibratory rollers, tampers) for granular soils, and equipment which penetrates the stratum (e.g., sheepsfoot roller) for the more cohesive material such as silt and/or clay. Many of these units, such as sheepsfoot rollers, vibratory rollers, and tampers, are common inventory of most of the larger construction firms or equipment dealers, who are generally also good sources for information regarding other types of equipment as well.

Performance Control

Compaction control and inspection are done by a qualified technican under the auspices of an engineer. Within the scope of duties, the technician performs field density tests, observes the placement of the fill (e.g., layer thickness, consistency of material), makes water content tests, and generally guides the contractor. Observations and density readings are made, usually a designated number for every lift (layer) or for a specified volume of fill placed. It is common to take such readings for each layer, for every 500 to 1000 m^2 depending on the importance of the site. Within the scope of inspection, when working with clay, compaction may result in a relatively smooth interface between lifts, and thereby create a potential seepage conduit— indeed undesirable in the case of a dam. This is less likely to be the case if sheepsfoot rollers are used in the process.

In-Place Field Tests

A number of methods for determining field densities are available. Among these are the nuclear method (ASTM D-2922-70), the rubber balloon method (ASTM D-2167-66-1977), the sand-cone method (ASTM D-1556-66-1974), and others. The author has developed a method more expedient and more accurate than others, which has been used extensively under his supervision during the past 20 years (6). Briefly the method consists of digging a hole in the compacted stratum, say 8 to 12 cm in diameter and for a similar depth. All the soil is carefully extracted from the hole and weighed via a portable field scale. A representative sample of the soil is preserved and dried, and subsequently the moisture content is determined. From this information both the dry and the wet states of the soil could be determined. Up to this point the procedure is similar to that followed in the case of the rubber balloon or the sand-cone methods. For the determination of volume, a glass or transparent plastic graduate is filled with a uniform-size dry sand to a point that would ensure an adequate volume of sand to fill the hole. With the initial level of the sand noted, the sand is poured into the hole until the hole is filled to the level judged by the technician to be that of the original ground surface. The new level of the sand in the graduate is now observed, with the difference being the volume of the hole. The unit weight of the soil may thus be determined by merely dividing the weight of the soil in the hole by the volume.

16-3 COMPACTED CLAYS

Compacted silts and clays are much less likely to fit into a well-defined category regarding their behavior and characteristics than granular soils. For example, while granular soils show gain in strength from compaction and subsequent increase in density, some silt and clays may, under certain conditions (e.g., method of compaction, molding water), display a decrease in strength past a certain increase in density. For this reason it is advisable to determine the characteristics of a given fine-grained soil via laboratory tests of that soil and not depend on general data available in the literature. Nevertheless much may be gained from the findings of others in this regard. Generally considerable experimental data show that the compaction of clay at various water contents results in:

1. Change in particle structure or grain arrangement
2. Change in engineering properties

   Figure 16-3 shows the effects of compaction on soil structure as presented by Lambe (14). A possible explanation for the change in structure is tied to the change in electrolyte concentration. At small water contents, such as points A or E located at dry of optimum, the concentration of electrolytes is relatively high; this impedes the diffuse double layer of ions surrounding each clay particle from full development. The result is low interparticle repulsion and subsequent flocculation of the colloids, and, thereby, a lack of significant particle orientation of the compacted clay. On the other hand as the water content is increased, say to point C or D located wet of optimum, the electrolyte concentration is reduced, and there is an increase in repulsion between clay particles, a reduction of flocculation, and thus an increase in particle orientation. Typical data illustrating this relationship between particle orientation and water content are given in Figs. 16-4 and 16-5. (See Sec. 2-4 for a more detailed discussion of clay minerals.)

   A pronounced change in engineering properties of a compacted clay results from a change in the structure of the soil. The shrinkage of samples compacted dry of

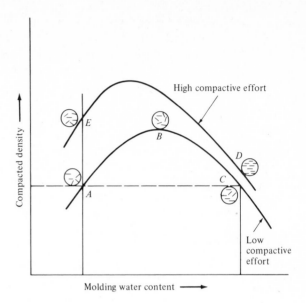

High compactive effort

Low compactive effort

Compacted density

Molding water content

FIGURE 16-3
Effect of compaction on soil
structure. [After Lambe (14).]

optimum is less than for those compacted wet of optimum, as shown in Fig. 16-6.
Conversely, samples compacted dry of optimum swell more than those compacted wet
of optimum. Likewise, as indicated by Fig. 16-7 for a kaolinite clay, molding water and
the corresponding particle orientation exert a significant influence on the stress–strain
relationships of compacted clay.

In general, the more flocculated samples develop their maximum strength at low
strains and exhibit much deeper stress–strain slopes, while the dispersed samples
display much flatter stress–strain curves, reaching maximum strength at much higher
strains.

Although not applicable to all clays, and depending on the particle content and
grain structure or arrangement, Fig. 16-7 depicts a general relationship between water
content and the density, particle orientation, and strength of samples of silty clay subjected to kneading compaction. One notes that a significant reduction in strength
results, when the strength is determined at relatively low shear strain (5 percent in this
case), beyond certain densities and certain water contents (e.g., compaction at wet of
optimum). Conversely, strength appears to increase somewhat with density when it
is determined at rather high strains, for various conditions of water content, and ir-
respective of flocculated or dispersed soil structure. Also, clays compacted dry of
optimum display much higher permeability than those compacted wet of optimum.

Generally the soil characteristics seem to be related to the method of compaction
(e.g., kneading, impact, laboratory, static). They are similar for samples compacted
dry of optimum but are different when the samples are compacted wet of optimum.
Furthermore, the strength determined at low strains tends to increase in the following
order of compaction: kneading, impact, vibratory, static (25).

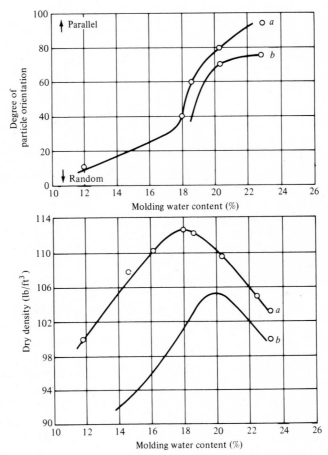

FIGURE 16-4
Influence of molding water content on particle orientation for compacted samples of Boston blue clay. [After Pacey (22).]

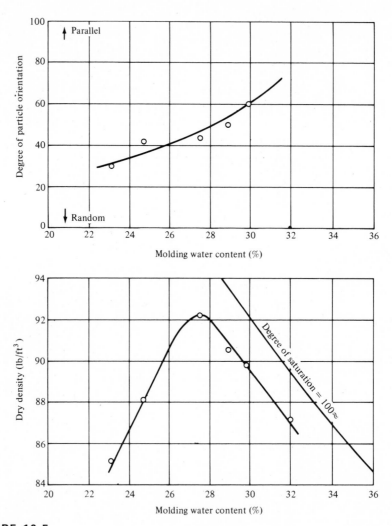

FIGURE 16-5
Influence of molding water content on particle orientation and axial shrinkage for compacted samples of kaolinite. [After Seed and Chan(25).]

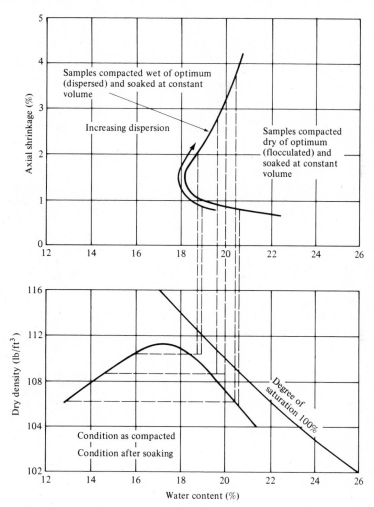

FIGURE 16-6
Influence of soil structure on shrinkage of silty clay. [After Seed and Chan (25).]

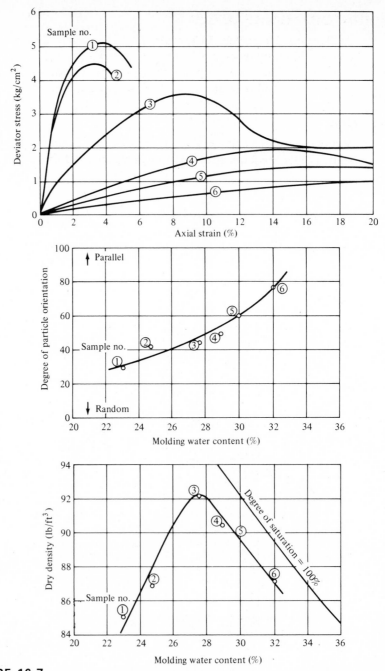

FIGURE 16-7
Influence of molding water content on structure and stress vs. strain relationship for compacted samples of kaolinite. [After Seed and Chan (25).]

16-4 VIBRATORY COMPACTION

As mentioned in Section 16-1, conventional surface-type compactors, both the static and the vibrating types, exert a compacting effect to only a relatively shallow depth, generally less than 2 m. For greater depths other procedures are sometimes used; some of these are briefly described in the following paragraphs.

Vibroflotation

Vibroflotation is a compaction technique which is employed mostly for cohesionless soils. The basic equipment used in conjunction with this method is shown in Fig. 16-8

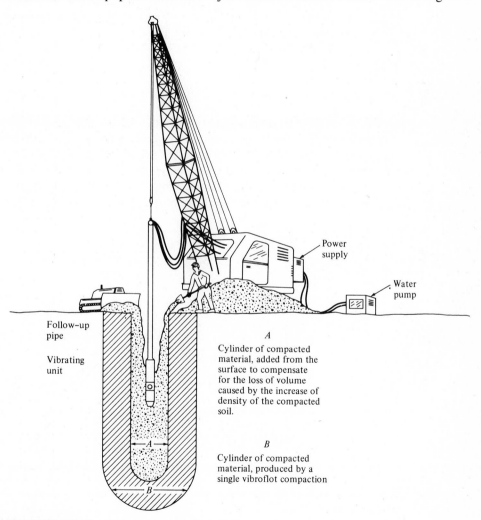

FIGURE 16-8
Vibroflotation equipment. [After Brown, (5).]

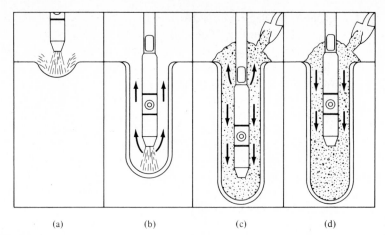

(a)   (b)   (c)   (d)

FIGURE 16-9
Vibroflotation compaction process. [After Brown (5).]

(5). Supplemented by some auxiliary units, the essence of the equipment is the vibro-flot probe which is freely suspended from a crane and equipped with water jets at the upper and lower ends. It consists of a cylindrical tubular section that houses eccentrically rotating weights which induce horizontal vibratory motion.

While vibrating and with water jetting, as shown in Fig. 16-9a, the probe is lowered under its own weight to the required penetration depth. When this depth is

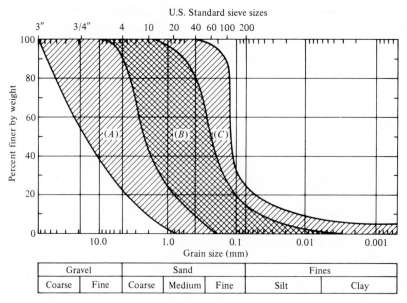

FIGURE 16-10
Soils suitable for vibroflotation. [After Brown (5).]

reached, the water flow is decreased and reversed from the lower to the upper interior jets, as shown in Fig. 16-9c. While still vibrating, the vibroflot unit is raised in incremental steps as a suitable granular material (e.g., river gravel) is shoved into the hole which, by now, is enlarged by the horizontal vibratory motion. This material is then compacted by vibrating the probe at each incremental position for a certain time. The final result is essentially a column of granular material, as shown in Fig. 16-9d. Simultaneously the adjoining soil formation is likewise denser than originally.

Vibroflotation is most effective for loose sands, particularly those below the water table. According to Brown (5), the range of the suitable size for most effective composition falls within zone B of Fig. 16-10, with zones A and C less desirable. Zone C appears appreciably more difficult to compact, while zone A, the zone containing coarse material such as gravels, may pose installation obstacles and subsequent economic problems if the penetration rate of the vibroflot is too low.

Terra Probe

The basic equipment employed for this method consists of a vibrodriver, a probe suspended from a crane, and an electric generator. The probe is an open-ended pipe which is vibrated down into the stratum with the aid of water jets. The weight, the vibration, and the water jets facilitate the penetration process. Once the depth of penetration has been reached, the still vibrating probe is extracted at a slow uniform rate and with continuous water jetting. This is repeated over the site to be compacted.

Although the terra-probe and vibroflotation methods are somewhat similar, a relative comparison of the results obtained via the two methods may be more problematical than unique. Soil types, mechanical installation procedures, and equipment are influential factors in this regard. Generally, however, (1) the extraction rate for the terra probe is higher than that for the vibroflot; (2) more probes may be needed for the terra probe than for vibroflotation to achieve equivalent results (4).

Pounding

This process consists of dropping a heavy weight from designated heights to achieve densification of loose cohesionless soils or fills. Successful applications of this method have been reported for various site conditions, including a former garbage dump (15, 17). In some cases the weight ranged up to 40 tons, and the dropping height exceeded 30 m. Cohesive soils may be liquefied by this process.

Menard and Broise (17) propose a formula relating the required energy to the required depth of densification:

$$W \times H = D^2 \tag{16-4}$$

where W = weight (metric tons)
$\quad\ H$ = height of drop (m)
$\quad\ D$ = densification (m)

As might be expected, the rather high intensity of vibration induced by the dropping weight may affect adjacent structures. Generally the effects are much greater than those from the vibroflotation or terra-probe methods, and they approach those from blasting—another method sometimes used for deep-soil densification. It should be used with caution in the proximity of adjacent buildings in order to minimize the potential dangers of damage to such buildings.

16-5 PRELOADING

The method consists of subjecting a building site to an artificial load, generally in the form of added fill or surcharge, prior to the building loads. The duration of the load may be several months or perhaps years, commensurate with the desired results. It is generally governed by an acceptable designated deformation or rate of settlement of the preloading. Sometimes a system of vertical sand and/or gravel drains is employed in conjunction with the preloading, in order to decrease the time of consolidation and, therefore, permit an earlier use of the site.

Preloading could be employed for virtually all sites with some beneficial results. However, it is generally limited to compressible strata, perhaps soft silt and clays, or sanitary landfills. It is not usually associated with granular formations where consolidation is a relatively insignificant factor. Furthermore, economic considerations may indeed be a governing factor. In this regard the availability of soil or some other material to be used for preloading may have a decisive bearing on the economic feasibility of the method.

16-6 DEWATERING

As we have noted in various sections of this text, water can have a pronounced influence on the stability and/or behavior of a soil formation. In other words, the change in effective stress and subsequently the strength and stability of the soil mass may be directly tied to the change in pore-water pressure. For example, an increase in pore-water pressure results in a decrease in the effective stress, a reduction of soil strength, and perhaps a reduction in stability. Conversely, a reduction in water levels may have just the opposite effect—an increase in intergranular pressure, strength, and stability. Dewatering is a method of improving the soil properties by reducing the water content and the pore-water pressure.

Vertical sand and/or gravel drains are commonly used to dewater and, thereby, increase the consolidation and settlement of soft saturated and compressible soils, such as under embankments or for stabilizing liquefiable soil deposits (26). Horizontal drains are common means of dewatering and subsequently stabilizing natural slopes against seepage and erosion. Ditches along highways may serve the prupose of not only channeling the surface runoff from the road but also dewatering the base and perhaps the subbase of the road. Wellpoints are sometimes used to dewater a soil formation to greater depths.

Electroosmosis is a technique sometimes employed during dewatering to facilitate the water movement. Electroosmosis develops a time-dependent imbalance in the pore-water pressure in a fine-grained soil, thereby expediting the water movement through the soil into an installed drain or wellpoint; see for example, references (1, 19, 28).

Routinely tied to considerations regarding the use of dewatering as a means of improving or stabilizing the site are factors such as cost vs. benefit, feasibility, and effectiveness, and the time schedule to achieve a designated objective. In addition, one should evaluate the effect (e.g., settlement) dewatering may have on adjacent structures.

16-7 CHEMICAL STABILIZATION

Various chemicals added to a soil may yield one, but more likely a number of changes in a soil formation: (1) reduce permeability of the soil (e.g., in dam construction, excavation infiltration); (2) increase soil strength; (3) increase bearing capacity; (4) decrease settlement; (5) produce a stiffening of loose sand formation and thus minimize undesirable effects, such as from vibrations.

The designated objective of the engineer in this regard becomes the central consideration point for the selection of the method and chemicals to be used. Numerous cases of successful chemical stabilization attempts have been described in the literature. Indeed an attempt to even briefly reflect on the very large number of methodologies and techniques and materials used in connection with this method of soil stabilization is perhaps unwarranted and may be even counterproductive. Instead, the discussion will be limited to a general overall view of the more common methods and chemicals.

Chemical stabilization may be employed for surface soils, for subsurface formations, or for both (29). Surface treatment, common in connection with subgrades or bases for pavement construction, generally consists of mechanically mixing the soil with a chemical (e.g., cement, bitumen, lime, bentonite, or some other chemicals) in place or by a batch process. Lime is an effective agent to be mixed with fine-grained soils with high plasticity. Subsurface treatment generally entails the impregnation of the subsurface formation with the chemical. Some of the more common additives are cement or cement grout, sodium silicate (7), and calcium chloride solution.

Grouting generally refers to the process of injecting a stabilizing substance into the soil stratum under pressure. The substance might be a Portland cement, a cement–sand mixture, a bentonite, or a solution of sodium silicate or a number of other chemicals (2, 12, 21, 27, 30). Some of these solutions will penetrate the subsurface formation more easily than others, depending on the viscosity and the rate of chemical reaction and/or hardening. Some of these form a gel within a matter of seconds (a potential advantage when unstable conditions are critical), while others take perhaps hours to set or harden. The setting can be and often is controlled. Furthermore the distance of penetration is not only related to the applied pressure, but it is also closely tied to the size of the soil particles and the permeability of the stratum. The coarser grain soils will normally permit much easier penetration for longer distances than silt or clay formations. Grouting with either cement or chemicals is a relatively expensive form of soil stabilization. Thus except for special instances where the desired effect cannot be obtained by more conventional stabilization techniques, its use is closely scrutinized from an economical point of view. In many instances an alternate method may be more economical. In addition, the degree of consistency and effectiveness over a layered system may require close monitoring of the installation effort as well as of the final results, particularly when subsurface conditions are not uniform or well defined (e.g., erratic layer formation, crevices in rock). In fact, a detailed evaluation of the soil formation, a careful evaluation of the method of injection, a close scrutiny of the type and desired characteristics of the stabilizing agent, and an economic assessment of the benefit are recommended prerequisites to the use of this method.

BIBLIOGRAPHY

1. Banarjee, S., and J. K. Mitchel, "In-Situ Volume-Change Properties by Electro-Osmosis—Theory," *ASCE J. Geotech. Eng. Div.*, Apr. 1980.
2. Benzekri, M., and R. J. Marchand, "Foundation Grouting of Moulay Youssef Dam," *ASCE J. Geotech. Eng. Div.*, Sept. 1978.
3. Bowles, J. E., *Engineering Properties of Soils and Their Measurements*, McGraw-Hill, New York, 1970.
4. Brown, R. E., and A. J. Glenn, "Vibroflotation and Terra-Probe Comparison," *ASCE J. Geotech. Eng. Div.*, Oct. 1976.
5. Brown, R. E., "Vibroflotation Compaction of Cohesionless Soils," *ASCE J. Geotech. Eng. Div.*, Dec. 1977.
6. Cernica, J. N., "Proposed New Method for the Determination of Density of Soil in Place—Proposed New Technique," *ASTM Geotech. Test. J.*, Sept. 1980.
7. Clough, G. W., W. M. Kuck, and G. Kasali, "Silicate-Stabilized Sands," *ASCE J. Geotech. Eng. Div.*, Jan. 1979.
8. D'Appolonia, D. J., R. V. Whitman, and E. D'Appolonia, "Sand Compaction with Vibratory Rollers," *ASCE Specialty Conf. Placement and Improvement of Soil to Support Structure*, 1968.
9. D'Appolonia, D. J., "Soil–Bentonite Slurry Trench Cutoffs," *ASCE J. Geotech. Eng. Div.*, Apr. 1980.
10. Foster, C. R., "Reduction in Soil Strength with Increase in Density," *Trans. ASCE*, vol. 120, 1955.
11. Holtz, W. G., and H. J. Gibbs, "Engineering Characteristics of Expansive Clays," *Trans. ASCE*, 1956.
12. Houlsby, A. C., "Engineering of Grout Curtains to Standards," *ASCE J. Geotech. Eng. Div.*, Sept. 1977.
13. Lambe, T. W., *Soil Testing for Engineers*, Wiley, New York, 1951.
14. Lambe, T. W., "The Structure of Compacted Clay," *ASCE Soil Mech. Found. Div.*, vol. 84, no. SM2, May 1958.
15. Lukas, R. G., "Densification of Loose Deposits by Pounding," *ASCE J. Geotech. Eng. Div.*, Apr. 1980.
16. Majumdar, D. K., "Effect of Catalyst in Soil–Resin Stabilization," *ASCE J. Geotech. Eng. Div.*, June 1975.
17. Menard, L., and Y. Broise, "Theoretical and Practical Aspects of Dynamic Consolidation," *Geotechnique*, vol. 25, 1975.

18. Mitchell, J. K., "The Fabric of Natural Clays and Its Relation to Engineering Properties," *Proc. Highw. Res. Board*, vol. 35, 1956.

19. Mitchell, J. K., and S. Banarjee, "In-Situ Volume-Change Properties by Electro-Osmosis—Evaluation," *ASCE J. Geotech. Eng. Div.*, Apr. 1980.

20. Moorhouse, D. C., and G. I. Baker, "Sand Densification by Heavy Vibratory Compactor," *3rd Proc. Soil Mech. Found. Div.*, ASCE 1968.

21. O'Rourke, J. E., "Structural Grouting of Oroville Dam Coreblock," *ASCE J. Geotech. Eng. Div.*, May 1977.

22. Pacey, J. G., Jr., "The Structure of Compacted Soils," M.S. thesis, Mass. Inst. Technol., Cambridge, Mass., 1956.

23. Proctor, R. R., "Fundamental Principles of Soil Compaction," *Eng. News Rec.*, Aug. 31, Sept. 7, 21, 28, 1933.

24. Seed, H. B., R. L. McNeill, and J. DeGuenin, "Increased Resistance to Deformation of Clay Caused by Repeated Loading," *ASCE J. Soil Mech. Found. Div.*, vol. 84, no. SM2, May 1958.

25. Seed, H. B., and C. K. Chan, "Structure and Strength Characteristics of Compact Clays," *ASCE J. Soil Mech. Found. Div.*, Oct. 1959.

26. Seed, H. B., and J. R. Booker, "Stabilization of Potentially Liquefiable Sand Deposits Using Gravel Drains," *ASCE J. Geotech. Eng. Div.*, July 1971.

27. Vinson, T. S., and J. K. Mitchell, "Polyurethane Foamed Plastics in Soil Grouting," *ASCE J. Geotech. Eng. Div.*, June 1972.

28. Wan, T. Y., and J. K. Mitchell, "Electro-Osmotic Consolidation of Soils," *ASCE J. Geotech. Eng. Div.*, May 1976.

29. Warner, J., "Strength Properties of Chemically Solidified Soils," *ASCE J. Geotech. Eng. Div.*, Nov. 1972.

30. Warner, J., and D. R. Brown, "Planning and Performing Compaction Grouting," *ASCE J. Geotech. Eng. Div.*, June 1974.

31. Wilson, S. D., "Small Soil Compaction Apparatus Duplicates Field Results Closely," *Eng. News Rec.*, vol. 145, no. 18, Nov. 1950.

Appendix A

Westergaard Equations

Point Load P

$$\sigma_z = \frac{Q_P}{z^2} \frac{1}{2\pi} \frac{\sqrt{(1 - 2\mu)/(2 - 2\mu)}}{\{(1 - 2\mu)/(2 - 2\mu) + (r/z)^2\}^{3/2}} = \frac{Q}{z^2} A_w \qquad \text{(A-1)}$$

Line Load q

$$\sigma_z = \frac{q}{z} \frac{k}{2\pi} \left[\frac{n}{m^2 + k^2} \cdot \frac{1}{(m^2 + n^2 + k^2)^{1/2}} \right] = \frac{q}{z} P_0 \qquad \text{(A-2)}$$

Rectangular Load q

$$\sigma_z = q \frac{1}{2\pi} \left[\cot^{-1} \sqrt{\left(\frac{1 - 2\mu}{2 - 2\mu} \right) \left(\frac{1}{m^2} + \frac{1}{n^2} \right) + \left(\frac{1 - 2\mu}{2 - 2\mu} \right)^2 \left(\frac{1}{m^2 n^2} \right)} \right] = q f(m, n) \quad \text{(A-3)}$$

where μ = Poisson ratio
k, m, n = constants for Boussinesq's equations

Circular Load q

$$\sigma_z = q \left\{ 1 - \frac{k}{[k^2 + (r/z)^2]^{1/2}} \right\} = q W_0 \qquad \text{(A-4)}$$

TABLE A-1
**Single Concentrated (Point) Load for $\mu = 0.2$ Values of A_w
for Various r/z Ratios for Use in the Westergaard Equation
$\sigma_z = (Q/z^2)A_w$**

r/z	A_w	r/z	A_w	r/z	A_w	r/z	A_w
0.00	0.31832	2.45	0.00679	4.90	0.00093	7.35	0.00028
0.05	0.31595	2.50	0.00642	4.95	0.00090	7.40	0.00027
0.10	0.30900	2.55	0.00607	5.00	0.00087	7.45	0.00027
0.15	0.29798	2.60	0.00575	5.05	0.00085	7.50	0.00026
0.20	0.28361	2.65	0.00546	5.10	0.00082	7.55	0.00026
0.25	0.26677	2.70	0.00518	5.15	0.00080	7.60	0.00025
0.30	0.24834	2.75	0.00492	5.20	0.00078	7.65	0.00025
0.35	0.22914	2.80	0.00467	5.25	0.00076	7.70	0.00024
0.40	0.20990	2.85	0.00445	5.30	0.00074	7.75	0.00024
0.45	0.19114	2.90	0.00423	5.35	0.00072	7.80	0.00023
0.50	0.17327	2.95	0.00403	5.40	0.00070	7.85	0.00023
0.55	0.15655	3.00	0.00384	5.45	0.00068	7.90	0.00023
0.60	0.14111	3.05	0.00367	5.50	0.00066	7.95	0.00022
0.65	0.12702	3.10	0.00350	5.55	0.00064	8.00	0.00022
0.70	0.11425	3.15	0.00335	5.60	0.00063	8.05	0.00021
0.75	0.10276	3.20	0.00320	5.65	0.00061	8.10	0.00021
0.80	0.09246	3.25	0.00306	5.70	0.00059	8.15	0.00021
0.85	0.08326	3.30	0.00293	5.75	0.00058	8.20	0.00020
0.90	0.07506	3.35	0.00280	5.80	0.00056	8.25	0.00020
0.95	0.06776	3.40	0.00269	5.85	0.00055	8.30	0.00019
1.00	0.06126	3.45	0.00258	5.90	0.00054	8.35	0.00019
1.05	0.05548	3.50	0.00247	5.95	0.00052	8.40	0.00019
1.10	0.05033	3.55	0.00237	6.00	0.00051	8.45	0.00018
1.15	0.04574	3.60	0.00228	6.05	0.00050	8.50	0.00018
1.20	0.04165	3.65	0.00219	6.10	0.00049	8.55	0.00018
1.25	0.03800	3.70	0.00211	6.15	0.00047	8.60	0.00018
1.30	0.03473	3.75	0.00203	6.20	0.00046	8.65	0.00017
1.35	0.03180	3.80	0.00195	6.25	0.00045	8.70	0.00017
1.40	0.02917	3.85	0.00188	6.30	0.00044	8.75	0.00017
1.45	0.02681	3.90	0.00181	6.35	0.00043	8.80	0.00016
1.50	0.02468	3.95	0.00174	6.40	0.00042	8.85	0.00016
1.55	0.02276	4.00	0.00168	6.45	0.00041	8.90	0.00016
1.60	0.02103	4.05	0.00162	6.50	0.00040	8.95	0.00016
1.65	0.01946	4.10	0.00156	6.55	0.00039	9.00	0.00015
1.70	0.01803	4.15	0.00151	6.60	0.00038	9.05	0.00015
1.75	0.01674	4.20	0.00146	6.65	0.00038	9.10	0.00015
1.80	0.01556	4.25	0.00141	6.70	0.00037	9.15	0.00015
1.85	0.01449	4.30	0.00136	6.75	0.00036	9.20	0.00014
1.90	0.01351	4.35	0.00132	6.80	0.00035	9.25	0.00014
1.95	0.01261	4.40	0.00127	6.85	0.00034	9.30	0.00014
2.00	0.01179	4.45	0.00123	6.90	0.00034	9.35	0.00014
2.05	0.01104	4.50	0.00119	6.95	0.00033	9.40	0.00013
2.10	0.01035	4.55	0.00115	7.00	0.00032	9.45	0.00013
2.15	0.00971	4.60	0.00112	7.05	0.00032	9.50	0.00013
2.20	0.00912	4.65	0.00108	7.10	0.00031	9.60	0.00013
2.25	0.00858	4.70	0.00105	7.15	0.00030	9.70	0.00012
2.30	0.00808	4.75	0.00102	7.20	0.00030	9.80	0.00012
2.35	0.00762	4.80	0.00099	7.25	0.00029	9.90	0.00012
2.40	0.00719	4.85	0.00096	7.30	0.00029	10.00	0.00011

TABLE A-2

Single Concentrated (Point) Load for $\mu = 0.20$ **Values of** A_w
for Various r/z **Ratios for Use in the Westergaard Equation**
$\sigma_z = (Q/z^2)A_w$

r/z	A_w	r/z	A_w	r/z	A_w	r/z	A_w
0.00	0.42443	2.45	0.00605	4.90	0.00081	7.35	0.00024
0.05	0.42002	2.50.	0.00572	4.95	0.00079	7.40	0.00024
0.10	0.40800	2.55	0.00540	5.00	0.00076	7.45	0.00023
0.15	0.38890	2.60	0.00511	5.05	0.00074	7.50	0.00023
0.20	0.36457	2.65	0.00484	5.10	0.00072	7.55	0.00022
0.25	0.33681	2.70	0.00459	5.15	0.00070	7.60	0.00022
0.30	0.30738	2.75	0.00436	5.20	0.00068	7.65	0.00022
0.35	0.27775	2.80	0.00414	5.25	0.00066	7.70	0.00021
0.40	0.24907	2.85	0.00394	5.30	0.00064	7.75	0.00021
0.45	0.22209	2.90	0.00374	5.35	0.00062	7.80	0.00020
0.50	0.19726	2.95	0.00356	5.40	0.00061	7.85	0.00020
0.55	0.17478	3.00	0.00340	5.45	0.00059	7.90	0.00020
0.60	0.15467	3.05	0.00324	5.50	0.00058	7.95	0.00019
0.65	0.13685	3.10	0.00309	5.55	0.00056	8.00	0.00019
0.70	0.12115	3.15	0.00295	5.60	0.00055	8.05	0.00019
0.75	0.10737	3.20	0.00282	5.65	0.00053	8.10	0.00018
0.80	0.09531	3.25	0.00269	5.70	0.00052	8.15	0.00018
0.85	0.08477	3.30	0.00258	5.75	0.00050	8.20	0.00018
0.90	0.07556	3.35	0.00247	5.80	0.00049	8.25	0.00017
0.95	0.06750	3.40	0.00236	5.85	0.00048	8.30	0.00017
1.00	0.06045	3.45	0.00227	5.90	0.00047	8.35	0.00017
1.05	0.05427	3.50	0.00217	5.95	0.00046	8.40	0.00016
1.10	0.04884	3.55	0.00209	6.00	0.00044	8.45	0.00016
1.15	0.04407	3.60	0.00200	6.05	0.00043	8.50	0.00016
1.20	0.03986	3.65	0.00192	6.10	0.00042	8.55	0.00015
1.25	0.03614	3.70	0.00185	6.15	0.00041	8.60	0.00015
1.30	0.03285	3.75	0.00178	6.20	0.00040	8.65	0.00015
1.35	0.02992	3.80	0.00171	6.25	0.00039	8.70	0.00015
1.40	0.02732	3.85	0.00165	6.30	0.00038	8.75	0.00014
1.45	0.02500	3.90	0.00158	6.35	0.00038	8.80	0.00014
1.50	0.02292	3.95	0.00153	6.40	0.00037	8.85	0.00014
1.55	0.02106	4.00	0.00147	6.45	0.00036	8.90	0.00014
1.60	0.01935	4.05	0.00142	6.50	0.00035	8.95	0.00014
1.65	0.01788	4.10	0.00137	6.55	0.00034	9.00	0.00013
1.70	0.01652	4.15	0.00132	6.60	0.00033	9.05	0.00013
1.75	0.01529	4.20	0.00127	6.65	0.00033	9.10	0.00013
1.80	0.01418	4.25	0.00123	6.70	0.00032	9.15	0.00013
1.85	0.01317	4.30	0.00119	6.75	0.00031	9.20	0.00013
1.90	0.01225	4.35	0.00115	6.80	0.00031	9.25	0.00012
1.95	0.01142	4.40	0.00111	6.85	0.00030	9.30	0.00012
2.00	0.01065	4.45	0.00108	6.90	0.00029	9.35	0.00012
2.05	0.00995	4.50	0.00104	6.95	0.00029	9.40	0.00012
2.10	0.00931	4.55	0.00101	7.00	0.00028	9.45	0.00011
2.15	0.00873	4.60	0.00098	7.05	0.00028	9.50	0.00011
2.20	0.00818	4.65	0.00094	7.10	0.00027	9.60	0.00011
2.25	0.00769	4.70	0.00092	7.15	0.00026	9.70	0.00011
2.30	0.00723	4.75	0.00089	7.20	0.00026	9.80	0.00010
2.35	0.00681	4.80	0.00086	7.25	0.00025	9.90	0.00010
2.40	0.00641	4.85	0.00083	7.30	0.00025	10.00	0.00010

TABLE A-3

Single Concentrated (Point) Load for $\mu = 0.40$ **Values of** A_w
for Various r/z **Ratios for Use in the Westergaard Equation**
$\sigma_z = (Q/z^2)A_w$

r/z	A_w	r/z	A_w	r/z	A_w	r/z	A_w
0.00	0.95496	2.45	0.00424	4.90	0.00055	7.35	0.00016
0.05	0.93387	2.50	0.00400	4.95	0.00053	7.40	0.00016
0.10	0.87503	2.55	0.00377	5.00	0.00051	7.45	0.00016
0.15	0.78975	2.60	0.00356	5.05	0.00050	7.50	0.00015
0.20	0.69159	2.65	0.00337	5.10	0.00049	7.55	0.00015
0.25	0.59228	2.70	0.00319	5.15	0.00047	7.60	0.00015
0.30	0.49969	2.75	0.00302	5.20	0.00046	7.65	0.00014
0.35	0.41786	2.80	0.00287	5.25	0.00045	7.70	0.00014
0.40	0.34802	2.85	0.00272	5.30	0.00043	7.75	0.00014
0.45	0.28968	2.90	0.00259	5.35	0.00042	7.80	0.00014
0.50	0.24159	2.95	0.00246	5.40	0.00041	7.85	0.00013
0.55	0.20219	3.00	0.00234	5.45	0.00040	7.90	0.00013
0.60	0.17000	3.05	0.00223	5.50	0.00039	7.95	0.00013
0.65	0.14368	3.10	0.00211	5.55	0.00038	8.00	0.00013
0.70	0.12211	3.15	0.00203	5.60	0.00037	8.05	0.00012
0.75	0.10436	3.20	0.00194	5.65	0.00036	8.10	0.00012
0.80	0.08968	3.25	0.00185	5.70	0.00035	8.15	0.00012
0.85	0.07750	3.30	0.00177	5.75	0.00034	8.20	0.00012
0.90	0.06732	3.35	0.00169	5.80	0.00033	8.25	0.00012
0.95	0.05877	3.40	0.00162	5.85	0.00032	8.30	0.00011
1.00	0.05156	3.45	0.00155	5.90	0.00031	8.35	0.00011
1.05	0.04544	3.50	0.00149	5.95	0.00031	8.40	0.00011
1.10	0.04023	3.55	0.00142	6.00	0.00030	8.45	0.00011
1.15	0.03576	3.60	0.00137	6.05	0.00029	8.50	0.00011
1.20	0.03191	3.65	0.00131	6.10	0.00028	8.55	0.00010
1.25	0.02858	3.70	0.00126	6.15	0.00028	8.60	0.00010
1.30	0.02568	3.75	0.00121	6.20	0.00027	8.65	0.00010
1.35	0.02316	3.80	0.00116	6.25	0.00026	8.70	0.00010
1.40	0.02095	3.85	0.00112	6.30	0.00026	8.75	0.00010
1.45	0.01901	3.90	0.00108	6.35	0.00025	8.80	0.00010
1.50	0.01730	3.95	0.00104	6.40	0.00025	8.85	0.00009
1.55	0.01578	4.00	0.00100	6.45	0.00024	8.90	0.00009
1.60	0.01443	4.05	0.00096	6.50	0.00024	8.95	0.00009
1.65	0.01323	4.10	0.00093	6.55	0.00023	9.00	0.00009
1.70	0.01216	4.15	0.00090	6.60	0.00022	9.05	0.00009
1.75	0.01120	4.20	0.00086	6.65	0.00022	9.10	0.00009
1.80	0.01033	4.25	0.00083	6.70	0.00021	9.15	0.00008
1.85	0.00956	4.30	0.00081	6.75	0.00021	9.20	0.00008
1.90	0.00885	4.35	0.00078	6.80	0.00021	9.25	0.00008
1.95	0.00322	4.40	0.00075	6.85	0.00020	9.30	0.00008
2.00	0.00764	4.45	0.00073	6.90	0.00020	9.35	0.00008
2.05	0.00712	4.50	0.00070	6.95	0.00019	9.40	0.00008
2.10	0.00664	4.55	0.00068	7.00	0.00019	9.45	0.00008
2.15	0.00620	4.60	0.00066	7.05	0.00018	9.50	0.00008
2.20	0.00580	4.65	0.00064	7.10	0.00018	9.60	0.00007
2.25	0.00543	4.70	0.00062	7.15	0.00018	9.70	0.00007
2.30	0.00510	4.75	0.00060	7.20	0.00017	9.80	0.00007
2.35	0.00479	4.80	0.00058	7.25	0.00017	9.90	0.00007
2.40	0.00450	4.85	0.00056	7.30	0.00017	10.00	0.00006

TABLE A-4
Influence Values P_0 for Case of Line Load of Finite Length Uniformly Loaded

m	0.1	0.2	0.3	0.4	0.5	n 0.6	0.7	0.8	0.9	1.0	1.2
0.0	0.03152	0.06126	0.08791	0.11082	0.12995	0.14563	0.15835	0.16865	0.17699	0.18378	0.19392
0.1	0.03060	0.05951	0.08546	0.10783	0.12656	0.14195	0.15447	0.16462	0.17286	0.17958	0.18963
0.2	0.02810	0.05473	0.07877	0.09964	0.11724	0.13181	0.14374	0.15348	0.16143	0.16794	0.17773
0.3	0.02463	0.04806	0.06939	0.08810	0.10406	0.11742	0.12848	0.13759	0.14509	0.15127	0.16065
0.4	0.02083	0.04076	0.05907	0.07532	0.08937	0.10130	0.11130	0.11964	0.12657	0.13235	0.14120
0.5	0.01721	0.03376	0.04921	0.06292	0.07503	0.08545	0.09433	0.10182	0.10812	0.11343	0.12168
0.6	0.01403	0.02759	0.04028	0.05183	0.06210	0.07109	0.07884	0.08548	0.09114	0.09595	0.10364
0.7	0.01137	0.02240	0.03282	0.04240	0.05104	0.05870	0.06541	0.07123	0.07626	0.08058	0.08751
0.8	0.00921	0.01818	0.02670	0.03463	0.04187	0.04836	0.05413	0.05919	0.06362	0.06748	0.07375
0.9	0.00748	0.01479	0.02178	0.02834	0.03439	0.03989	0.04482	0.04922	0.05310	0.05652	0.06217
1.0	0.00611	0.01209	0.01785	0.02329	0.02836	0.03301	0.03723	0.04103	0.04442	0.04745	0.05251
1.2	0.00415	0.00825	0.01221	0.01601	0.01960	0.02295	0.02605	0.02889	0.03148	0.03383	0.03786
1.4	0.00291	0.00579	0.00859	0.01131	0.01389	0.01635	0.01864	0.02079	0.02277	0.02459	0.02780
1.6	0.00210	0.00418	0.00622	0.00820	0.01011	0.01193	0.01366	0.01530	0.01683	0.01825	0.02080
1.8	0.00155	0.00310	0.00461	0.00609	0.00753	0.00892	0.01024	0.01150	0.01270	0.01382	0.01587
2.0	0.00118	0.00235	0.00350	0.00463	0.00573	0.00681	0.00784	0.00882	0.00977	0.01066	0.01231
2.5	0.00064	0.00128	0.00191	0.00254	0.00315	0.00375	0.00434	0.00491	0.00546	0.00599	0.00699
3.0	0.00038	0.00077	0.00115	0.00152	0.00190	0.00226	0.00262	0.00298	0.00332	0.00366	0.00430
4.0	0.00017	0.00034	0.00050	0.00067	0.00083	0.00100	0.00116	0.00132	0.00148	0.00163	0.00193
5.0	0.00009	0.00017	0.00026	0.00035	0.00043	0.00052	0.00061	0.00069	0.00077	0.00086	0.00102
6.0	0.00005	0.00010	0.00015	0.00020	0.00025	0.00030	0.00035	0.00040	0.00045	0.00050	0.00060
8.0	0.00002	0.00004	0.00007	0.00009	0.00011	0.00013	0.00015	0.00017	0.00019	0.00022	0.00026
10.0	0.00001	0.00002	0.00003	0.00004	0.00006	0.00007	0.00008	0.00009	0.00010	0.00011	0.00013

TABLE A-5
Influence Values P_0 for Case of Line Load of Finite Length Uniformly Loaded

m	\multicolumn{12}{c}{n}
	1.4	1.6	1.8	2.0	2.5	3.0	4.0	5.0	6.0	8.0	10.0	∞
0.0	0.20091	0.20587	0.20950	0.21221	0.21658	0.21908	0.22164	0.22286	0.22353	0.22421	0.22452	0.22508
0.1	0.19657	0.20151	0.20511	0.20782	0.21218	0.21467	0.21723	0.21845	0.21912	0.21979	0.22011	0.22057
0.2	0.18453	0.18939	0.19295	0.19562	0.19995	0.20242	0.20498	0.20619	0.20686	0.20753	0.20785	0.20841
0.3	0.16723	0.17196	0.17544	0.17806	0.18233	0.18478	0.18732	0.18853	0.18920	0.18987	0.19019	0.19075
0.4	0.14748	0.15204	0.15542	0.15798	0.16217	0.16459	0.16710	0.16831	0.16897	0.16964	0.16996	0.17052
0.5	0.12761	0.13196	0.13522	0.13770	0.14179	0.14417	0.14666	0.14785	0.14851	0.14918	0.14949	0.15005
0.6	0.10910	0.11322	0.11633	0.11872	0.12269	0.12502	0.12748	0.12867	0.12932	0.12999	0.13030	0.13086
0.7	0.09266	0.09653	0.09949	0.10173	0.10562	0.10790	0.11032	0.11149	0.11215	0.11281	0.11312	0.11363
0.8	0.07850	0.08211	0.08491	0.08709	0.09079	0.09300	0.09538	0.09654	0.09719	0.09785	0.09816	0.09872
0.9	0.06651	0.06987	0.07249	0.07256	0.07811	0.08026	0.08259	0.08374	0.08439	0.08504	0.08535	0.08591
1.0	0.05047	0.05959	0.06203	0.06398	0.06736	0.06946	0.07174	0.07287	0.07351	0.07416	0.07447	0.07503
1.2	0.04112	0.04375	0.04583	0.04760	0.05068	0.05262	0.05478	0.05588	0.05651	0.05715	0.05745	0.05801
1.4	0.03046	0.03267	0.03449	0.03600	0.03875	0.04054	0.04259	0.04365	0.04426	0.04489	0.04520	0.04575
1.6	0.02298	0.02482	0.02637	0.02768	0.03013	0.03177	0.03370	0.03471	0.03531	0.03593	0.03623	0.03678
1.8	0.01765	0.01918	0.02050	0.02163	0.02380	0.02529	0.02709	0.02806	0.02864	0.02925	0.02954	0.03009
2.0	0.01378	0.01506	0.01618	0.01716	0.01907	0.02042	0.02209	0.02302	0.02358	0.02417	0.02446	0.02501
2.5	0.00791	0.00874	0.00949	0.01017	0.01156	0.01260	0.01398	0.01479	0.01530	0.01586	0.01614	0.01667
3.0	0.00490	0.00546	0.00597	0.00645	0.00746	0.00826	0.00938	0.01008	0.01054	0.01105	0.01132	0.01185
4.0	0.00222	0.00250	0.00276	0.00301	0.00357	0.00405	0.00479	0.00529	0.00565	0.00608	0.00632	0.00682
5.0	0.00118	0.00138	0.00148	0.00163	0.00196	0.00225	0.00274	0.00311	0.00338	0.00373	0.00394	0.00441
6.0	0.00070	0.00079	0.00088	0.00097	0.00118	0.00137	0.00170	0.00197	0.00217	0.00246	0.00264	0.00308
8.0	0.00030	0.00034	0.00038	0.00042	0.00052	0.00061	0.00078	0.00092	0.00104	0.00123	0.00136	0.00174
10.0	0.00015	0.00018	0.00020	0.00022	0.00027	0.00032	0.00041	0.00050	0.00058	0.00070	0.00079	0.00112

TABLE A-6
Influence Values $f(m, n)$ **for Case of Rectangular Area Uniformly Loaded** [Westergaard Solution, $\sigma_z = q \cdot f(m, n)$]

m	n 0.1	0.2	0.3	0.4	0.5	0.6	0.7	0.8	0.9	1.0	1.2
0.1	0.00312	0.00607	0.00871	0.01098	0.01288	0.01444	0.01570	0.01673	0.01756	0.01824	0.01925
0.2	0.00607	0.01180	0.01695	0.02139	0.02511	0.02817	0.03066	0.03269	0.03433	0.03567	0.03767
0.3	0.00871	0.01695	0.02437	0.03080	0.03620	0.04066	0.04430	0.04727	0.04968	0.05166	0.05462
0.4	0.01098	0.02139	0.03083	0.03897	0.04588	0.05160	0.05630	0.06014	0.06327	0.06584	0.06972
0.5	0.01288	0.02511	0.03620	0.04588	0.05409	0.06093	0.06657	0.07120	0.07500	0.07813	0.08286
0.6	0.01444	0.02817	0.04066	0.05160	0.06093	0.06874	0.07521	0.08055	0.08495	0.08858	0.09410
0.7	0.01570	0.03066	0.04430	0.05630	0.06657	0.07521	0.08241	0.08837	0.09330	0.09739	0.10364
0.8	0.01673	0.03269	0.04727	0.06014	0.07120	0.08055	0.08837	0.09487	0.10027	0.10477	0.11168
0.9	0.01756	0.03433	0.04968	0.06327	0.07500	0.08495	0.09330	0.10027	0.10609	0.11096	0.11846
1.0	0.01824	0.03567	0.05166	0.06584	0.07813	0.08858	0.09739	0.10477	0.11006	0.11614	0.12418
1.2	0.01925	0.03767	0.05462	0.06972	0.08286	0.09410	0.10364	0.11168	0.11846	0.12418	0.13313
1.4	0.01995	0.03906	0.05668	0.07242	0.08617	0.09799	0.10806	0.11660	0.12383	0.12996	0.13963
1.6	0.02044	0.04004	0.05815	0.07431	0.08854	0.10079	0.11126	0.12017	0.12775	0.13421	0.14445
1.8	0.02080	0.04076	0.05921	0.07577	0.09029	0.10258	0.11363	0.12283	0.13068	0.13739	0.14809
2.0	0.02107	0.04130	0.06002	0.07683	0.09161	0.10441	0.11542	0.12484	0.13291	0.13982	0.15089
2.5	0.02151	0.04217	0.06132	0.07856	0.09375	0.10696	0.11835	0.12816	0.13658	0.14384	0.15555
3.0	0.02176	0.04267	0.06206	0.07954	0.09497	0.10842	0.12005	0.13007	0.13872	0.14619	0.15830
4.0	0.02202	0.04318	0.06283	0.08056	0.09624	0.10994	0.12181	0.13207	0.14096	0.14866	0.16122
5.0	0.02214	0.04345	0.06320	0.08105	0.09685	0.11066	0.12265	0.13303	0.14203	0.14985	0.16263
6.0	0.02221	0.04356	0.06310	0.08132	0.09718	0.11106	0.12312	0.13357	0.14263	0.15051	0.16341
8.0	0.02227	0.04370	0.06360	0.08158	0.09752	0.11146	0.12359	0.13410	0.14323	0.15117	0.16421
10.0	0.02230	0.04376	0.06369	0.08171	0.09768	0.11165	0.12380	0.13435	0.14351	0.15148	0.16459
∞	0.02236	0.04387	0.06386	0.08193	0.09796	0.11199	0.12420	0.13480	0.14401	0.15204	0.16525

TABLE A-7
Influence Values for Case of Rectangular Area Uniformly Loaded [Westergaard Solution,
$\sigma_z = q \cdot f(m, n)]$

m	1.4	1.6	1.8	2.0	2.5	3.0	4.0	5.0	6.0	8.0	10.0	∞
0.1	0.01995	0.02044	0.02080	0.02107	0.02151	0.02176	0.02207	0.02214	0.02221	0.02227	0.02230	0.02236
0.2	0.03906	0.04004	0.04076	0.04130	0.04217	0.04267	0.04318	0.04343	0.04356	0.04370	0.04376	0.04387
0.3	0.05668	0.05814	0.05921	0.06002	0.06132	0.06206	0.06283	0.06320	0.06340	0.06360	0.06369	0.06386
0.4	0.07242	0.07434	0.07577	0.07683	0.07856	0.07954	0.08056	0.08105	0.08132	0.08158	0.08171	0.08193
0.5	0.08617	0.08854	0.09029	0.09161	0.09375	0.09497	0.09624	0.09685	0.09718	0.09752	0.09768	0.09796
0.6	0.09799	0.10079	0.10285	0.10441	0.10696	0.10842	0.10994	0.11066	0.11106	0.11146	0.11165	0.11199
0.7	0.10806	0.11127	0.11363	0.11542	0.11835	0.12005	0.12184	0.12265	0.12312	0.12359	0.12380	0.12420
0.8	0.11660	0.12017	0.12283	0.12484	0.12816	0.13007	0.13207	0.13303	0.13357	0.13410	0.13435	0.13480
0.9	0.12383	0.12775	0.13068	0.13291	0.13658	0.13872	0.14096	0.14203	0.14263	0.14323	0.14351	0.14401
1.0	0.12996	0.13421	0.13739	0.13982	0.14384	0.14619	0.14866	0.14905	0.15051	0.15117	0.15148	0.15204
1.2	0.13963	0.14445	0.14809	0.15089	0.15555	0.15830	0.16122	0.16263	0.16341	0.16421	0.16458	0.16525
1.4	0.14672	0.15203	0.15606	0.15918	0.16443	0.16755	0.17089	0.17252	0.17343	0.17435	0.17478	0.17556
1.6	0.15203	0.15773	0.16210	0.16550	0.17127	0.17474	0.17847	0.18031	0.18134	0.18239	0.18288	0.18377
1.8	0.15606	0.16210	0.16676	0.17040	0.17663	0.18041	0.18452	0.18655	0.18770	0.18887	0.18942	0.19043
2.0	0.15918	0.16550	0.17040	0.17426	0.18090	0.18496	0.18941	0.19164	0.19290	0.19419	0.19480	0.19591
2.5	0.16443	0.17127	0.17663	0.18090	0.18836	0.19302	0.19823	0.20089	0.20242	0.20400	0.20475	0.20613
3.0	0.16755	0.17474	0.18041	0.18496	0.19302	0.19813	0.20397	0.20701	0.20877	0.21062	0.21151	0.21316
4.0	0.17089	0.17847	0.18452	0.18941	0.19823	0.20397	0.21072	0.21436	0.21653	0.21885	0.21999	0.22215
5.0	0.17252	0.18031	0.18655	0.19164	0.20089	0.20701	0.21436	0.21843	0.22001	0.22363	0.22499	0.22764
6.0	0.17343	0.18134	0.18770	0.19290	0.20242	0.20877	0.21653	0.22091	0.22863	0.22666	0.22822	0.23133
8.0	0.17435	0.18239	0.18887	0.19419	0.20400	0.21062	0.21885	0.22863	0.22666	0.23017	0.23208	0.23597
10.0	0.17478	0.18288	0.18942	0.19480	0.20475	0.21151	0.21999	0.22499	0.22822	0.23203	0.23412	0.23876
∞	0.17556	0.18377	0.19043	0.19591	0.20613	0.21316	0.22215	0.22764	0.23133	0.23597	0.23876	0.25000

TABLE A-8
Influence Values W_0 for Case of Circular Area
Uniformly Loaded (Westergaard Solution,
$\sigma_z = q \cdot W_0$)

r/z	W_0	r/z	W_0	r/z	W_0
0.00	0.00000	0.50	0.18351	1.00	0.42265
0.01	0.00010	0.51	0.18895	1.01	0.42648
0.02	0.00040	0.52	0.19439	1.02	0.43027
0.03	0.00090	0.53	0.19982	1.03	0.43403
0.04	0.00160	0.54	0.20525	1.04	0.43774
0.05	0.00249	0.55	0.21067	1.05	0.44142
0.06	0.00358	0.56	0.21607	1.06	0.44506
0.07	0.00486	0.57	0.22146	1.07	0.44867
0.08	0.00634	0.58	0.22683	1.08	0.45223
0.09	0.00800	0.59	0.23218	1.09	0.45577
0.10	0.00985	0.60	0.23751	1.10	0.45926
0.11	0.01188	0.61	0.24282	1.11	0.46272
0.12	0.01410	0.62	0.24810	1.12	0.46615
0.13	0.01648	0.63	0.25336	1.13	0.46954
0.14	0.01904	0.64	0.25859	1.14	0.47290
0.15	0.02177	0.65	0.26379	1.15	0.47622
0.16	0.02466	0.66	0.26896	1.16	0.47950
0.17	0.02770	0.67	0.27411	1.17	0.48276
0.18	0.03091	0.68	0.27922	1.18	0.48598
0.19	0.03426	0.69	0.28429	1.19	0.48917
0.20	0.03775	0.70	0.28933	1.20	0.49233
0.21	0.04138	0.71	0.29434	1.21	0.49545
0.22	0.04515	0.72	0.29931	1.22	0.49854
0.23	0.04904	0.73	0.30425	1.23	0.50160
0.24	0.05306	0.74	0.30915	1.24	0.50463
0.25	0.05719	0.75	0.31401	1.25	0.50763
0.26	0.06144	0.76	0.31883	1.26	0.51060
0.27	0.06579	0.77	0.32362	1.27	0.51354
0.28	0.07024	0.78	0.32836	1.28	0.51645
0.29	0.07479	0.79	0.33307	1.29	0.51933
0.30	0.07943	0.80	0.33774	1.30	0.52218
0.31	0.08415	0.81	0.34236	1.31	0.52500
0.32	0.08895	0.82	0.34695	1.32	0.52779
0.33	0.09383	0.83	0.35150	1.33	0.53056
0.34	0.09877	0.84	0.35601	1.34	0.53330
0.35	0.10378	0.85	0.36047	1.35	0.53601
0.36	0.10885	0.86	0.36490	1.36	0.53869
0.37	0.11397	0.87	0.36929	1.37	0.54135
0.38	0.11914	0.88	0.37363	1.38	0.54398
0.39	0.12436	0.89	0.37794	1.39	0.54658
0.40	0.12961	0.90	0.38220	1.40	0.54916
0.41	0.13491	0.91	0.38643	1.41	0.55171
0.42	0.14023	0.92	0.39061	1.42	0.55424
0.43	0.14558	0.93	0.39475	1.43	0.55674
0.44	0.15096	0.94	0.39886	1.44	0.55922
0.45	0.15635	0.95	0.40292	1.45	0.56168
0.46	0.16176	0.96	0.40695	1.46	0.56411
0.47	0.16719	0.97	0.41093	1.47	0.56651
0.48	0.17262	0.98	0.41483	1.48	0.56890
0.49	0.17806	0.99	0.41878	1.49	0.57126

TABLE A-9
Influence Values W_0 for Case of Circular Area
Uniformly Loaded (Westergaard Solution Continued, $\sigma_z = q \cdot W_0$)

r/z	W_0	r/z	W_0	r/z	W_0
1.50	0.57359	2.00	0.66666	7.50	0.90673
1.51	0.57591	2.02	0.66960	7.80	0.90971
1.52	0.57820	2.04	0.67249	8.00	0.91195
1.53	0.58047	2.06	0.67533	8.20	0.91408
1.54	0.58272	2.08	0.67813	8.40	0.91611
1.55	0.58495	2.10	0.68088	8.60	0.91805
1.56	0.58715	2.15	0.68757	8.80	0.91990
1.57	0.58934	2.20	0.69400	9.00	0.92167
1.58	0.59150	2.25	0.70018	9.20	0.92336
1.59	0.59364	2.30	0.70613	9.40	0.92498
1.60	0.59577	2.35	0.71186	9.60	0.92654
1.61	0.59787	2.40	0.71737	9.80	0.92803
1.62	0.59996	2.45	0.72269	10.00	0.92946
1.63	0.60202	2.50	0.72783	10.20	0.93084
1.64	0.60406	2.55	0.73278	10.40	0.93216
1.65	0.60609	2.60	0.73756	10.60	0.93344
1.66	0.60810	2.65	0.74218	10.80	0.93466
1.67	0.61009	2.70	0.74664	11.00	0.93585
1.68	0.61206	2.75	0.75096	11.20	0.93699
1.69	0.61401	2.80	0.75514	11.50	0.93862
1.70	0.61595	2.85	0.75918	11.80	0.94018
1.71	0.61786	2.90	0.76310	12.00	0.94117
1.72	0.61976	2.95	0.76690	12.20	0.94213
1.73	0.62164	3.00	0.77058	12.50	0.94352
1.74	0.62351	3.10	0.77760	12.80	0.94484
1.75	0.62536	3.20	0.78423	13.00	0.94568
1.76	0.62719	3.30	0.79047	13.20	0.94650
1.77	0.62901	3.40	0.79638	13.50	0.94769
1.78	0.63081	3.50	0.80196	13.80	0.94882
1.79	0.63259	3.60	0.80726	14.00	0.94955
1.80	0.63436	3.70	0.81228	14.20	0.95026
1.81	0.63611	3.80	0.81705	14.50	0.95129
1.82	0.63784	3.90	0.82159	14.80	0.95227
1.83	0.63957	4.00	0.82591	15.00	0.95291
1.84	0.64127	4.20	0.83397	15.40	0.95413
1.85	0.64296	4.40	0.84132	15.80	0.95529
1.86	0.64464	4.60	0.84806	16.00	0.95585
1.87	0.64630	4.80	0.85425	16.40	0.95692
1.88	0.64795	5.00	0.85997	16.60	0.95744
1.89	0.64958	5.20	0.86525	17.00	0.95844
1.90	0.65120	5.40	0.87106	17.40	0.95939
1.91	0.65281	5.60	0.87427	17.80	0.96030
1.92	0.65440	5.80	0.87898	18.00	0.96074
1.93	0.65598	6.00	0.88295	20.00	0.96467
1.94	0.65754	6.20	0.88668	25.00	0.97173
1.95	0.65909	6.40	0.89018	30.00	0.97644
1.96	0.66063	6.60	0.89347	40.00	0.98233
1.97	0.66216	6.80	0.89657	50.00	0.98586
1.98	0.66367	7.00	0.89949	100.00	0.99293
1.99	0.66517	7.30	0.90358	∞	1.00000

Index